Advanced Macromolecular
and Supramolecular Materials
and Processes

Advanced Macromolecular and Supramolecular Materials and Processes

Edited by

Kurt E. Geckeler

Kwangju Institute of Science and Technology
Kwangju, South Korea

Kluwer Academic / Plenum Publishers
New York, Boston, Dordrecht, London, Moscow

Library of Congress Cataloging-in-Publication Data

Advanced macromolecular and supramolecular materials and processes/edited by Kurt
E. Geckeler
     p.   cm.
Includes bibliographical references and index.
ISBN 0-306-47405-0
   1. Macromolecules   2. Supramolecular chemistry.   I. Geckeler, Kurt E.

QD381 .A385 2002
547'.7—dc21

2002027466

ISBN 0-306-47405-0

©2003 Kluwer Academic / Plenum Publishers, New York
233 Spring Street, New York, New York 10013

http://www.wkap.nl/

10 9 8 7 6 5 4 3 2 1

A C.I.P. record for this book is available from the Library of Congress

Printed in the United States of America

LIST OF AUTHORS

Seiji Akimoto
Department of Molecular Chemistry, Graduate School of Engineering, Hokkaido
University, Sapporo 060-8628, Japan

Natalia I. Akritskaya
Polymer Chemistry Department, Faculty of Chemistry, Moscow State University V-234,
Moscow 119899, Russia

Naoki Aratani
Department of Chemistry, Graduate School of Science, Kyoto University, Sakyo-ku,
Kyoto 606-8502, Japan

Christian Bailly
General Electric, Plastics BV, Box 117, 4600 AC Bergen op Zoom, The Netherlands

Larisa A. Bimendina
Institute of Polymer Materials and Technology, Satpaev Str.18a, 480013 Almaty,
Kazakhstan

Christoph Burdack
Technical University of Munich, Department of Organic Chemistry, Chair 1, Lichtenbergstr. 4, D-85747 Garching, Germany

Robert P. Burford
Centre of Polymer Science and Engineering, School of Chemical Engineering and
Industrial Chemistry, University of New South Wales, Sydney - NSW 2052, Australia

K. P. Chaudhari
PPV Division, Department of Chemical Technology, University of Mumbai (U.D.C.T.),
Matunga, Mumbai 400 019, India

Alexander G. Didukh
Institute of Polymer Materials and Technology, Satpaev Str.18a, 480013 Almaty, Kazakhstan

Andrzej Dworak
University of Opole, Institute of Chemistry, 54-052 Opole, Poland; Polish Academy of Sciences, Institute of Coal Chemistry, 44-121 Gliwice, Poland

Sven Eggerstedt
Institut für Technische und Makromolekulare Chemie, Universität Hamburg, Bundesstr. 45, D-20146 Hamburg, Germany

Björn Fechner
Institut für Technische und Makromolekulare Chemie, Universität Hamburg, Bundesstr. 45, D-20146 Hamburg, Germany

Justyna Filak
University of Opole, Institute of Chemistry, 54-052 Opole, Poland

Kurt E. Geckeler
Laboratory of Applied Macromolecular Chemistry, Department of Materials Science and Engineering, 1 Oryong-dong, Buk-gu, Kwangju 500-712, South Korea

Yasuhiro Hatanaka
Department of Applied Chemistry, Faculty of Engineering, Kyushu University, Hakozaki 6-10-1, Fukuoka 812-8581, Japan

Gil Tae Hwang
Department of Chemistry, Center for Intergrated Molecular Systems, Pohang University of Science and Technology, Pohang 790-784, South Korea

Kohzo Ito
Graduate School of Frontier Sciences, University of Tokyo, 7-3-1 Hongo, Bunkyo-ka, Tokyo, 113-8656, Japan

Yuji Iwaki
Research Institute for Electronic Science (RIES), Hokkaido University, Sapporo 060-0812, Japan; and Graduate School of Environmental Earth Science, Hokkaido University, Sapporo 060-0810, Japan

Vladimir A. Izumrudov
Polymer Chemistry Department, Faculty of Chemistry, Moscow State University V-234, Moscow 119899, Russia

D. D. Kale
PPV Division, Department of Chemical Technology, University of Mumbai (U.D.C.T.), Matunga, Mumbai 400 019, India

Eugenii Katz
Institute of Chemistry, The Hebrew University of Jerusalem, Jerusalem 91904, Israel

Byeang Hyean Kim
Department of Chemistry, Center for Intergrated Molecular Systems, Pohang University of Science and Technology, Pohang 790-784, South Korea

Su Jeong Kim
Department of Chemistry, Center for Intergrated Molecular Systems, Pohang University of Science and Technology, Pohang 790-784, South Korea

Nobuo Kimizuka
Department of Applied Chemistry, Faculty of Engineering, Kyushu University, Hakozaki 6-10-1, Fukuoka 812-8581, Japan

Agnieszka Kowalczuk
Silesian University of Technology, Faculty of Chemistry, 44-121 Gliwice, Poland

Hans R. Kricheldorf
Institut für Technische und Makromolekulare Chemie, Universität Hamburg, Bundesstr. 45, D-20146 Hamburg, Germany

Sarkyt E. Kudaibergenov
Institute of Polymer Materials and Technology, Satpaev Str.18a, 480013 Almaty, Kazakhstan

Toyoki Kunitake
Frontier Research System, RIKEN, Wako, Saitama, 351-0198, Japan

Dennis Langanke
Institut für Technische und Makromolekulare Chemie, Universität Hamburg, Bundesstr. 45, D-20146 Hamburg, Germany

Gulmira Sh. Makysh
Institute of Polymer Materials and Technology, Satpaev Str.18a, 480013 Almaty, Kazakhstan

Chivikula N. Murthy
Laboratory of Applied Macromolecular Chemistry, Department of Materials Science and Engineering, 1 Oryong-dong, Buk-gu, Kwangju, 500-712, South Korea

Aiko Nakano
Department of Chemistry, Graduate School of Science, Kyoto University, Sakyo-ku, Kyoto 606-8502, Japan

Marzena Nowicka
Silesian University of Technology, Faculty of Chemistry, 44-121 Gliwice, Poland

Nobuhiro Ohta,
Research Institute for Electronic Science (RIES), Hokkaido University, Sapporo 060-0812, Japan; and Graduate School of Environmental Earth Science, Hokkaido University, Sapporo 060-0810, Japan

Yasushi Okumura
Graduate School of Frontier Sciences, University of Tokyo, 7-3-1 Hongo, Bunkyo-ka, Tokyo, 113-8656, Japan

Atsuhiro Osuka
Department of Chemistry, Graduate School of Science, Kyoto University, Sakyo-ku, Kyoto 606-8502, Japan

Alvise Perosa
Dipartimento di Scienze Ambientali, Università Ca' Foscari, Dorsoduro 2137, 30123 Venezia, Italy

Houssain Qariouh
Université de Montpellier II, Science et Techniques de Languedoc, Laboratoire de Chimie Macromoléculaire, Place E. Bataillon, 34095 Montpellier Cedex 5, France

Nabil Raklaoui
Université de Montpellier II, Science et Techniques de Languedoc, Laboratoire de Chimie Macromoléculaire, Place E. Bataillon, 34095 Montpellier Cedex 5, France

Helmut Ritter
Institute of Organic Chemistry and Macromolecular Chemistry, Heinrich-Heine University Düsseldorf, Universitätsstr. 1, 40225 Düsseldorf, Germany

Bernabé L. Rivas
Polymer Department, Faculty of Chemistry, University of Concepcion, Casilla 160-C, Concepción, Chile

Evelyn M. Rodrigues
Centre of Polymer Science and Engineering, School of Chemical Engineering and Industrial Chemistry, University of New South Wales, Sydney - NSW 2052, Australia

Günther Roß
Technical University of Munich, Department of Organic Chemistry, Chair 1, Lichtenbergstr. 4, D-85747 Garching, Germany

Yoshiteru Sakata
The Institute of Scientific and Industrial Research, Osaka University, Mihoga-oka, Ibaraki, Osaka 567-0047, Japan

Shashadhar Samal
Laboratory of Applied Macromolecular Chemistry, Department of Materials Science and Engineering, 1 Oryong-dong, Buk-gu, Kwangju 500-712, South Korea

Shin-ichiro Sato
Department of Molecular Chemistry, Graduate School of Engineering, Hokkaido University, Sapporo 060-8628, Japan

François Schué
Université de Montpellier II, Science et Techniques de Languedoc, Laboratoire de Chimie Macromoléculaire, Place E. Bataillon, 34095 Montpellier Cedex 5, France

Rossitza Schué
Université de Montpellier II, Science et Techniques de Languedoc, Laboratoire de Chimie Macromoléculaire, Place E. Bataillon, 34095 Montpellier Cedex 5, France

Takeshi Shimomura
Graduate School of Frontier Sciences, University of Tokyo, 7-3-1 Hongo, Bunkyo-ka, Tokyo, 113-8656, Japan

Andrea Stricker
Institut für Technische und Makromolekulare Chemie, Universität Hamburg, Bundesstr. 45, D-20146 Hamburg, Germany

Audist I. Subekti
Centre of Polymer Science and Engineering, School of Chemical Engineering and Industrial Chemistry, University of New South Wales, Sydney - NSW 2052, Australia

Monir Tabatabai
Institute of Organic Chemistry and Macromolecular Chemistry, Heinrich-Heine University Düsseldorf, Universität Str. 1, 40225 Düsseldorf, Germany

Barbara Trzebicka
Polish Academy of Sciences, Institute of Coal Chemistry, 44-121 Gliwice, Poland

Akihiko Tsuda
Department of Chemistry, Graduate School of Science, Kyoto University, Sakyo-ku, Kyoto 606-8502, Japan

Pietro Tundo
Dipartimento di Scienze Ambientali, Università Ca' Foscari, Dorsoduro 2137, 30123 Venezia, Italy

Ivar Ugi
Technical University of Munich, Department of Organic Chemistry, Chair 1, Lichtenbergstr. 4, D-85747 Garching, Germany

Ludovico Valli
Dipartimento di Ingegneria dell'Innovazione, Università degli Studi di Lecce, Via Monteroni, 73100 Lecce, Italy

Wojciech Walach
Polish Academy of Sciences, Institute of Coal Chemistry, 44-121 Gliwice, Poland

Itamar Willner
Institute of Chemistry, The Hebrew University of Jerusalem, Jerusalem 91904, Israel

Iwao Yamazaki
Department of Molecular Chemistry, Graduate School of Engineering, Hokkaido University, Sapporo 060-8628, Japan

Tomoko Yamazaki
Department of Molecular Chemistry, Graduate School of Engineering, Hokkaido University, Sapporo 060-8628, Japan

Naoya Yoshida
Department of Chemistry, Graduate School of Science, Kyoto University, Sakyo-ku, Kyoto 606-8502, Japan

Marina V. Zhiryakova
Polymer Chemistry Department, Faculty of Chemistry, Moscow State University V-234, Moscow 119899, Russia

Gulmira T. Zhumadilova
Institute of Polymer Materials and Technology, Satpaev Str.18a, 480013 Almaty, Kazakhstan

PREFACE

The area of macromolecular and supramolecular science and engineering has gained substantial interest and importance during the last decade and many applications can be envisioned in the future. The rapid developments in this interdisciplinary area justify a snapshot of the state-of-the-art in the research of materials and processes that is given in this monograph. It goes without saying that this book cannot be a comprehensive survey but is rather an attempt to highlight some important fields and exciting developments. It summarizes the advances in the area of macromolecular and supramolecular materials and processes made by world-leading research groups over the past years.

This monograph is based primarily on synthetic architectures and systems covered by the contents of selected plenary and invited lectures delivered at the 1st International Symposium on Macro- and Supramolecular Architectures and Materials (MAM-01): Biological and Synthetic Systems, which was held from 11-14 April, 2001 on the international campus of the Kwangju Institute of Science and Technology (K-JIST) in Kwangju, South Korea. In addition, it contains several complementing contributions in this novel field of science dealing with synthetic architectures and represents a unique compilation of reviewed research accounts of the in-depth knowledge of macromolecular and supramolecular materials and processes. It comprises 22 pioneering chapters written by 64 renowned experts from 13 different countries.

On behalf of all contributors I thank all publishers and authors for granting copyright permissions to use their illustrations in this book. I am also thankful to all contributing authors who devoted their precious time and efforts to write these chapters. I am grateful to the President, faculty, and staff of K-JIST for their kind and outstanding support during the preparation and in the course of the development of this work. Lastly, the continuous cooperation, patience, support, and understanding of my family during many evenings, nights, and weekends is highly appreciated. I hope that this book will attract more scientists to this interesting area.

Kurt E. Geckeler, M.D., Ph.D.
Professor of Materials Science
and Engineering

CONTENTS

POLYMERIZATION IN AQUEOUS MEDIUM USING CYCLODEXTRIN
    AS HOST COMPONENT

Helmut Ritter and Monir Tabatabai

SUPRAMOLECULAR COMPOUNDS OF CYCLODEXTRINS WITH
    [60]FULLERENE

Chivikula N. Murthy and Kurt E. Geckeler

STRUCTURE AND FUNCTION OF POLYMERIC INCLUSION COMPLEX
    OF MOLECULAR NANOTUBES AND POLYMER CHAINS

Kohzo Ito, Takeshi Shimomura, and Yasushi Okumura

DESIGN, STRUCTURE, AND BEHAVIOR OF INTERPOLYMER COMPLEX MEMBRANES 139

Sarkyt E. Kudaibergenov, Larisa A. Bimendina, and Gulmira T. Zhumadilova

OPTIMISATION OF MEMBRANE MATERIALS FOR PERVAPORATION 155

François Schué, Houssain Qariouh, Rossitza Schué, Nabil Raklaoui,
and Christian Bailly

Eugenii Katz and Itamar Willner

Iwao Yamazaki, Seiji Akimoto, Tomoko Yamazaki, Shin-ichiro Sato, and
Yoshiteru Sakata

BIODEGRADABLE BLOCK COPOLYMERS, STAR-SHAPED POLYMERS, AND NETWORKS *VIA* RING-EXPANSION POLYMERIZATION

Hans R. Kricheldorf, Sven Eggerstedt, Dennis Langanke, Andrea Stricker, and Björn Fechner

AMPHIPHILIC POLYETHERS OF CONTROLLED CHAIN ARCHITECTURE

Andrzej Dworak, Wojciech Walach, Barbara Trzebicka, Agnieszka Kowalczuk, Marzena Nowicka, and Justyna Filak

BLENDS OF WASTE POLY(ETHYLENE TEREPHTHALATE) WITH POLYSTYRENE AND POLYOLEFINS

K. P. Chaudhari and D. D. Kale

THE CHEMICAL PROGRESS OF MULTICOMPONENT REACTIONS (MCRs)

Ivar Ugi, Günther Roß, and Christoph Burdack

WATER-SOLUBLE POLYMER-METAL INTERACTION

Bernabé L. Rivas

BOVINE SERUM ALBUMINE COMPLEXATION WITH SOME POLYAMPHOLYTES

Alexander G.. Didukh, Gulmira Sh. Makysh, Larisa A. Bimendina, and Sarkyt E. Kudaibergenov

MACROMOLECULAR AND SUPRAMOLECULAR ARCHITECTURES BASED ON FULLERENES

Shashadhar Samal and Kurt E. Geckeler[*]

1. INTRODUCTION

The fullerenes with the π-electron orbitals extending outward provide an excellent opportunity for a variety of orbital-orbital interactions with molecules appropriately positioned close to their surface. The extent of these interactions is dependent on the π-electron availability, shape, and dimension of the interacting molecules. The typical arrangement of the five- and the six-membered rings in C_{60} lead to a spherical shape. When five-membered ring is attached to six-membered rings on each side, each carrying three alternate π-bonds in such away that none of the π-bonds are in the 5,6-ring junctions, and the structure is energy-minimized, the result is a curved surface (Figure 1a). On the curved exterior of such a surface, the π-bonds are bent (Figure 1b).

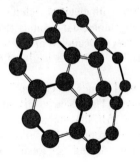 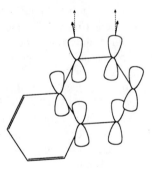

Figure 1. (a) An energy-minimized π-surface built out of a pentagon and five hexagons showing the curved face, and (b) bent π-bonds at 6,6-ring junctions of C_{60}.

[*] Laboratory of Applied Macromolecular Chemistry, Department of Materials Science and Engineering, Kwangju Institute of Science and Technology, 1 Oryong-dong, Buk-gu, Kwangju 500-712, South Korea.

Advanced Macromolecular and Supramolecular Materials and Processes
Edited by K. Geckeler, Kluwer Academic/Plenum Publishers, 2003

All the double bonds in fullerenes are between the 6, 6-ring junctions. The double bonds are exocyclic to the 5-membered rings, i.e., all the 5,6-ring junctions are single bonds, because a double bond on a 5-membered ring is energetically unfavorable. Thus, all the 6-membered rings have three double bonds each and delocalization is unfavorable. On the other hand, each 5-membered ring has to contend with 5-electron each, one short of aromatization, and hence these rings are highly electron-deficient. In C_{60}, for example, the 12 pentagons are arranged in the form of six pyracylene units, and addition of an electron would bring in aromaticity to a pentagonal ring (Figure 2).

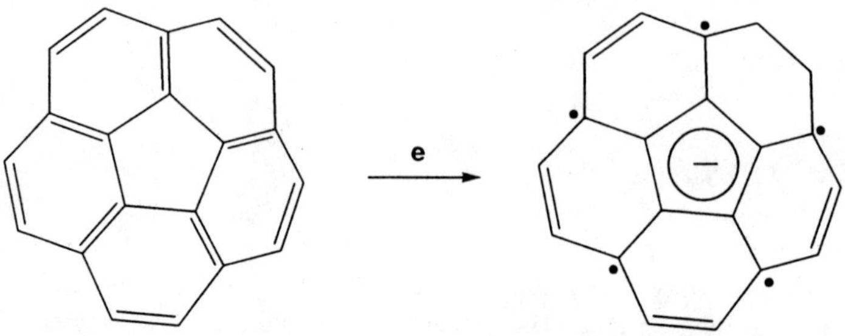

Figure 2. Addition of an electron brings in aromaticity to a pentagonal ring.

Thus, it should be possible to reversibly add six electrons to C_{60}. Hence, the electron-deficient C_{60} surface combined with its symmetrical spherical shape provides the most suitable π-face for uniform interaction in three-dimensions with electron-rich molecules.

The 60 π-electrons of C_{60} are present in 30 bonding molecular orbitals (MO). The highest occupied molecular orbital (HOMO) and the lowest unoccupied molecular orbital (LUMO) are separated by an energy gap of 1.5-2.0 eV. The LUMO is triply degenerated and energetically low lying.

The estimated electron affinity and ionization potential values for C_{60} are 2.7 eV (2.8 eV for C_{70}) and 7.8 eV (7.3 eV for C_{70}), respectively.[1] The electron affinity values indicate that fullerenes should easily accept electrons. All the six successive, fully reversible, one-electron reductions have been observed from cyclic voltametry and differential pulse voltametry studies. The added electrons do not get localized on any specific carbon atom.

In higher fullerenes the structure is built from a higher number of 6-membered rings whereas the 5-membered rings remain at 12. The number of 6-membered rings, m, in a fullerene (C_n) is given by the Euler's theorem, $m = (n - 20)/2$. Thus C_{70} contains 12 pentagons and 25 hexagons. In C_{70} the LUMO is doubly degenerate and the energy separation between LUMO and LUMO+1 is small. For C_{70} also all the six reversible one-electron reductions are observed. Although reductions are relatively readily achieved, oxidations are more difficult.

Cyclic voltametry studies show that C_{60} and C_{70} undergo a single one-electron oxidation at 1.26 and 1.2 V, respectively, versus a ferro-cene/ferrocinium electrode. The electron-deficient nature and lack of delocalization of the π-bonds has led to a variety of chemical reactions and a rich chemistry.[2,3]

The strong electron-deficient character coupled with the three-dimensional shape of the fullerene sphere favors π-π and π-n interactions with electron donors. Due to its three-dimensional architecture, one fullerene sphere can interact with a number of donor molecules. If the donor has a large number of electron-rich centers, even one donor molecule can efficiently interact with a single fullerene molecule. If the electron centers are arranged in the donor molecule in a manner so as to uniformly cover a large portion of the curved exterior of the fullerene sphere, the interaction would be very strong leading to a stable 1:1 complex formation. The electron rich function of the donor tends to position it self so as to remain in close proximity to the pentagons. Also the donor molecule could interact with the fullerene acceptor, if covalently bound to each other, usually through a linker molecule.

These aspects of fullerene science have led to a broad area of supramolecular fullerene chemistry, which is interesting because of the diverse applications of the materials. The present chapter aims at discussing the various known fullerene supramolecular architectures and materials starting from small molecules such as simple aromatic donors to large macrocycles. Specific attention is given to those fullerene macrocycles having interesting applications.

2. FULLERENES AND SMALL MOLECULES

When C_{60} crystallizes from solutions of solvents, the solvents often are incorporated into the crystal as in the case of $C_{60} \cdot 4(C_6H_6)$. Both C_{60} and C_{70} have the remarkable ability to form inclusion complexes with a number of compounds (Table 1).

Table 1. Donor-acceptor complexes of C_{60} with some small molecules

Donor	Acceptor	Structure of Complex	Intermolecular distance (pm)
Hydroquinone (HQ)	C_{60}	$[C_{60}][HQ]_3$	310
Ferrocene (Fc)	C_{60}	$[C_{60}][Fc]_2$	330
Bis(ethylenedithio)tetrathiafulvalene (BEDT-TTF)	C_{60}	$[C_{60}][BEDT-TTF]_2$	345-357

A solution of hydroquinone and C_{60} in hot benzene on evaporation leads to the black $[C_{60}][HQ]_3$ crystal.

The hydroquinone molecules form a three-dimensional network through H-bonding creating a cage structure in the process, the center of which is occupied by C_{60}. Solutions of C_{60} (or C_{70}) and ferrocene deposit black crystals of $[C_{60}][Fc]_2$. Close-packed layers of fullerene molecules are stacked directly one above the other, the Fc molecules filling the voids. The fullerenes make close face-to-face contact with the ferrocene molecules through the five-membered rings of the fullerene. C_{60} could be co-crystallized from a CS_2 solution of bis(ethylenedithio)tetrathiafulvalene (BEDT-TTF), in which a C_{60} molecule is sandwiched by two BEDT-TTF molecules.

The fullerenes C_{60}, C_{70}, and C_{78} have been found to form co-crystals with cyclo-octa-sulfur (S_8). All the three fullerenes and S_8 yield crystals of $C_n \cdot 6(S_8)$. In the solid state one fullerene molecule is surrounded by 14 molecules of $S_8 \cdot C_{60}$ dissolved in toluene cocrystallizes with a CS_2 solution of white phosphorous (P_4). In the mass spectrum of the product, a prominent maximum was noticed for the composition $C_{60} \cdot 32(P_4)$. In these examples, the π-π interaction between donor and acceptor was the driving force for the complex formation. These interactions, however, are noticed in the solid state, though in solution, the interactions disappear.

3. FULLERENE MACROCYCLES

The interaction between the fullerenes and the macyclic compounds, on the other hand, is strong and even exists in solution, often leading to stable complexes. A large variety of macrocycles and fullerenes, bound noncovalently or covalently, have been extensively studied. The following is a concise account of the important class of fullerene-macrocycles.

3.1. Fullerene-Cyclotriveratrylenes

Cyclotriveratrylene (CTV), a nine-carbon macrocyclic compound, forms an inclusion complex with both C_{60} and C_{70}. From a toluene solution of C_{60} and cyclotriveratrylene, crystals of $(C_{60})_{1.5} \cdot (CTV)(PhMe)_{0.5}$ have been isolated (Figure 3).[4]

The macrocycle covers C_{60} in such a manner that three adjacent five membered rings of C_{60} are covered by the electron-rich aryl rings of CTV. The stability of the complex is an outcome of the π − π interactions. This interaction further leads to changes in the absorption spectra of CTV. In solution, the stability of the complexes is is due to *van der Waals* interactions.

Thin films containing fullerenes are interesting owing to the possibility of transferring the interesting fullerene properties to the bulk materials by simple surface coating. However, monolayers of pure C_{60} are difficult to obtain since C_{60} is not an ampiphilic molecule and easily aggregates to form multi-layer films due to its strong hydrophobic interactions.

3.2. Fullerene-Crown Ethers

Stable monolayers can be obtained by embedding the pristine C_{60} in ampiphilic monolayer matrices or by attachment of polar groups to C_{60}.

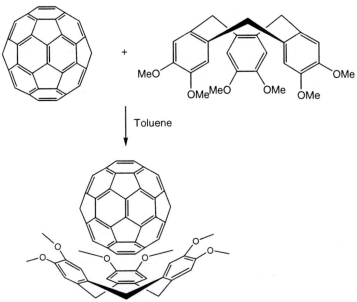

Figure 3. [60]Fullerene-cyclotriveratrylene complex showing a nesting position for the fullerene on the macrocycle.

Wang *et al.* reported the formation and characterization of Langmuir monolayers and Langmuir-Blodgett films of a [60]fullerene crown ether, which is a methanofullerene derivative with a benzo-18-crown-6 derivative.[5]

This derivative forms a thin film at the air/water interaface, especially when K^+ ions are present in the subphase interacting with the crown-moiety.[6] Monolayers of C_{60} derivatives such as studied a pseudo 15-crown-5 macro ring directly attached to C_{60} *via* a bis-aziridine linkage (Figure 4 b) and the subphase containing K^+, Ba^{2+}, and NH_4^+ ions were also studied. The choice of the cations was based on the ability of these cations to interact with the crown-ether moiety. The addition of these ions stabilized the monolayers and increased the limiting area of the monolayer.

In an interesting experiment, molecular recognition of a 18-crown-6 functionalized fullerene was used to induce the formation of molecular monolayers (Figure 5).[7] A gold surface was modified using a thiol-terminated ammonium salt. When the modified gold layer was immersed into a solution of the fullerene derivative in dichloromethane, a surface coverage was obtained by a compact monolayer of C_{60}. The attachment of the crown ether to the gold surface was demonstrated to be reversible.

Wilson and Lu[8] reported the results on a C_{60}-Diels-Alder reaction with a relatively unreactive crown ether diene (Figure 6) and followed the reaction course by the electrospray mass spectrometry method. The reaction was found not to be reversible. At room temperature a monoadduct was obtained and at higher temperature the reaction yielded a 1:1 mixture of the mono- and bis-adduct.

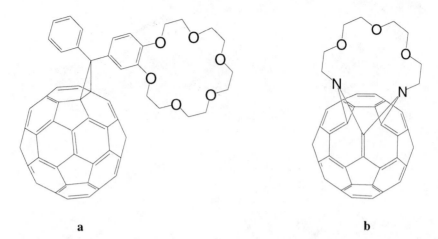

a   b

Figure 4. (a) A [60]fullerene-benzo-18-crown-6 derivative, and (b) a pseudo 15-crown-5 macro ring directly attached to C_{60} *via* a bis-aziridine-like linkage.

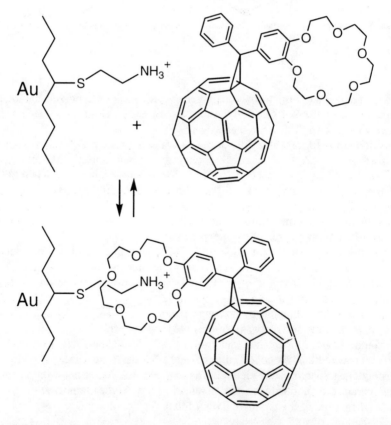

Figure 5. Molecular recognition of a 18-crown-6 functionalized fullerene on a gold surface modified using a thiol-terminated ammonium salt.

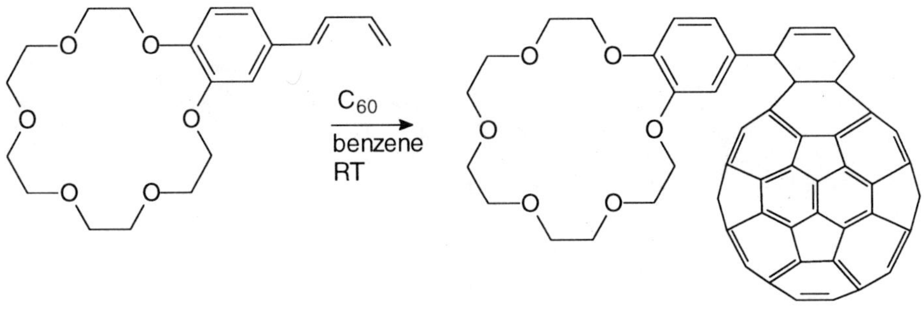

Figure 6. Crown-ether derivative of C_{60} obtained by a Diels-Alder reaction.

The crown ether can also be precisely positioned in close proximity to the fullerene surface (Figure 7).[9] In these molecules, when alkali metals are complexed in the crown ether moiety, which is positioned closely and tightly on the fullerene surface, a significant perturbation of the electronic structure of the fullerene is observed, indicating that placing the macrocycle at a precise position on the fullerene surface lead to more effective interaction between the metal ion and the fullerene sphere.

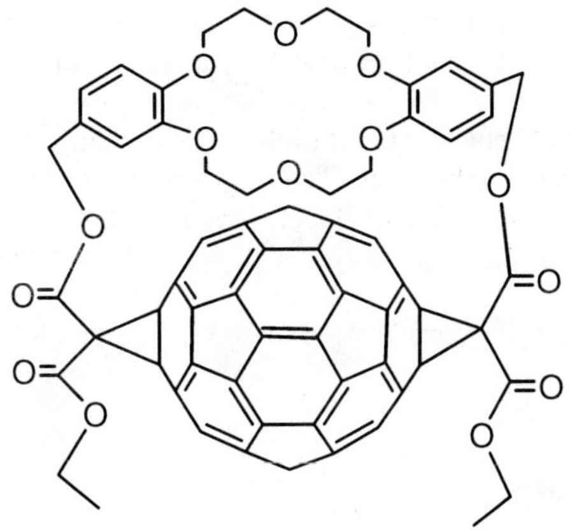

Figure 7. Precise positioning of the crown ether above the fullerene sphere.

In order to solve the aggregation problem in monolayer formation, a series of amphiphilic, covalent C_{60} derivatives such as the fullerene crown ether conjugate and the fullerene-cryptate conjugate were prepared (Figure 8).[10]

The compounds should lead to monomolecular layer formation at the air-water interface. However, the compression-expansion cycles were irreversible, indicating that aggregation leading to multi-layer formation could not be avoided. It is likely that the hydrophobic head groups were not bulky enough to prevent aggregation of the carbon spheres.

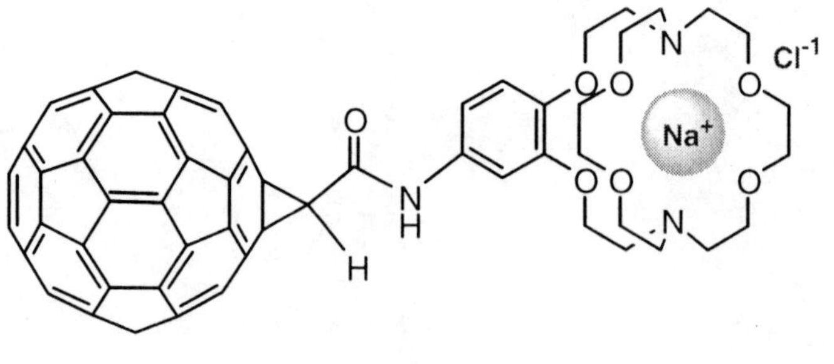

Figure 8. A fullerene-cryptate conjugate.

Therefore, the synthesis of amphiphilic fullerene derivatives with hydrophilic head groups, large enough to keep the fullerene units apart in the absence of pressure, were targeted. In order to prevent fullerene aggregation, carbohydrate- containing dendrons as bulky hydrophilic head groups were attached to the carbon cores and the resulting fullerene glycodendron conjugates were shown to give layers at the air-water interface that are monomolecular, stable, and reversible.[11]

Another possible approach to avoid the aggregation of the fullerene moieties at the air-water interface is to surround the carbon spheres by a hydrophobic environment, which will prevent a close contact among the fullerene molecules. Mono- and multilayers of 1:1 mixtures of C_{60} and C_{70} and the ampiphilic azacrown derivatives were spread on the air-water interface. The binding between the fullerene molecule and the azacrown was a noncovalent interaction. The macrocycle and the fullerenes could stay in close proximity because the substituents on the nitrogen atoms wrapped around the fullerenes like lipophilic arms, thus creating a hydrophobic envelope around the fullerene sphere (Figure 9).[12]

The monolayers could be transferred onto solid supports by the Langmuir-Blodgett technique.

3.3. Fullerene-Calixarenes

Calix[n]arenes are macrocyclic arrays of aromatic rings, where 'n' represents the number of aromatic rings (Figure 10). These rings often carry substitutents and the phenolic groups at the lower rim form a circular H-bonding network giving the molecule a typical cone shape with a hydrophobic cavity.

The groups of Atwood[13] and Shinkai[14] independently reported that the *p-tert*-butyl-calix[8]arene forms a complex with C_{60}. Most of the reported complexes are insoluble in water.[15,16] However, a water-soluble calix[8]arene with a hydrophilic $-SO_3^- Na^+$ function was reported to form a water-soluble complex.[17] Atwood et al. demonstrated that the

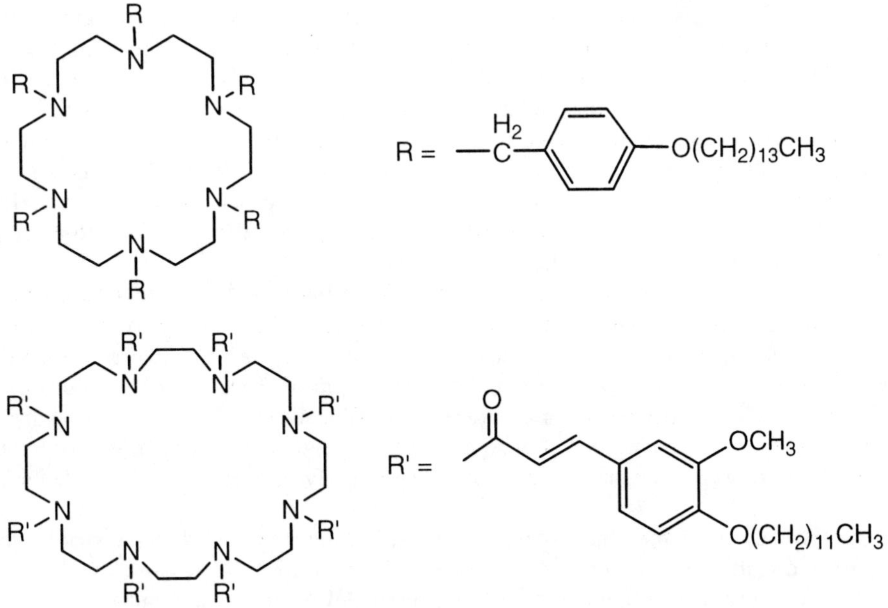

R =
H_2
C
O(CH2)13CH3
R' =
O
OCH3
O(CH2)11CH3

Figure 9. Azacrown ethers with lipophilic arms.

R
OX
n
But
tButBu
tButBu
tBu

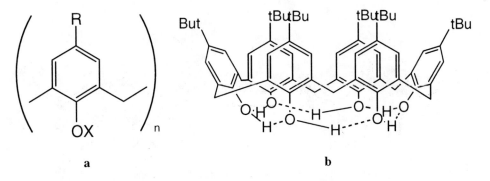

a

b

Figure 10. (a) General representation of a calix[n]arene, and (b) the conical shape of the molecule showing the OH groups forming a H-bond belt at the lower rim.

reaction of toluene solutions of C_{60} with *p-tert*-butylcalix[8]arene gave a sparingly soluble yellow-brown precipitate which was identified as a 1:1 complex.

When carbon soot was allowed to react with *p-tert*-butylcalix[8]arene, the crude product consisted of 89% C_{60} and 11% C_{70} complex. This corresponds to a 90% extraction of the C_{60} content of the soot. It has been believed that the origin of this selective inclusion stems from the conformity of the C_{60} size with the calix[8]arene cavity. Treatment of this *p-tert*-butylcalix[8]arene:fullerene complex with chloroform dissolved the host that led the fullerenes to precipitate. This paved the way for a substantial enrichment of the C_{60} content over the C_{70} in the fullerene mixture and a repetition of the process led to the effective separation C_{60} from C_{70}. Shinkai and coworkers also described the selectivity of the calix[8]arene derivatives over calix[6]-arene and calix[4]arene. They could isolate 102 mg of 99.8% pure C_{60} from 200 mg of carbon soot containing 72 wt% of C_{60}.

The nature of bonding between the guest and the fullerene host appear to originate from the π-π interaction between of the aromatic surface of the host and the π-surface of the guest. ^{13}C-NMR spectroscopic analysis of the 1:1 *p-tert*-butylcalix[8]arene-C_{60} complex revealed a significant up-field shift of 1.4 ppm of the C_{60} carbon resonance.[18] This shift was accompanied by a sharpening of all calixarene signals, suggesting a complex-induced conformational change of the calixarene. The hydrogen bonds among the OH groups in the lower rim of the cone were slightly weakened, as concluded from IR spectra.

X-ray analysis of the calix[8]arene-fullerene crystal revealed that the macrocycle adopted a double cone conformation to host two fullerene balls (Figure 11).[19] In the double cone conformation, the hydrogen bonds in the lower rim remain intact.

Raston *et al.*[20] showed that the 1:1 complexes between *p-tert*-butylcalix[8]arene and C_{60} is micelle-like with trimeric aggregates of fullerenes as the core surrounded by three host molecules, each in double cone conformation and spanning two fullerenes along a triangular edge (Figure 12).

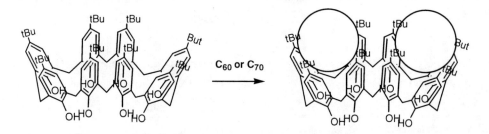

Figure 11. A double cone conformation of calix[8]arene.

The *p-tert*-butylcalix[8]arene-C_{60} complex breaks into its components in solution; no spectroscopic indication for the existence of a complex could be found. Shinkai and coworkers[21,22] introduced electron-rich aniline and 1,3-diaminobenzene derivatives into calix[8]arene, expecting that the macrocycle would create a new C_{60} receptor.

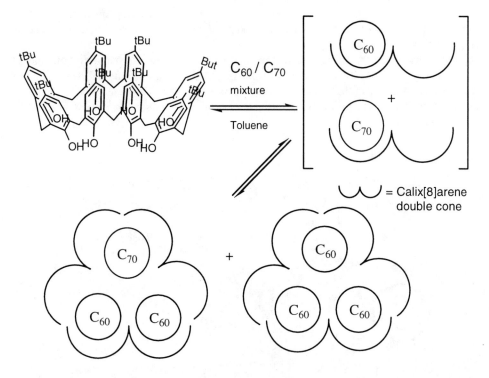

Figure 12. Trimeric aggregate of fullerenes as the core surrounded by three host molecules, each in double cone conformation.

However, this compound did not give any indication of complex formation. They demonstrated that such calix[n]arenes that have a conformation and a proper inclination of the benzene rings can interact with C_{60} even in solution. After examination of 28 calixarene derivatives, the authors eventually discovered that calix[5]arene, homocalix[3]arene, and to some extent calix[6]arene do interact with C_{60} in toluene. This means that deep inclusion of C_{60} into calix[n]arene cavity, as expected for calix[8]arenes, is not a prerequisite for C_{60} inclusion.

Fullerenes with well-characterized ionophoric cavities interact with C_{60} in the solid state and in solution. Highly ion-selective and highly ionophoric cavities can be created by the appropriate modification of the lower rim OH groups while the upper rim π-basic cavities of certain calixarenes can accept [60]fullerenes even in solution. It has been demonstrated that certain metal cations, which appropriately preorganize the calix[n]aryl ester derivatives into their cone conformers, can alter them into excellent [60]fullerene receptors in solution (Figure 13).[23]

Conformational freedom remaining in calix[n]aryl esters can be frozen to cone conformation by complexation with appropriate metal cations. The authors used a calix[6]aryl ester and observed that the conformationally-mobile ester derivative did not react with C_{60}.

The phenol units in the calix[6]aryl ester are forced to stand up by introduction of bulky ester groups to OH groups, which makes the cavity edge smaller and the benzene ring inclination improper.

However, when the calixarene was reacted with a Cs^+ salt, the metal complex could easily include C_{60} into the cavity. When a large metal cation, such as Cs^+ was added, it formed a complex with the phenolic oxygen converting the calixarene to a symmetrical cone conformation. The Cs^+-binding to the lower rim facilitates the [60]fullerene-binding to the upper rim.

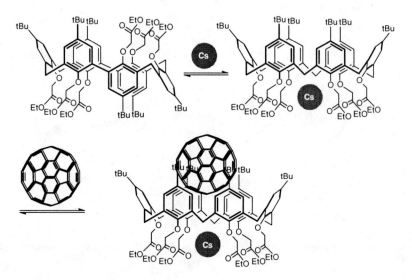

Figure 13. A Cs^+ ion preorganizes appropriately the calix[n]aryl ester derivatives into their cone conformers, altering them into excellent [60]fullerene receptors.

Fukazawa and coworkers linked two calix[5]arene macrocycles.[24] The bridged calix[5]arene macrocycle led to a shape-selective receptor with a well-defined cavity size. A complex with C_{60} was formed as well as with C_{70} in toluene with very high association constants (K_{ass} 76,000 M^{-1} for C_{60} and 163,000 M^{-1} for C_{70}, 298 K).

Noncovalently-bound calixarene-fullerene complexes suffer from the disadvantage of low stability in solution, and hence the interactions between the two molecules could not be fully understood. Therefore, a number of covalently-bound calixarene-fullerenes were synthesized. C_{60}-functionalized calix[8]arene connected by a polyether chain has been designed (Figure 14).[25]

An azido-terminated polyether chain attached to the rim of the calix[8]arene was reacted with C_{60} leading to the formation of the fulleroid. From a series of UV-Vis and fluorescence studies, the authors proved that the covalently-bound fullerene still interacts with the calixarene in solution.

In another approach, two calixarene-fullerenes were synthesized, one of which had two fullerene rings bonded to the lower rim of the same calixarene (Figure 15).[26]

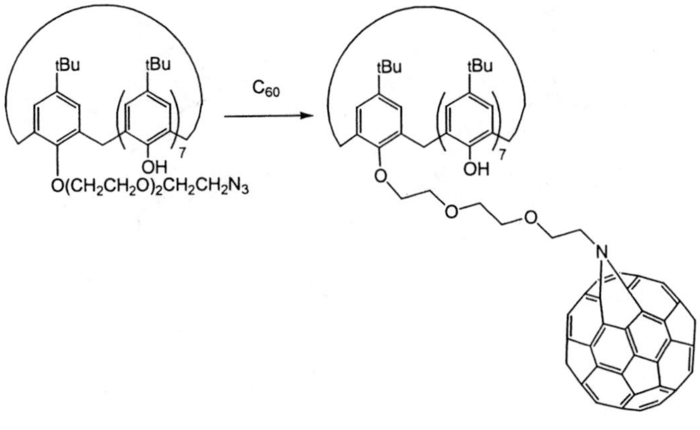

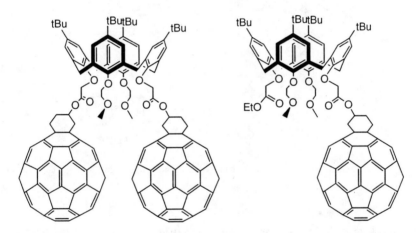

Figure 14. A covalently linked fullerene-calixarene *via* a polyether chain.

Figure 15. Two calixarene-fullerenes, one of which has two fullerenes bonded to the lower rim of the same calixarene.

The one in which there is one fullerene was expected to aggregate less strongly in polar solvents than the one in which there are two fullerenes in close proximity. From light scattering measurement, the authors observed that in chloroform both compounds did not aggregate, whereas in chloroform-acetonitrile and chloroform-methanol the compound with two fullerenes tends to aggregate strongly with particle diameters of 240-390 nm. No aggregation was observed for compound with a single fullerene.

Also the synthesis of fullerenocalixarenes, in which the calix[4]arene and [60]fullerene were included in the same ionophoric ring, was reported (Figure 16).[27] Two novel compounds, in which C_{60} is linked to a cone-calix[4]arene or to a 1,3-alternate-calix[4]arene, were synthesized. Fullerene was used as a 'lid' for the ionophoric cavity.

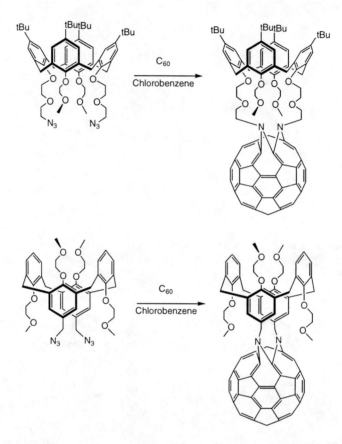

Figure 16. Fullerenocalixarenes, in which the calix[4]arene and [60]fullerene were included in the same ionophoric ring.

Several metal cations (Li⁺, Na⁺, Ag⁺) added from outside influenced the electronic spectra of C_{60} in cone-calix[4]arene, indicating the metal cation bound therein interact with the [60]fullerene surface. On the other hand, no spectral change could be noticed when 1,3-alternate-calix[4]arene was similarly treated with the metal ions.

3.4. Fullerene-Prophyrins

[60]Fullerene has a rich electronic and electrochemical behavior for which it has been considered as the ideal molecule in photo-induced processes.[28-30] The absorption and electron-accepting properties of C_{60} and the properties of electro- and photoactive molecules can be combined by covalently linking these two types of molecules. The objective is to synthesize donor-acceptor (D-A) systems with long-lived charge-separated state for several applications such as artificial photosynthesis or conversion of solar light to electric current.

Many classes of donor units such as ferrocene, electron-rich aromatics, ruthenium(III) complex, tetrathiafulvalene, benzoquinone have been attached to C_{60}. The porphyrins are ideal donors, and hence a number of molecules have been synthesized to covalently link the fullerene sphere with the macrocycle.[31-36]

A fullerene-porphyrin conjugate, covalently linked *via* a pyrrolidine linkage, was prepared by treating β-formyltetraphenyl porphyrin with C_{60} and $MeNHCH_2COOH$ (Figure 17).[37] The Ni(II) complex was similarly prepared from the Ni(II) complex of the porphyrin. A porphyrin-linked fullerene, where a C_{60} moiety is covalently linked to the meso position of 5,15-diarylporphyrin with a pyrrolidine spacer, was prepared.[38] Sun *et al.*[39] investigated three pyrrolidine-functionalized C_{60}. A simple monomeric methyl-pyrrolidine derivative, a pyrrolidine-linked tetraphenyl porphyrin-C_{60} dyad, and a pair of bifullerenes have been investigated.

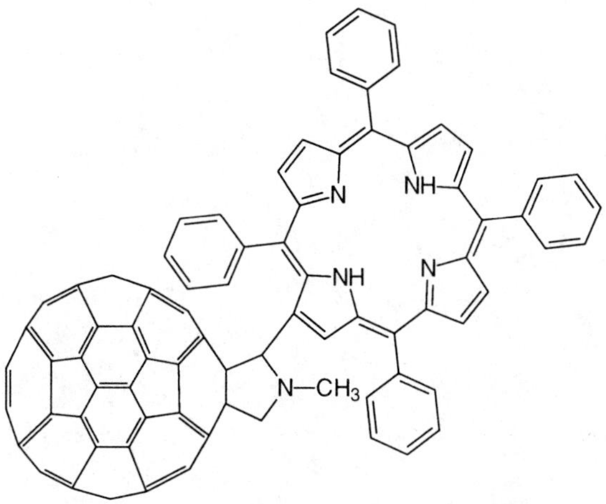

Figure 17. A fullerene-porphyrin conjugate, covalently linked *via* a pyrrolidine linker.

A Diels-Alder reaction has been used to prepare porphyrin-C_{60} dyads.[40] In addition to their potential activity as photosynthetic model systems, the compounds are potential anti-tumor agents due to their activity as photosensitizers for the generation of singlet oxygen. The first two of such Diels-Alder-derived pophyrin-C_{60} dyads are shown in Figure 18.[41,42]

C_{60} can accelerate photo-induced charge separation and decelerate charge recombination. The extent and rate of recombination could be decreased by modulating the separation or spatial orientation of the donor and acceptor by means of flexible or rigid bridges. By changing the bond connectivity in their spacer, Imahori and coworkers systematically changed the spacing and relative orientation of the porphyrin and fullerene moieties in dyads (Figure 19).[43]

In these and related compounds, a 3,5-di-tert-butylphenyl group is used instead of a simple phenyl substituent on the porphyrin to improve solubility characteristics.

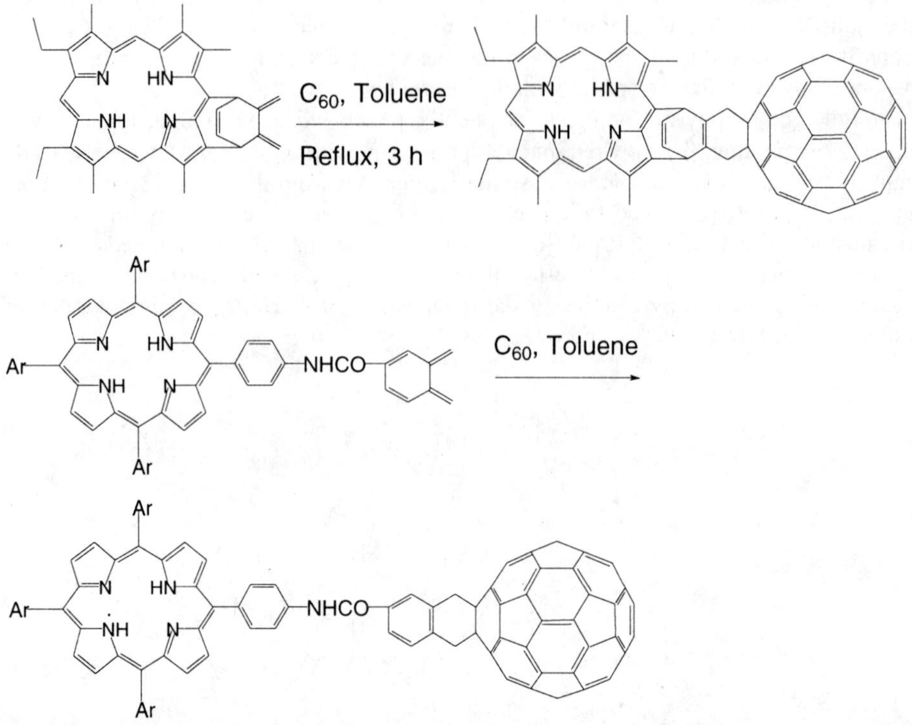

Figure 18. Porphyrin-C$_{60}$ dyads prepared by Diels-Alder reactions.

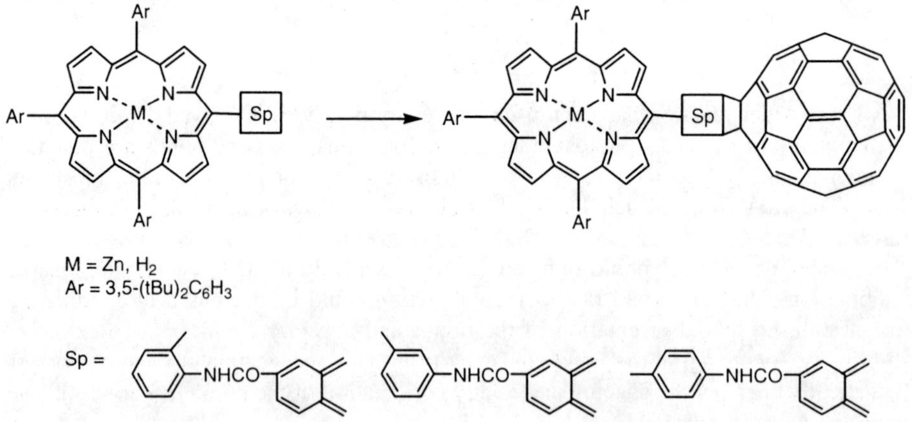

M = Zn, H$_2$
Ar = 3,5-(tBu)$_2$C$_6$H$_3$

Sp =

Figure 19. Spacing and relative orientation of the porphyrin and fullerene moieties in dyads.

The ability of these dyads, as well as related porphyrin-substituted fullero-pyrrolidines to undergo photo-induced intermolecular electron transfer has been examined as a function of spatial disposition of the two chromophores. It is clear that such dyads prefer to adopt a conformation in which the porphyrin and fullerene moieties are as close as possible in space as allowed by the structural constraints of the spacer (Figure 20).[41-43]

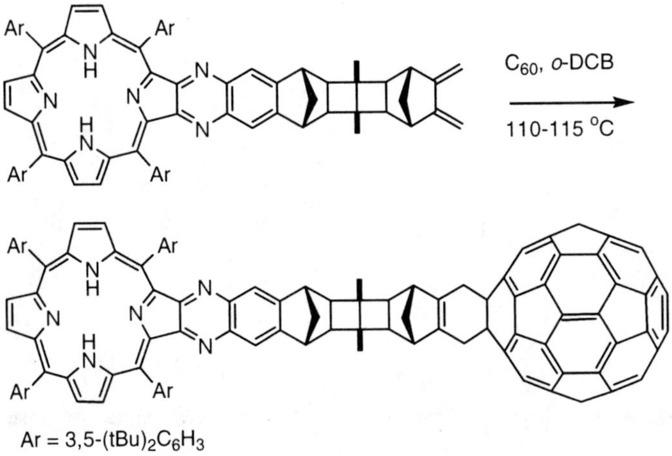

Ar = 3,5-(tBu)$_2$C$_6$H$_3$

Figure 20. Synthesis of fullerene-porphyrin dyads; o-DCB = o-dichlorobenzene.

A few triads have been synthesized and studied in terms of electron and energy transfer (Figure 21).[44,45] The results have confirmed the ability of C$_{60}$ to slow down back-electron transfer in the charge separated states.

An interesting case of a C$_{60}$-porphyrin interaction was reported by Diederich and coworkers, in which the macrocycle is not covalently bound to C$_{60}$ leading to a true supramolecular dyad (Figure 22).[46] In this approach, ligation was used to associate the donor and the acceptor. A pyridine-functionalized fullero-pyrrolidone was linked to the zinc(II) ion of a tetraphenylphorphyrin by coordination with high affinity.

The macrocyclization between C$_{60}$ and bismalonate derivatives provides the most versatile and simple method for the preparation of covalent bisadducts of C$_{60}$ with high regio- and diastereoselectivity. Using this principle it was possible to synthesize a product in which the spatial position of the porphyrin ring is fixed above the fullerene (Figure 23).

In a similar approach, a molecule with four appended fullerenes was synthesized (Figure 24).[47] The organic chromophore could be precisely positioned in close proximity to the fullerene surface.

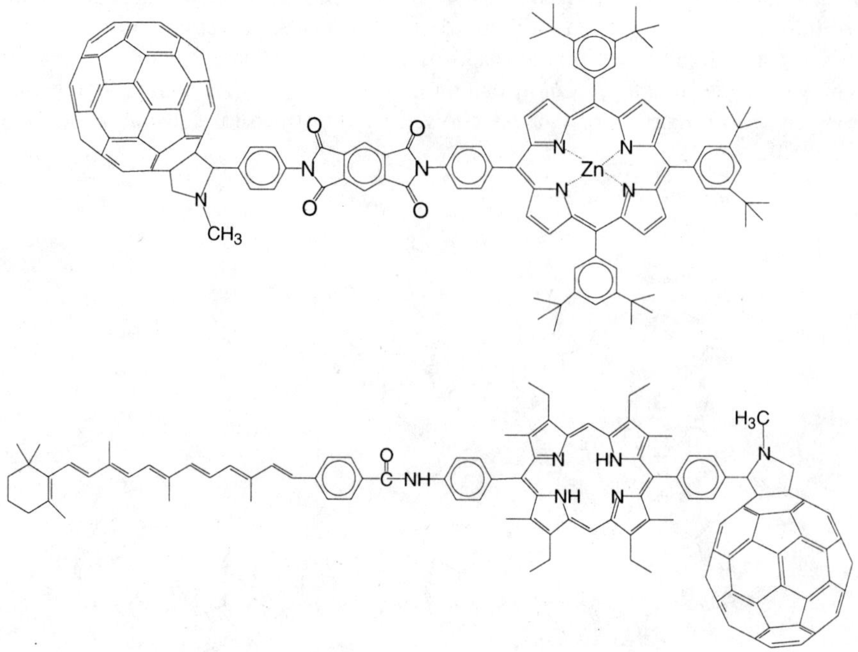

Figure 21. Synthesis of triads to slow down the ability of C$_{60}$ back-electron transfer in the charge-separated states.

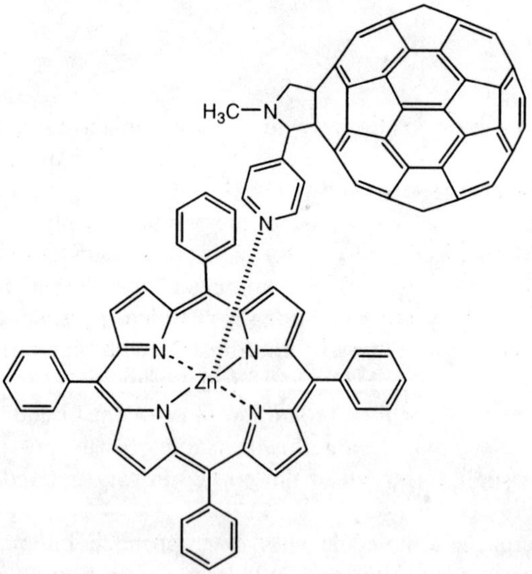

Figure 22. A supramolecular dyad in which ligation was used to associate the donor and the acceptor.

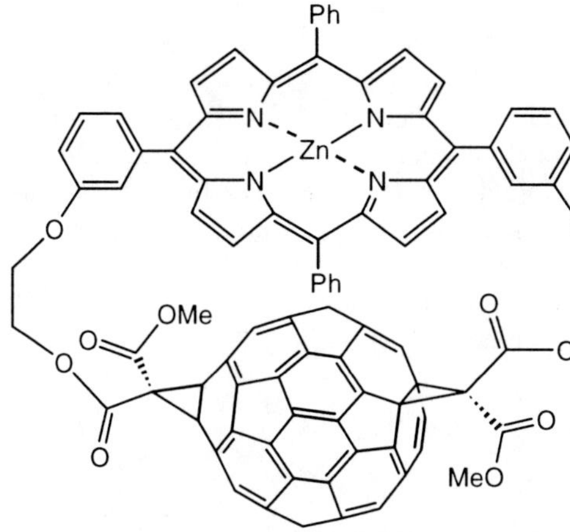

Figure 23. Noncovalent interaction between fullerene and porphyrin, in which the porphyrin is at a fixed distance in space from the fullerene sphere.

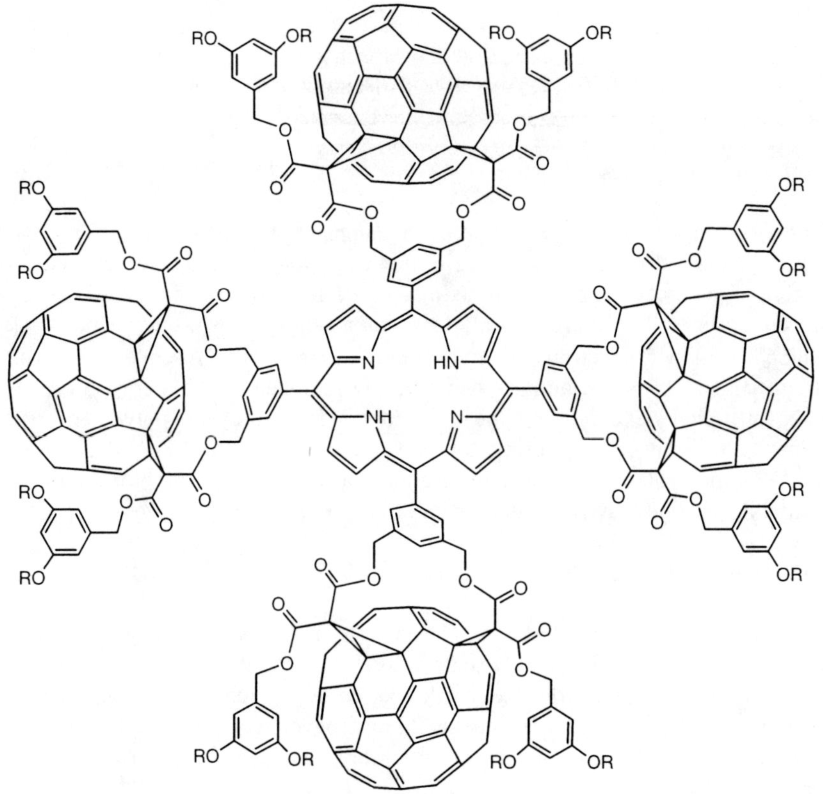

Figure 24. Spatial disposition of a porphyrin appended to four fullerene molecules.

3.5. Fullerene-Phthalocyanines

The synthesis and electrochemistry of a Diels-Alder adduct of C_{60} with a nickel phthalocyanine was described_(Figure 25).[48] The nickel phthalocyanine was synthesized from 3,6-diheptylphthalodinitrile and 1,2,3,4-tetrahydro-2,3-dimethyl ene-1,4-epoxy-naphthalene-6,7-dicarbonitrile to react with nickel acetate in the presence of a catalytic amount of 1,8-diazabicyclo[5.4.0]undec-7-ene. The fullerene-phthalocyanine could be an interesting molecule, because the phthalocyanines themselves are known for their remarkable electronic and optical behavior. From cyclovoltammetry studies it was found that the reduction of the fullerene moiety has a pronounced influence on the optical properties of the phthalocyanine.

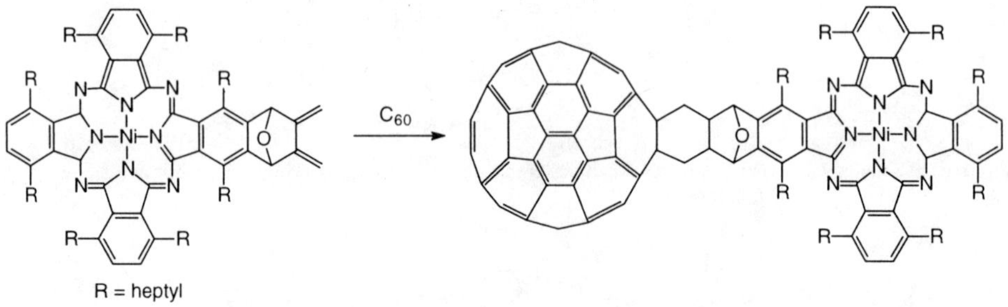

R = heptyl

Figure 25. A Diels-Alder adduct of C_{60} with a nickel phthalocyanine.

3.6. Fullerene-Catenanes

The first report of a C_{60} containing a catenane was reported by Diederich and coworkers (Figure 26).[49] Based on a regio- and diastereoselective bisfunctionalization on cyclization of the carbon sphere with bis(malonate) derivatives, the method was extended for the synthesis of the fullerene-containing catenane. The bis(p-phenylene)-[34]crown-10, which has two π–electron-rich hydroquinone rings, was fused to C_{60} in a regio-selective manner. Then the authors describe the template-directed formation of cyclo-bis(paraquat-p-phenylene) around the C_{60}-appended macrocyclic polyether to give the [2]catenane, a compound featuring an unusual topology. An ^{1}H-NMR investigation indicated the molecule to be an intramolecular stacks of donor-acceptor-donor-acceptor, and spatial interaction between fullerene and bipyridinium units.

3.7. Fullerene-Rotaxanes

Diederich *et al.* reported the synthesis of a rotaxane assembled *via* the Cu(I)-templated approach, bearing two C_{60} units as stoppers (Figure 27).[50] The synthesis of this novel structure was based on the oxidative coupling reaction of terminal alkynes functionalized on C_{60} and a Cu(I) rotaxane, a macrocycle containing a 1,10-phenan-thro-line-coordinating ring. The molecular assembly displays interesting photo-physical properties.

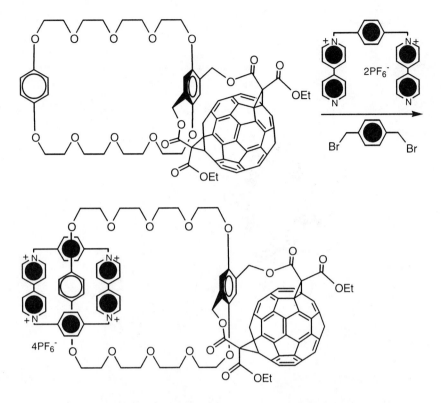

Figure 26. A catenane containing C$_{60}$ eading to inter-locking molecules.

3.8. Fullerene-Cyclophanes

An structure built out of a [2.2]cyclophane and C$_{60}$ was reported (Figure 28).[51] The [2.2]cyclophane had a diazoalkane moiety, through which reaction took place at the [6,6]-ring junction, leading to a methanofullerene derivative. From UV-vis spectra, the authors concluded an intramolecular charge transfer interaction between the cyclophane and fullerene.

3.9. Fullerene-Cyclodextrins

The preparation of a water-soluble C$_{60}$ complex has been achieved by generating a C$_{60}$/γ-cyclodextrin (γ-CD) host-guest complex, where the fullerene core is embedded between the cavities of two γ-CD molecules. With a boiling aqueous solution of γ-CD it was possible to extract C$_{60}$ from a mixture of C$_{60}$ and C$_{70}$ in solid state. However, in similar experiments α- and β-CD could not extract C$_{60}$. This was ascribed to the lower cavity diameter of α- and β-CD as compared to γ-CD. In this complexation process, n-π donor-acceptor interactions between the n-orbitals of the sugar O-atoms and the π-system of the fullerene contribute to the stability of the complex.[52]

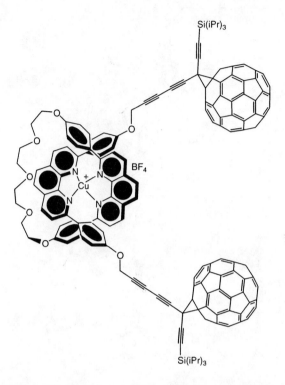

Figure 27. A rotaxane assembled *via* the Cu(I)-templated approach, bearing two C_{60} un
its as stoppers.

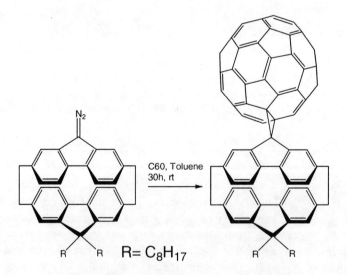

Figure 28. Covalent binding between a cyclophane and C_{60}.

In an extraction process involving γ-CD and a C_{60}-C_{70} solid mixture, the dissolution of C_{60} in the aqueous medium must be a consequence of matching sizes. Still the cone-shaped cavity of γ-CD ranging from 0.75 – 0.83 nm, and the cavity depth of 0.8 nm could not completely encapsulate the C_{60} ball of an der Waals diameter of ~ 1 nm. Hence a complex with a 2:1 (γ-CD:C_{60}) stoichiometry was proposed to explain the selective extraction of C_{60}. In Laser excitation (7 ns pulse, λ_{max} = 355 nm), the quenching rate of the transient fullerene state by dioxygen was slower than for pristine C_{60} by a factor of 2 indicating that γ-CD is covering a large portion of the carbon sphere in the complex.[53]

A recent report showed that ß-CD does indeed form an inclusion complex of C_{60}, when a reaction was carried out between the macrocycle dissolved in DMF and C_{60} in toluene.[54] Presumably, in such complexation processes, the polar solvent molecules play a definite role for complete encapsulation of the C_{60} in aqueous media.

Although the reactivity of the fullerene in this complex is noticeably lower than in the unprotected form, γ-CD/C_{60} is still susceptible for easy reduction by radicals generated in the aqueous phase.[55] Although the fullerene moiety is embedded between two host molecules of γ-CD, the formation of donor-acceptor complexes with electron-rich amines, such as triethylamine, pyridine, hexamethylene tetramine, acrylamide, thioglycolic acid and 2,2'-thiodiethanol could still be observed.[56]

Most of the fullerene-macrocycles reported, however, were insoluble in water (Table 2), restricting their applications in biological and biomedical fields.[57] The biomedical application of water-soluble γ-CD-C_{60} complex has been scarcely explored[58], presumably because of the low fullerene content of the fullerene in solution and the stability concerns of such complexes in different bioenvironments. This limitation can be circumvented if the cyclodextrin and fullerene could be covalently attached.

The first synthesis of a recently water-soluble covalently-bound fullerene-cyclo-dextrin conjugate was recently reported (Figure 29).[59] The experiment involved converting the cyclodextrin monotosylate to a monoamine derivative reacting the tosylate with a number of aliphatic and aromatic diamines. The aminocyclodextrins were reacted with C_{60} following a titration technique,[60] fullerene could be monofunctionalized.

In the synthesis of fullerenols, or fullerene carboxylates, at least twelve OH or six COOH groups, respectively, need to be attached to the C_{60} for the products to be water-soluble. In this process, a number of π-bonds of C_{60} need to be broken. It is now known that the larger the number of C=C bonds retained in the functionalized water-soluble fullerene, the better is its free radical scavenging property. This property is of vital importance in evaluating the application potential of the water-soluble fullerene derivative in the fields of biology and medicine. In the covalently bound monocyclodextrin-fullerene conjugate only one π-bond of fullerene moiety has been used up. Further, cyclodextrins being bioamenable, thus the product has an edge over other monofunctional water-soluble fullerenes.

The synthesis of water-soluble cyclodextrin-fullerene conjugates has been further amended when the linking molecule between the macrocycle and the fullerene moiety has been completely eliminated. By using azidocyclodextrins in place of amino cyclodextrins, it was possible to attach cyclodextrin with fullerene *via* a bridging nitrogen atom.

TABLE 2. Important macrocycles used in covalent linking to fullerenes

Macrocycle		Macrofullerene	
Name	Structure	Water-solubility	Year reported
Crown ether		–	1993
Calixarene		–	1994
Porphyrin		–	1995
Phthalocyanine		–	1995
Catenane		–	1997
Fluorenophane		–	1999
Cyclodextrin		+	2000

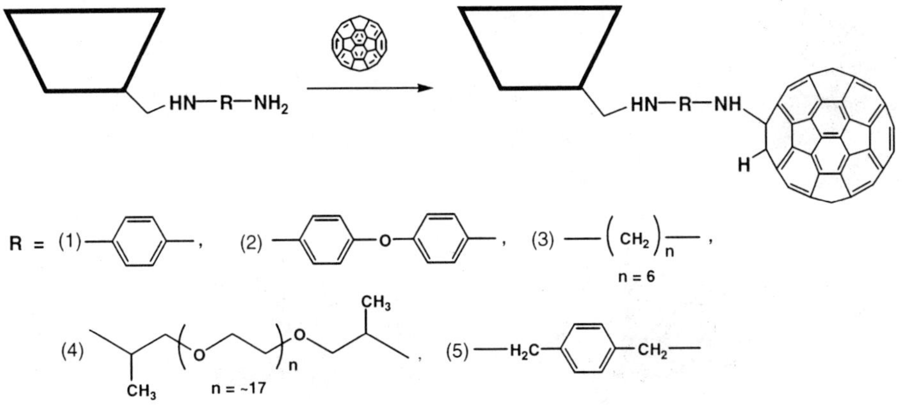

Figure 29. Water-soluble covalently-bound fullerene-cyclodextrin conjugates.

The cyclodextrinazafullerenes were highly soluble in water. The extent of aqueous solubility was again dependent on the solubility of the macrocycle. Since α-CD is much more soluble in water than β-CD, the solubility of α-cyclodextrin-azafullerene was much higher when compared to the product from the β-CD.

To prove whether these cyclodextrin-fullerene conjugates are potential candidates for biomedical applications, their radical scavenging property was evaluated using a simple experiment between the fullerene derivative and the stable free radical 1,1'-diphenyl-2-picryl hydrazil (DPPH), and monitoring the reaction progress by UV-Vis spectroscopy. A control reaction was also carried out between C_{60} dissolved in dichloromethane and DPPH in ethanol.

This color reaction, as a manifestation of the fullerene core property, was conclusively demonstrated recently.[61] Water-soluble fullerene cyclodextrins were reacted with DPPH, and it was observed that the materials strongly scavenge the living free radical, indicating that the fullerene component is as active as the pristine fullerene itself.

In another novel approach, we synthesized several inclusion complexes of aliphatic and aromatic diamines and cyclodextrins, and used these complexes for a polyreaction between the complex and fullerene (Figure 30).[62]

The resulting poly(fullerocyclodextrins) had no direct covalent bond between the macrocycle and the fullerene moieties. The diamine, which is the linking molecule between the fullerenes, is masked by the macrocycle, and in the process, the resulting polyfullerene was highly water-soluble. This was the first approach towards a water-soluble fullerene main-chain polymer, an important class of the macromolecular fullerene derivatives. The poly(fullerocyclodextrin) could scavenge DPPH even more strongly than C_{60}, itself, indicating its good application potential.

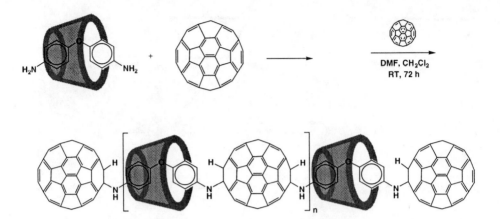

Figure 30. The polyfullerocyclodextrin, the first water-soluble fullerene main-chain polymer.

Both the fullerene-cyclodextrins and the poly(fullerocyclodextrin)s were also proved to very efficiently cleave a DNA oligonucleotine in presence of light. Thus, a small quantity of a poly(fullerocyclodextrin) could cleave into half above 80% of a large quantity of the DNA. [63] This means that the materials could be used in diverse applications such as photodynamic cancer therapy.

Another advantage of the fullerene-cyclodextrin conjugates over other water-soluble fullerenes is that the macrocycle has its cavity free (see Figure 31), thus being able to accomodate useful molecules such as drugs. To prove this, we monitored a complexation reaction between a fullerene-cyclodextrin conjugate and 4-nitrophenol. A control reaction was carried out with cyclodextrin. It was found that the complexation ability of the macrocycle component in the fullerene derivative is fully retained.

Figure 31. Space-filling models of fullerene-cyclodextrin conjugate, one structure showing the vacant macrocycle cavity for complexation of guest molecules.

Since it is known that water-soluble fullerenes are absorbed preferentially in cancer bearing tissue more effectively than in normal tissue, the macrocycle covalently bound to the fullerene can be used to carry a drug in addition. Thus, while the fullerene component acts as an anti-tumor agent in the presence of light, the included guest in the macrocycle cavity would provide an additional source for a known drug.

4. ACKNOWLEDGEMENTS

The authors gratefully acknowledge the generous financial support from the Korean Federation of Science and Technology, the Ministry of Science and Technology, the Ministry of Education, and the Kwangju Institute of Science and Technology.

5. REFERENCES

1. L. Echegoyen, F. Diederich, and L. E. Echegoyen, in *Fullerenes: Chemistry, Physics and Technology,* Eds. K. M. Kadish and R. S. Ruoff, Wiley-Interscience, New York, 1-52 (2000).
2. K. E. Geckeler, *Trends Polym. Sci.* **2**, 355 (1994).
3. S. Samal and K. E. Geckeler, *in Advanced Functional Molecules and Polymers*, Ed. H. S. Nalwa, Gordon and Breach Publishing, Vol. 1, Chapter 1, 1-85 (2001).
4. J. W. Steed, P. C. Junk, J. L. Atwood, M. J. Barnes, C. L. Raston, and R. S. Bulkhalter, *J. Am. Chem. Soc.* **116**, 10346 (1994).
5. S. Wang, R. M. Leblan, F. Arias, and I. Echegoyen, *Langmuir* **13**, 1672 (1997).
6. C. J. Pedersen and H. K. Freosdorff, *Angew. Chem., Int. Ed. Engl.* **11**, 16 (1972).
7. F. Arias, L. A. Godinez, S. R. Wilson, A. E. Kaifer, and L. Echegoyen, *J. Am. Chem. Soc.* **118**, 6086 (1996).
8. S. R. Wilson and Q. Lu, *Tetrahedron Lett.* **34**, 8043 (1993).
9. F. Diederich and R. Kessinger, *Acc. Chem. Res.* **32**, 537 (1999).
10. U. Jonas, F. Cardullo, P. Belik, F. Diederich, A. Gügel, E. Hart, A. Hermann, L. Isaacs, K. Müllen, H. Ringsdorf, C. Thilgen, P. Uhlmann, A. Vasella, C. A. A. Waldraff, and M. Walter, *Chem. Eur. J.* **1**, 243 (1995).
11. F. Cardullo, F. Diederich, L. Echegoyen, T. Habicher, N. Jayaraman, R. M. Leblane, J. F. Stoddart, and S. Wang, *Langmuir* **14**, 1995 (1998).
12. F. Diederich, J. Effing, U. Jonas, L. Jullien, T. Plesnivy, H. Ringsdorf, C. Thilgen, and D. Weinstein, *Angew. Chem., Int. Ed. Engl.* **31**, 1599 (1992).
13. J. L. Atwood, G. A. Koutsantonis, and C. L. Raston, *Nature* **368**, 229 (1994).
14. T. Suzuki, K. Nakashima, and S. Shinkai, *Chem. Lett.* 699 (1994).
15. E. C. Constable, *Angew. Chem., Int. Ed. Engl.* **33**, 3269 (1994).
16. S. Shinkai and A. Ikeda, *Gazz. Chim. Ital.* **127**, 657 (1997).
17. R. M. Williams and J. W. Verhoeven, *Recl. Trav, Chim. Pays-Bas* **111**, 531 (1992).
18. R. M. Williams, J. M. Zwier, J. W. Verhoeven, G. H. Nachtegaal, and A. P. M. Kentgens, *J. Am. Chem. Soc.* **116**, 6965 (1994)
19. J. L. Atwood, L. J. Barbour, C. L. Raston, and I. B. N. Sudria, *Angew. Chem., Int. Ed. Engl.* **37**, 981 (1998).
20. C. L. Raston, J. L. Atwood, B. J. Nichols, and I. B. N. Sudria, *Chem. Commun.* 2615 (1996).
21. K. Araki, K. Akao, A. Ikeda, T. Suzuki, and S. Shinkai, *Tetrahedron Lett.* **37**, 73 (1996).
22. A. Ikeda, M. Yoshimura, and S. Shinkai, *Tetrahedron Lett.* **38**, 2107 (1997).
23. A. Ikeda, Y. Suzuki, M. Yoshimura, and S. Shinkai, *Tetrahedron* **54**, 2497 (1998).
24. T. Haino, M. Yanase, and Y. Fukazawa, *Angew. Chem,. Int. Ed. Engl.* **36**, 259 (1997).
25. S. Shinkai, H. Adams, and J. M. Stirling, *J. Chem, Soc., Chem. Commn.* 2527 (1994).
26. A. Ikeda and S. Shinkai, *Chem. Lett.* 803 (1996).
27. M. Kawaguchi, A. Ikeda, and S. Shinkai, *J. Chem. Soc., Perkin Trans. I*, **179** (1998).
28. M. Prato, *Mater. Chem.* **7**, 1097 (1997).
29. H. Imahori and Y. Sakata, *Adv. Mater.* **9**, 537 (1997).
30. N. Martin, J. Segura, and C. Seoane, Mater. *Chem.* **7**, 1661 (1997).
31. P. A. Liddell, J. P. Sumida, A. N. Macpherson, L. Noss, G. R. Seely, K. N. Clark, A. L. Moore, T. A. Moore, and D. Gust, *Photochem. Photobiol.* **60**, 537 (1994).
32. H. Imahori, T. Hagiwara, T. Akiyama, S. Taniguchi, T. Okada, and Y. Sakata, *Chem. Lett.* 265 (1995).
33. M. G. Ranasinghe, A. M. Oliver, D. F. Ruthenfluh, A. Salek, and M. N. Paden-Row, *Tetrahedron Lett.* **37**, 4797 (1996).
34. T. Bell, T. Smith, K. Shiggino, M. Ranasinghe, M. Shephard, and M. Paddon-Row, *Chem. Phys. Lett.* **268**, 223 (1997).

35. D. Kuciauskas, S. Lin, G. R. Seely, A. L. Moore, T. A. Moore, D. Gust, T. Drovetskaya, C. A. Reed, and
     P. D. W. Boyd, *J. Phys. Chem.* **100**, 15926 (1996) (Check page, it is given as 15, 926 in the review of
     Prato).
36. H. Imahori, S. Cardoso, D. Tatman, S. Lin, L. Noss, G. Seely, L. Sereno, J. Chessa de Silber, T. A. Moore,
     A. L. Moore, and D. Gust, *Photochem. Photobiol.* **62**, 1009 (1995).
37. T. Drovetskaya, C. A. Reed, and P. Boyd, *Tetrahedron Lett.* **36**, 7971 (1995).
38. H. Imahori and Y. Sakata, *Chem. Lett.* **3**, 199 (1996).
39. Y. Sun, T. Drovetskaya, R. D. Bolskar, R. Bau, P. D. W. Boyd, and C. A. Reed, *J. Org. Chem.* **62**, 3642
     (1997).
40. N. Martin, L. Sanchez, B. Illescâs, and I. Pérez, *Chem. Rev.* **98**, 2527 (1998).
41. P. A. Liddel, J. P. Sumida, A. N. MacPherson, L. Noss, G. R. Seely, K. N. Clark, A. L. Moore, T. A.
     Moore, and D. Gust, *Photochem. Photobiol.* **60**, 537 (1994).
42. H. Imahori, K. Hagiwara, T. Akiyama, S. Taniguchi, T. Okada, and Y. Sakata, *Chem. Lett.* 265 (1995).
43. H. Imahori, K. Hagiwara, M. Aoki, T. Akiyama, S. Taniguchi, T. Okada, M. Shirakawa, and Y. Sakata, *J.
     Am. Chem. Soc.* **118**, 11771 (1996).
44. R. M. Williams, M. Koeberg, J. M. Lawson, Y. –Z. An, Y. Rubin, M. N. Paddon-Row, and J. W.
     Verhoeven *J. Org. Chem.* **61**, 5055 (1996).
45. P. A. Liddell, D. Kuciauskas, J. P. Sumida, B. Nash, D. Nguyen, A. L. Moore, T. A. Moore, and D. Gust,
     J. Am. Chem. Soc. **119**, 1400 (1997).
46. N. Armaroli, F. Diederich, C. O. Dietrich-Buchecker, L. Flamigni, G. Marconi, and J. F. Nierengarten,
     New J. Chem. **77** (1999).
47. F. Diederich, and R. Kissinger, *Acc. Chem. Res.* **32**, 537 (1999).
48. T. G. Linssen, K. Durr, M. Hanack, and A. Hirsch, *J. Chem. Soc, Chem. Commn*, 103 (1995).
49. P. R. Ashton, F. Diederich, M. Gomez-Lopez, J.F. Nierengarten, J. A. Preece, F. M. Raymo, and J. F.
     Stoddart, *Angew. Chem., Int. Ed. Engl.* **36**, 1448 (1997).
50. F. Diederich, C. Dietrich-Buchecker, J.-F. Nierengarten, and J.-P. Sauvage, *J. Chem. Soc., Chem.
     Commun.* 781 (1995).
51. K-Y. Kay and I. C. Oh, *Tetrahedron Lett.* **40**, 1709 (1999).
52. Z. Yoshida, H. Takekuma, S-I. Takekuma, and Y. Matsubara, *Angew. Chem., Int. Ed. Engl.* **33**, 1597.
     (1994).
53. T. Anderson, K. Nilson, M. Sundhal, G. Westman, and O. Wennerström, *J. Chem. Soc., Chem. Commun.*
     604 (1992).
54. C. N. Murthy and K. E. Geckeler, *Chem. Commun.* 1194 (2001).
55. V. Ohlendorf, A. Willnow, H. Hungerbuhler, D. M. Guldi, and K.-D. Asmus, *J. Chem. Soc. Chem.
     Commun.* 759 (1995).
56. K. I. Priyadarsini, H. Mohan, and J. P. Mittal, *J. Photochem. Photobiol. A*, 63 (1995).
57. A. Ikeda, T. Hatano, M. Kawaguchi, M. Suenaga, and H. Shinkai, S. *J. Chem. Soc., Chem. Commun.* **15**,
     1403 (1999).
58. J. P Kamat, T. P. A Devasagayam, K. I Priyadarsini, H Mohan, and J. P. Mittal, *Chem.-Biol. Interact.* **114**,
     145 (1998).
59. S. Samal and K. E. Geckeler, *Chem. Commun.* 1101 (2000).
60. K. E. Geckeler and A. Hirsch, *J. Am. Chem. Soc.* **115**, 3850 (1993).
61. K. E. Geckeler and S. Samal, *Fullerene Sci. Technol.* **9**, 17 (2001).
62. S. Samal, B-J. Choi, and K. E. Geckeler, *Chem. Commun.* 1373 (2000).
63. S. Samal and K. E. Geckeler, *Macromol. Biosci.* **1**, 329 (2001).

RECENT DEVELOPMENTS IN LANGMUIR-BLODGETT FILMS CONTAINING FULLERENE DERIVATIVES

Peptide-derivatised fullerenes

Ludovico Valli, Alvise Perosa, and Pietro Tundo[*]

1. INTRODUCTION: PREVIOUS ACHIEVEMENTS, RELATED PROBLEMS, AND NEW PERSPECTIVES

As a consequence of the detection of the fullerene series by Kroto et al.[1] and their consequential production in large amounts and by an easy procedure at the beginning of the nineties thanks to Krätschmer contribution,[2] great interest has been concentrated on the investigations of the physical and chemical characteristics of these innovative materials. A considerable impulse to these examinations was given by the discovery of the superconductivity at high temperature which occurs when fullerenes are doped with alkali metals.[3]

Fullerenes constitute a newborn and uncommon class of allotropes of carbon (the so-called third allotropic form of carbon). They and their functionalised derivatives are stimulating from both the basic and application perspectives owing to their appealing characteristics, such as superconductivity, ferromagnetism, nonlinear optical properties, charge-transfer behavior.[4] Practicable functions range from carbon composites and molecular sieves to catalysts, lubricants, optical limiters, superconducting and nanoscale devices. Anyhow, there are substantial requirements to promote processing procedures starting from fullerene and producing fullerene-derived materials that are the active constituents in thin films. In fact, in many cases, an essential requirement for the systematic investigation of the above-mentioned properties is the incorporation of fullerenes in organised 2D-arrays and 3D-networks.

[*] Alvise Perosa and Pietro Tundo, Dipartimento di Scienze Ambientali, Università Ca' Foscari, Dorsoduro 2137, 30123 Venezia, Italy; Ludovico Valli, Dipartimento di Ingegneria dell'Innovazione, Università degli Studi di Lecce, Via Monteroni, 73100 Lecce, Italy.

Advanced Macromolecular and Supramolecular Materials and Processes
Edited by K. Geckeler, Kluwer Academic/Plenum Publishers, 2003

In this connection, there have been few reports on interfacial properties of C_{60} and its derivatives, especially in the field of floating films assembled at the air-water interface.[5-7]

The consequential utilisation of the Langmuir-Blodgett (LB) technique allows to build, through the transfer of just one monolayer at a time, multilayers with very appealing characteristics: well-known composition and number of layers, pre-determined thickness and architecture, possibility to create alternate or more complex structure.[8-10] Even though this deposition method was introduced in the thirties,[11] recently it has experienced a novel attention because of its potentialities and possible applications. Among the deposition methods of organic substances, the LB technique represents one of the most attractive tools currently available for the arrangement of molecules in an ordered state. Controlling the orientation of active molecules and the architecture of the multilayer has become a matter of general interest from the point of view of controlling the functions of the films. Moreover, the employment of the LB technique for the deposition of highly organised fullerene thin films is thought to lead to the discovery of applications.[8] But, the ball-like C_{60} molecules are very rigid and hydrophobic, thus evidencing characteristics intrinsically dissimilar from those of rod-like self-assembling amphiphilic molecules generally appropriate for the LB experiments. Thus, the deposition of LB films of these compounds has confirmed to be distinctly problematic and lacks reproducibility. Their high cohesive energy of more than 30 kcal/mol[5,12,13] demonstrates the existence of strong intermolecular attractive π-π interactions, and, in turn, C_{60} aggregates spontaneously and assembles very stable Van der Waals crystals. As a direct repercussion, [60] fullerene is only sparingly soluble in most organic solvents.[14]

But notwithstanding this phenomenon, the preparation of monocomponent, uniform Langmuir monolayers is all the same possible: the devised stratagem is the functionalisation of C_{60} by polar addends.[15,16] In fact, with a view to preparing monomolecular condensed layers, we have pursued the strategy of producing floating films containing novel materials that in their structure evidence the contemporary presence of the lipophilic buckyball and a hydrophilic termination, covalently connected to C_{60}. In the light of this procedure, an essential advancement in the comprehension of factors governing fullerene LB film organisation is the methodical investigation of the film properties as a function of the substituent structure and characteristics.

It has been recognized that unsubstituted fullerene[17-19] assembles condensed layers on the water surface. In fact, the measured areas per molecule of C_{60} at zero pressure ($A_{\Pi \to 0}$), are considerably smaller than the minimum theoretical value of about 90 Å^2/molecule, already reported for an ideal close-packed monolayer[7,20,21] The thickness of the floating layers results critically from the characteristics of the spreading solution. Concentration of the spreading solution, nature, and volume of spreading solvent were all ascertained to influence the film thickness. The LB method has sometimes been used effectively to deposit the floating films onto solid supports. On the other hand, transfer ratios, considerably less than unity, have been observed.[22] Therefore, a typical procedure has been usually utilised: to employ a mixture of a film-promoting substance and the fullerene C_{60}.[23-25]

A different procedure involves the application of derivatised fullerenes bearing substituents capable of producing an amphiphilic character in these new compounds and therefore to induce a film forming capability. These materials exhibit much enhanced potentials in LB deposition with respect to pristine C_{60}.[26-29]

However, an elegant and novel approach implies the utilisation of amino acid derivatives of C_{60} in association with a subphase containing an oligopeptide.[30,31] The strategy is to take advantage of the formation of hydrogen bonding interactions between oligopeptide units in the aqueous subphase and the pendant group attached to the buckyball. A few examples concerning poly(L-glutamic acid)[32] and long-chain fatty acids and amines[33,34] have been reported in literature. Therefore, a novel amphiphilic fullerene bearing an oligopeptidic tail, and at the same time able to form a stable floating film, is here reported. This compound was expected to form a stabilised, regular, and homogeneous Langmuir film at the air-water interface through the formation of a hydrogen-bonding network with a dipeptide dissolved in the aqueous phase.

In such a way it is possible to fill the gap between the oligopeptide chains attached to adjacent fullerenes thus strongly anchoring the amphiphilic C_{60}'s to the subphase and to preclude the well-known aggregation of C_{60} in clusters at the air-water interface. In an ideal case we could represent the proposed arrangement as illustrated in Figure 1.

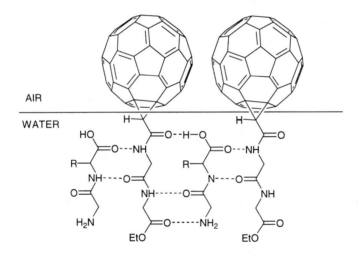

Figure 1

2. RESULTS AND DISCUSSION

2.1. Synthesis of the Amphiphilic Fullerene

Diglycine ethyl ester, -[NHCH$_2$C(O)]$_2$OEt (Gly$_2$OEt), was utilised to derivatize the buckyball. (1,2-Methanofullerene C_{60})-61-carboxylic acid **1**, prepared by the addition of (ethoxycarbonyl)-methyl diazoacetate to C_{60}, and successive hydrolysis[35,36] was coupled with the oligopeptide by the traditional peptide coupling procedures. The method used is analogous to the one previously reported for the synthesis of C_{60} linked to peptide T (where a few milligrams were prepared and tested for biological activity).[37]

Accordingly, reaction of **1** with Gly$_2$OEt in the presence of dicyclohexylcarbodiimide (DCC), hydroxybenzotriazole (HOBT), and bromobenzene as the solvent, proceeded at room temperature for 24 hours and afforded amphiphilic fullerene **2** in a 42% yield, based on recovered C$_{60}$-carboxylic acid **1** (Eq. 1).

The product appeared like a stable brown powder, which could be purified by silica gel column chromatography, by eluting with large volumes of a mixture of toluene and petroleum ether (due to its low solubility in commonly used solvents).

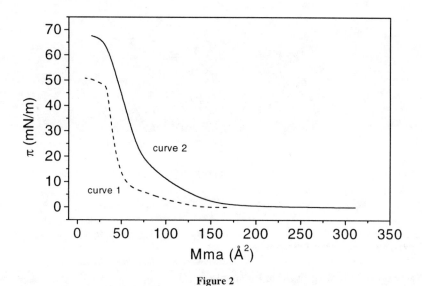

Figure 2

2.2. Langmuir Experiments

The surface pressure Π vs. area per molecule curve for derivative **2**, when were spread 400 μL of a 1.1 x 10^{-4} M chloroform solution of **2** on pure water as the subphase is illustrated in Figure 2 (curve 1, dotted line) together with the Langmuir isotherm obtained with a water subphase containing Gly-L-Leu (curve 2, solid line).

Chloroform was used as the spreading solvent instead of toluene or other aromatic solvents, since it is well known that the interpretation of surface pressure-area isotherms is complicated not only by aggregation phenomena, but also by solvent related effects. In fact, it has been reported that aromatic solvents remain trapped in the spread films, thus indicating a probable complexation of the solvent molecules with the fullerenes moieties.[20] The value of the limiting area per molecule, $A_{\pi \to 0}$, was extrapolated to zero pressure for each curve from the steepest portion of the Langmuir isotherm. In the presence of Gly-L-Leu in the water subphase (curve 2), the curve has manifested much better interfacial properties than when pure water is used. In fact, limiting area has a value of 89.1 Å^2 which is in close accordance with that reported (ca. 90 Å^2) for a two-dimensional close-packed monolayer of the buckyball with the hydrophilic tail oriented toward the water phase.

When pure water (curve 1) is used as the subphase, $A_{\pi \to 0}$ is significantly smaller, assuming, when dilute solutions and small volumes of spreading solutions are utilised,[38] the best value of. This suggests that fullerene π-π interactions overcame the weak hydrophilic-hydrophilic interactions (between the peptidic pendant group and water molecules) that control the construction process of the floating monolayer. An area value of 55 Å^2 is almost equivalent to the formation of a bilayer. The aggregates so formed are typical of molecular crystals. The interatomic separation for adjacent C_{60} balls is larger than 3 Å.[39] This means that of course there is not the formation of covalent bonds between contiguous buckyballs and the character of the inter-C_{60} interaction is Van der Waals binding. In order to overcome such adhesion forces, we have used hydrogen bonding interactions brought about by the presence of Gly-L-Leu in the subphase. $A_{\pi \to 0}$ data suggest that they are effective to this purpose.

In addition, another difference between curve 1 and curve 2 is that in the presence of Gly-L-Leu the Langmuir isotherm exhibits a slope larger than when pure water is used as the subphase. For example, on dissolution of the dipeptide in water Π vs. A curve evidences the transition from a liquid expanded phase to the condensed phase at about Π = 20 mN/m, while in the absence of Gly-L-Leu the curve flattens and the same transition appears at about Π = 10 mN/m. Moreover, also the collapse pressure is significantly different: while the Langmuir film of **2** in the presence of Gly-L-Leu (curve 2) collapsed at Π > 60 m/Nm, on pure water (curve 1) this occured at a lower pressure: Π < 50 mN/m. The presence of Gly-L-Leu helps in anchoring the derivatised buckyballs to the subphase and consequently the floating layer supports larger pressures before collapse. So, it is possible to suggest on analogy of the formation of self-assembled monolayers of alkyl-oligopeptide amphiphiles[30,40] and of the results acquired with dendritic amphiphilic fullerenes,[41] that all these experimental data indicate that the stabilisation, homogeneity and persistence of the monolayer are induced by the presence of the dipeptide in the subphase through hydrogen bonding.

In addition, the fact that the presence of Gly-L-Leu in the subphase assists the formation of a monolayer is also indirectly confirmed from another consideration: for curve 2, the extrapolated value of the limiting area per molecule is independent on the concentration of the spreading solution, as determined by spreading solutions of **2** with different concentrations.

This implies that the monotone expansion of the $A_{\pi \to 0}$ values upon lowering the concentration of the spreading solution, a general drawback typical of other fullerene derivatives onto pure water, does not occur for compound **2**.

This collaborative effect between the dipeptide in the subphase and the functional group covalently attached to the buckyball is corroborated also by hysteresis cycles. Successive compression and expansion cycles were carried out on the floating film of **2** with Gly-L-Leu in the subphase, in order to explore the reversibility of its formation. The response to such cycles is that compound **2** could be relaxed and recompressed without significant hysteresis, provided Gly-L-Leu is present in the aqueous subphase: during the expansion following the first compression, the surface pressure describes a curve close to the one registered while the floating layer is compressed.

In addition, the floating film of compound **2** with Gly-L-Leu can be maintained overnight at a pressure of 35 mN/m with an area loss of just 1%. On the other hand, in the absence of the dipeptide the limiting area per molecule evidenced a large loss. Concluding, the high collapse pressure, the response to the compression/expansion cycles, and the resistance over time of the film all leads to the conclusion that Gly-L-Leu inserted in the aqueous subphase has a strong anchoring effect on the film, rendering it rigid, organized, and stable.

The Langmuir floating film of **2** can be transferred at 20 °C onto hydrophobic quartz, glass and silicon slides, at a surface pressure maintained at 30 mN/m. After each spreading, a series of two down- and two up-strokes was reproducibly carried out, and thereafter up to at least 200 strokes could be performed. For the first down-stroke (D) and up-stroke (U), the transfer ratios were 1.0, while it was 0 for the second downward passage, and 1.0 again for the second up-stroke. This pattern for the deposition ratio (D = 1.0, U = 1.0, D = 0, U = 1.0) was then constantly repeated reproducibly for each following series of four strokes, the *repeat unit*, yielding 150 effectively deposited layers. In this way, the global sequence of the successful transfer was D, U, U, (cleaning and spreading), D, U, U,... and so on (D = down and U = up), by moving the substrate: D, U, D, U, (cleaning and spreading), D, U,....

The necessity of interrupting the deposition after only four dippings arose from the fact that the two moving barriers of the trough became too close one another and to the Wilhelmy plate, thus modifying the measurement of the surface pressure. The quality of the transfer was also checked by measuring the absorbance at a fixed wavelength for the films deposited onto a quartz substrate vs. the number of transferred layers. Figure 3 illustrates the plot of absorbance at 257 nm, on an absorption maximum. It is linear, thus substantiating that the transfer is reproducible cycle after cycle and a constant amount of the fullerene derivative is on the average picked up during every cycle.

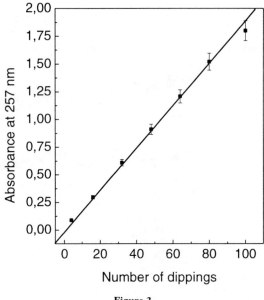

Figure 3

2.3. X-Ray Diffraction

Figure 4 shows the diffraction spectrum of the thin film obtained by 150 strokes through the monolayer of derivative **2**, in the presence of Gly-L-Leu, deposited onto a silicon substrate (curve a), in comparison with the x-ray diffraction profile collected on the pure silicon substrate (curve b). The two diffraction profiles differ only for one peak at 19.31° (see arrow), which was attributed to the film structure. A Montecarlo method of simulation was applied to fit this peak.[42] In this way it was possible to obtain the distribution of the structure dimensions, which cause the scattering. This is depicted in Figure 5.

Throughout the bulk of the LB film, 80% of the *repeat units* have a thickness ranging from 20 to 60 Å, with an average of 40 Å. This is consistent with the deposition pattern and with the size of compound **2**: by considering that the length of the tail attached to the buckyball is roughly 10 Å, and that the diameter of the C_{60} sphere is also about 10 Å, the maximum thickness-per-monolayer of the LB film should therefore be of about 20 Å (layer thickness). In the ideal case of a perfect vertical stacking of the multilayer, the maximum thickness of the three layers deposited during each cycle (assumed as the repeat unit) is 60 Å.

The smaller average thickness (40 Å) of layers with the same quality and the same transfer ratios, can be explained by the uncertainty on the orientation of **2** within the three layers respect to the plane of the silicon slide, by the possible interdigitation of the chains, and by the texture of the layers.

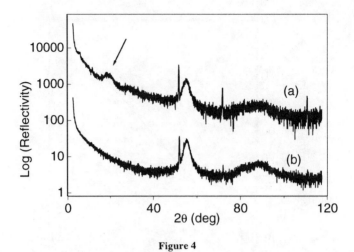

Figure 4

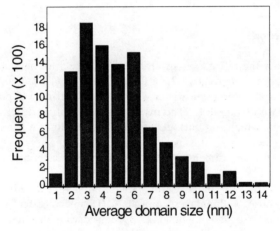

Figure 5

2.4. LB Film Thickness

An average thickness of 100 ± 15 nm was measured by a Tencor computerized surface profiler (Alpha-Step 200 Stylus Profilometer) for an LB film 25 repeat units (100 total strokes) thick, deposited onto hydrophobic Corning glass. This value is in agreement with the average domain size (4 nm) obtained by the diffraction spectrum (in fact: 4 nm x 25 units = 100 nm). The measurement was carried out after the deposition of a thin gold layer, of known thickness, both onto a clean region of the glass substrate and onto the LB film itself.

3. EXPERIMENTAL

3.1. Synthesis of the Amphiphilic Fullerene N-(1,2-methanofullerene[60]-61-carbonyl)-glycylglycyl ethyl ester 2.

A 50 mL flask was loaded with 20.0 mg (0.026 mmol) of (1,2-methanofullerene C_{60})-61-carboxylic acid 2, 7.0 mg (0.052 mmol) of HOBT, 5.5 mg (0.026 mmol) of DCC, and 15 mL of a 1 : 6 mixture of DMSO ad PhBr. The reaction was degassed by bubbling nitrogen, and cooled in an ice-water bath, with stirring, for 30 min. To this solution was added, *via* syringe, a solution made of 10.2 mg (0.052 mmol) of gly-gly ethyl ester and 5.3 mg (0.052 mmol) of triethyl amine in 10 mL of PhBr. The mixture was warmed to room temperature, and stirred under N_2 for 24 h. Formation of a brown precipitate was observed. PhBr was removed under reduced pressure, the product was adsorbed on a small portion of silica gel, and chromatographed over silica gel eluting first with toluene and then with a 1/1 mixture of toluene and ethyl acetate (R_f = 0.27). Yield (based on recovered C_{60}) = 30 mg (42%).

^{1}H-NMR (DMSO-d_6/CS$_2$/C$_6$D$_6$, 4 : 1 : 2) δ ppm 9.31 (t, 1H, J = 5.80 Hz), 8.47 (t, 1H, J = 5.80 Hz), 5.40 (s, 1H), 4.12 (q,2H, J = 7.02 Hz), 4.03 (d, 2H, J = 5.80 Hz), 3.92 (d, 2H, J = 5.80 Hz), 1.20 (t, 3H, J = 7.02 Hz). λ_{max} (*n*-hexane)/nm 257, 328, 437. MALDI/MS, matrix: 2,5-hydroxybenzoic acid, 720 (C_{60}), 921 (M+1), 944 (M+1+Na), 960 (M+1+K).

3.2. Langmuir Experiments

In the Langmuir experiments, compound 2 was dissolved in CHCl$_3$ (concentration: 1.1 x 10^{-4} M). The resulting solution was uniformly spreaded over the subphase by adding small drops (approximately 5 µL) at different locations on the water surface (850 cm^2, KSV5000 System 3 LB apparatus). After 15 min the floating film was slowly compressed by the use of two mobile Teflon barriers at a rate of 2 x 10^{-2} nm^2 sec^{-1} molecule^{-1} and the surface pressure vs. area per molecule curve was recorded. The Wilhelmy plate was placed perpendicularly to the direction of motion of barriers; it was burnt to redness and quenched in methanol between runs. During a series of Langmuir film formation experiments, ultrapure water with resistivity greater than 18 MΩ cm and pH = 5.9 ca. was produced by a Millipore Milli Elix3-MilliQ system and used as the subphase after filtration through a 0.5 µm nylon disk.

In a second series of experiments and during the multilayer depositions, a solution containing 1.0 g L^{-1} of glycyl-L-leucine was used as the subphase. The subphase was in both cases thermostated at 20 °C by a Haake GH-D8 apparatus.

Depositions of compound 2 was carried out at a surface pressure of 30 mN/m, using various substrates, such as hydrophobized quartz, glass and silicon having a surface area of 1.1 cm^2.

The LB film was successfully transferred onto various solid supports by up to 150 successive dipping/withdrawal cycles at a dipping speed of 1-2 mm/min for the downstrokes and 6 mm/min for the upstrokes.

4. CONCLUSIONS

The main problem connected to the deposition of ultrathin films containing fullerene derivatives by the Langmuir-Blodgett technique is the great tendency of the buckyball to form 3D aggregates on the water surface. This phenomenon of course does not allow the formation of monomolecular, homogeneous, and organised floating films at the air water interface and therefore prevents the fabrication of ordered multilayers onto solid substrates, as usual for typical amphiphilic substances.

The strategy followed by us and other researchers permits to assert that the modification of the buckyball by the attachment of strongly hydrophilic groups improves their amphiphilic behavior at the air-water interface. However, this notwithstanding, in many cases, the propensity of the buckyball to develop aggregates still manifests, unless very diluted spreading solutions and small volumes of them are used.[38] This confirms that the so-called hydrophile-lipophile balance, HLB[43] still leans towards the hydrophobic counterpart. Various solutions have been proposed in order to minimise the cohesive energy among C_{60} spheres and at the same time to promote the anchorage at the water surface. For example, an effective solution is to introduce large and bulky hydrophilic head groups, like in fullerene-glycodendron conjugates or in fullerene-aza-crown ethers.[5,44]

In this way the aggregation phenomenon has been prevented even though the buckyball are still in contact to form homogeneous monolayers. Moreover, this way provides an excellent balance between the driving force of the multilayer growth (intermolecular interactions) and the condition for well-behaved Langmuir films (intermolecular repulsion among the large hydrophilic groups). Of course, also the insertion of two or more suitable hydrophilic head groups in an appropriate sterical configuration permits a better anchorage at the water subphase and hence helps the transfer onto solid supports.[45-47] For example, fullerenes functionalised with malonic acid esters have allowed to inhibit the formation of three dimensional aggregates even when concentrated spreading solutions were used (2×10^{-3} M). Other practical examples reported in literature describe the assistance of monolayer formation through an amphiphilic carboxylic acid-terminated fullerene.[48]

In our research, we have followed another different approach: we have modified the fullerene by insertion of a new pendant group, but have also anchored it to the water subphase by dissolution of Gly-L-Leu. The Langmuir films on the water surface are probably monomolecular, reproducible, and stable. Moreover, the transfer onto various solid substrates has been successful and also reproducible. The regularity in the multi-layer structure has been checked also by X-ray diffraction measurements.

5. ACKNOWLEDGEMENTS

The authors are grateful to Dr. Cinzia Gianni for XRD measurements and to Mr. L. Dimo, whose technical assistance with the trough experiments has permitted to speed up the conclusion of this work.

6. REFERENCES

1. H. W. Kroto, J. R. Heath, S. C. Brien, R. F. Curl, and R. E. Smalley, *Nature* **318**, 162 (1985).
2. W. Krätschmer, L. D. Lamb, K. Fostiropoulos, and D. R. Huffman, *Nature* **347**, 254 (1990).
3. J. H. Weaver and D. M. Poirier, in: Ehrenreich H., and Spaepen F. (eds.), *Solid State Physics* **48**, 1
4. G. S. Hammand, and. V. J Kuck, *Fullerenes: Synthesis, Properties and Chemistry of Large Carbon Cluster*, ACS Symposium Series, vol. 481, American Chemical Society, Washington, DC (1992).
5. F. Cardullo, F. Diederich, L. Echegoyen, T. Habicher, N. Jayaraman, R. M. Leblanc, J. F. Stoddart, and S.Wang, *Langmuir* **14**, 1955 (1998).
6. Maggini, Pasimeni, L., Prato, M., Scorrano, G., and Valli, L., *Langmuir* **10**, 4164 (1994).
7. P. Wang, M. Shamsuzzoha, X. L. Wu, W. J. Lee, and R. M. Metzger, *J. Phys. Chem.* **96**, 9025 (1992).
8. M. C. Petty, *Langmuir-Blodgett Films, an Introduction,* Cambridge University Press, Cambridge (1996).
9. G. Roberts, *Langmuir-Blodgett Films,* Plenum Press, New York (1990).
10. A. Ulman, *An Introduction to Ultrathin Organic Films from Langmuir-Blodgett to Self-Assembly*, Academic Press, Inc., San Diego (1991).
11. K. B. Blodgett, *J. Am. Chem. Soc.* **57**, 1007 (1935).
12. C. Pan, M. P. Sampson, Y. Chai, R. H. Hauge, and J. L.Margrave, *J. Phys. Chem.* **95**, 2944 (1991).
13. X. Shi, W.B. Caldwell, K. Chen, and C. A. Mirkin, *J. Am. Chem. Soc.* **116**, 11598 (1994).
14. R. S. Ruoff, D. S. Tse, R. Malhotra, and D. C. Lorents, *J. Phys. Chem.* **97**, 3379 (1993).
15. C. A. Mirkin, and W. B. Caldwell, *Tetrahedron* **52**, 5113 (1996).
16. M. Prato, *J. Mater. Chem.* **7**, 1097 (1997).
17. J. L. Brousseau, K. Tian, S. Gauvin, R.M. Leblanc, and P. Delhaès, *Chem. Phys. Lett.* **202**, 521 (1994).
18. M. Iwahashi, K. Kikuchi, Y. Achiba, I. Ikemoto, T. Araki, T. Mochida, S. Yokoi, A. Tanaka, and K. Iriyama, *Langmuir* **8**, 2980 (1992).
19. C. F. Long, Y. Xu, F. X. Guo, Y. L. Li, D. F. Xu, Y. X. Yao, and D. B. Zhu, *Solid State Commun.* **82**, 381 (1992).
20. P. A. Heiney, J. E. Fischer, A. R. McGhie, W.J. Romanow, A. M. Donenstein, J. P. McCauley, Jr. Smith, A. B., and D. E. Cox, *Phys. Rev. Lett.* **66**, 2911 (1991).
21. P. W. Stephens, L. Mihaly, P. L. Lee, R. L. Whetten, S. M. Huang, R. Kaner, F. Diederich, and K. Holczer, *Nature* **351**, 632 (1991).
22. Y. S. Obeng, and A. J. Bard, *J. Am. Chem. Soc.* **113**, 6279 (1991).
23. F. Diederich, J. Effing, U. Jonas, L. Jullien, T. Plesnivy, H. Ringsdorf, C. Thilgen, and D. Weinstein, *Angew. Chem., Int. Ed. Engl.* **31**, 1599 (1992).
24. Z. I. Kazantseva, N. V. Lavrik, A. V. Nabok, O. P. Dimitriev, B. A. Nesterenko, V. I. Kalchenko, S. V. Vysotsky, L. N. Markovskiy, and A. A. Marchenko, *Supramol. Sci.* **4**, 341 (1997).
25. R. Rella, P. Siciliano, and L. Valli, *Phys. Stat. Sol.* **143**, K129 (1994).
26. Y. Hirano, H. Sano, J. Shimada, H. Chiba, J. Kawata, Y. F. Miura, M. Sugi, and T. Ishii, *Mol. Cryst. Liq. Cryst. Sci. Technol., Sect. A* **294**, 161 (1997).
27. U. Jonas, F. Cardullo, P. Belik, F. Diederich, A. Gügel, E. Harth, A. Herrmann, L. Isaacs, K. Müllen, H. Ringsdorf, C. Thilgen, P. Uhlmann, A. Vasella, C. A. A. Waldraff, and M. Walter, *Chem. Eur. J.* **1**, 243 (1995).
28. S. Ravaine, F. Le Peq, C. Mingotaud, P. Delhaes, J. C. Hummelen, F. Wudl, and L. K. Patterson, *J. Phys. Chem.* **99**, 9551 (1995).
29. D. J. Zhou, L. B. Gan, C. P. Luo, H. S. Tan, C. H. Huang, Z. F. Liu, Z. Y. Wu, X. S. Zhao, X. H. Xia, F. Q. Sun, S. B. Zhang, Z. J. Xia, and Y. H. Zou, *Chem. Phys. Lett.* **235**, 548 (1995).
30. X. Cha, K. Ariga, M. Onda, and T. Kunitake, *J. Am. Chem. Soc.* **117**, 11833 (1995).

31. X. Cha, K. Ariga, and T. Kunitake, *Bull. Chem. Soc. Jpn.* **69**, 163 (1996).
32. N. Higashi, M. Saitou, T. Mihara, and M. Niwa, *J. Chem. Soc., Chem. Commum* 2119 (1995).
33. D. Gidalevitz, I. Weissbuch, K. Kjaer, J. Als-Nielsen, and L. Leiserowitz, *J. Am. Chem. Soc.* **116**, 3271 (1994).
34. E. M. Landau, S. Grayer Wolf, M. Levanon, L. Leiserowitz, M. Lahav, and J. Sagiv, *J. Am. Chem. Soc.* **111,** 1436 (1989).
35. L. Isaacs, and F. Diederich, *Helv. Chim. Acta* **76** 2454 (1993).
36. L. Isaacs, A. Wehrsig, and F. Diedrich, *Helv.Chim. Acta* **76** 1231 (1993).
37. C. Toniolo, A. Bianco, M. Maggini, G. Scorrano, M. Prato, M. Marastoni, R. Tomatis, S. Spisani, G. Palù, and E. D. Blair, *J. Med. Chem.* **37** 4558 (1994).
38. L. Leo, G. Mele, G. Rosso, L. Valli, G. Vasapollo, D. M. Guldi, G. Mascolo, *Langmuir* **16**, 4599 (2000).
39. P. R. Surján, L. Udvardi, and K. Németh, *Synth. Met.* **77**, 107 (1996).
40. X. Cha, K. Ariga, and T. Kunitake, *J. Am. Chem. Soc.* **118**, 9545 (1996).
41. D. Felder, J.-L. Gallani, D. Guillon, B. Nicoud, J.-F., and J.-F. Nierengarten, *Angew. Chem. Int. Ed. Engl.* **39**, 201 (2000).
42. P. E. Di Nunzio, S. Martelli, R. Ricci Bitti, *J. Appl. Cryst.* **28**, 146 (1995).
43. I. J. Lin, J. P. Friend, and Y. Zimmels, *J. Coll. Interface Sci.* **45**, 378 (1973).
44. Y. Xiao, Z. Yao, D. Jin, F. Yan, and Q. Xue, *J. Phys. Chem.* **97**, 7072 (1993).
45. D. M. Guldi, Y. Tian, J. H. Fendler, H. Hungerbühler, and K. D. Asmus, *J. Phys. Chem.* **99**, 17673 (1995).
46. D. M. Guldi, Y. Tian, J.H. Fendler, Y. Hungerbühler, H., and Asmus, K. D., *J. Phys. Chem.* **100**, 2753 (1996).
47. D. M. Guldi, K. D. Asmus, Y. Tian, and J. H. Fendler, *Proc.-Electrochem. Soc. (Recent Advances in the Chemistry and Physics of Fullerenes and Related Materials, vol. 3)* **96-10**, 501 (1996).
48. M. Matsumoto, H. Tachibana, R. Azumi, M. Tanaka, T. Nakamura, G. Yunome, M. Abe, S. Yamago, and E. Nakamura, *Langmuir* **11**, 660 (1995).

POLYMERIZATION IN AQUEOUS MEDIUM USING CYCLODEXTRIN AS HOST COMPONENT

Helmut Ritter[*] and Monir Tabatabai

1. INTRODUCTION

Since cyclodextrins (CDs) are commercially available on a large scale they attract rapidly increasing attention in many fields of practical applications because of their ability to include several types of hydrophobic guest molecules into their cavity. They are produced in technical plants by enzymatically catalyzed modification of starch simply via degradation of helical polysaccharide into ring shaped molecules. The chemical structures of CDs are therefore derived from the repeating structural units of starch macromolecules. Thus, CDs are cyclic oligoamyloses, consisting of 6 (α), 7 (β), 8 (γ), or 9 (δ) units of 1,4-linked glucose. They exhibit a torus-shaped structure with a hydrophobic cavity and a hydrophilic outer side.[1-5] The formation of an inclusion complex leads to a significant change of the solution properties and reactivities of the guest molecules. This opens the possibility to create for example main-chain and side-chain rotaxanes from CDs as generally illustrated in Scheme 1.[6-10]

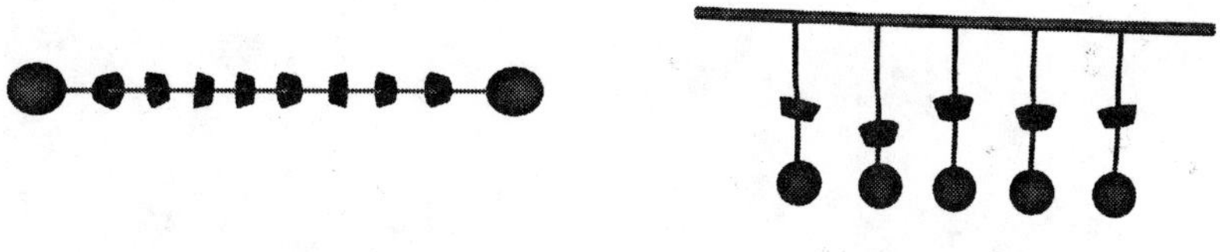

Scheme 1. General structure of main-chain and side-chain polyrotaxanes.

[*] Institut für Organische Chemie und Makromolekulare Chemie, Heinrich-Heine Universität Düsseldorf, Universitätsstr. 1, D-40225 Düsseldorf; Germany.

Advanced Macromolecular and Supramolecular Materials and Processes
Edited by K. Geckeler, Kluwer Academic/Plenum Publishers, 2003

By the way, water-insoluble molecules may become completely water-soluble simply on treatment with aqueous solutions of native CD or CD-derivatives, e.g., methylated or hydroxypropylated CD. It must be pointed out that no covalent bonds are formed by the interaction of the CD-host and the water-insoluble guest molecule. Only hydrogen bonds or hydrophobic interactions may stabilize the complex which is normally formed spontaneously in stirred water or organic solvent, e.g., tetrahydrofurane (THF), methanol, or methylene chloride.

2. CYCLODEXTRIN AS HOST COMPONENT

Water-soluble methylated β-CD complexes of phenyl methacrylate (**2h**) and cyclohexyl methacrylate (**2j**) were achieved by stirring an equimolar mixture of β-CD and the monomers **2h** or **2j**, respectively, in chloroform for six days. The complex can be isolated after evaporation of the solvent.[11] The thermodynamically stable CD-complexes are more or less dynamically mobile depending on the temperature and the shape of the included guest molecule.

The complex structure of n-butyl acrylate (**2b**)/methylated β-CD as an example, can be characterized by X-ray diffraction showing the monomeric guest fits perfectly into the cavity of β-cyclodextrin (Figure 1a).[12] Surprisingly, in the case of the 3,4-diethylenedioxythiophene (**5**)/α-CD complex, X-ray analysis showed the formation of a 1:1 non-inclusion channel-type packed supramolecular complex based on multiple hydrogen bounds (Figure 1b).[13]

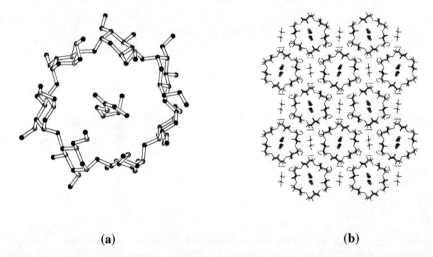

(a)   (b)

Figure 1. X-ray structure of a) n-butyl acrylate (**2b**)/methylated β-cyclodextrin; b) 3,4-diethylenedioxythiophene (**5**)/α-cyclodextrin complexes.

The formation of the complexes can be verified by ^{1}H NMR spectroscopy and 2D-ROESY-NMR. For example, in N-(4-hydroxyphenyl)-maleimide (**6**)/methylated β-CD complex, compared to the guest component, the signals of the aromatic protons in meta position to the OH-group and the signals of the maleimide double bound were shifted 0.06 ppm and 0.02 ppm, respectively, to a higher magnetic field, while no change occurred for the doublet assigned to the aromatic protons in ortho position to the OH-group.[14]

The 2D-ROESY-NMR spectrum of **6**/methylated β-CD complex showed interaction between the aromatic moieties of the monomer (at 7.12 ppm and 6.97 ppm) with the cyclodextrin cavity around 3.6 ppm (pointed out by circles in Figure 2).[14] In the case of cyclohexyl methacrylate (**2j**)/methylated β-CD complex a significant shift of the vinyl protons of the methacryl group to lower fields was observed as a result of complexation.

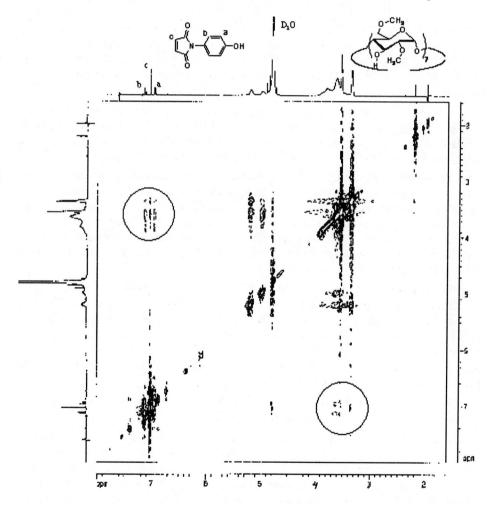

Figure 2. 2D-ROESY-NMR spectrum of N-(4-hydroxyphenyl)-maleimide (**6**)/β-cyclodextrin.

A further important method to prove the formation of a complex is thin layer chromatography. For example, the R_f-values of methylated β-cyclodextrin (R_f = 0.69) and uncomplexed phenyl methacrylate (R_f = 0.92) in methanol are significantly different from the values of the complex (R_f = 0.54).[11]

The FT-IR spectra of the complexed and uncomplexed guest molecule **1a** showed that the carbonyl bonds of the complexed monomer are shifted to higher frequencies from 1708 cm^{-1} to 1712 cm^{-1}. This clearly indicates the influence of the host ring molecule on the carbonyl vibration, as expected.[15]

3. POLYMERIZATION OF CYCLODEXTRIN COMPLEXES

Recently, we thus applied CDs to make different types of suitable vinyl monomers water-soluble. It turned out that it can be successfully extended to polymer synthesis via free radicals in water starting from CD-complexed hydrophobic vinyl monomers. The interest was focused on standard guest monomers without bearing barrier groups, which are given in Scheme 2.

Surprisingly, it was found that in nearly all cases, the resulting polymer precipitates rapidly in high yields and the cyclodextrin slips off step by step from the growing chain and thus remains nearly quantitatively in the water phase. The unthreaded cyclodextrin is soluble in water and thus can be reused to entrap new monomer (Scheme 3).

As an early example, the water-soluble complex consisting of the relatively hydrophobic monomer N-methacryloyl-11-aminoundecanoic acid (**1a**) and unmodified β-cyclodextrin was polymerized radically in an aqueous medium using the water-soluble azoinitiator 2,2'-azobis(N,N'-dimethylen isobutyramidine).[15] Furthermore, the monomers **1a** and **1b** were incorporated as a guest into the cavity of β-cyclodextrin and homopolymerized in water or in water/dimethyl sulfoxide (DMSO)-initiated by the free radical redox initiator system $K_2S_2O_8$/$KHSO_3$.[16]

The water-insoluble polymers obtained immediately precipitate after a few growing steps. At the end, the polymer contains only traces of cyclodextrin of about 5 mol% in the case of **1a** and 3 mol% in the case of **1b**. The amount of cyclodextrin can be reduced nearly quantitatively by extraction with water or alcohol. The kinetics of the polymerization reaction measured by the use of ^{1}H NMR spectroscopy shows relatively high polymerization rates of the complexed monomers **1a** and **1b** compared to the polymerization rates of the uncomplexed monomers in solution (Figure 3).

This demonstrated that precipitation polymerization reactions are faster than the corresponding polymerization in homogenous medium.

The functionalized monomers **1e** and **f** were prepared by the condensation of 4-aminophenyl-2-oxazoline with (meth)-acrylic acid in the presence of 1-ethoxycarbonyl-2-ethoxy-1,2-dihydroquinoline (EEDQ) in 1,4-dioxane. Subsequently, the monomers were quantitatively incorporated into the dimethylated β-cyclodextrin at room temperatur using methanol as solvent. After evaporating the organic solvent, the residual complex was completely water-soluble.

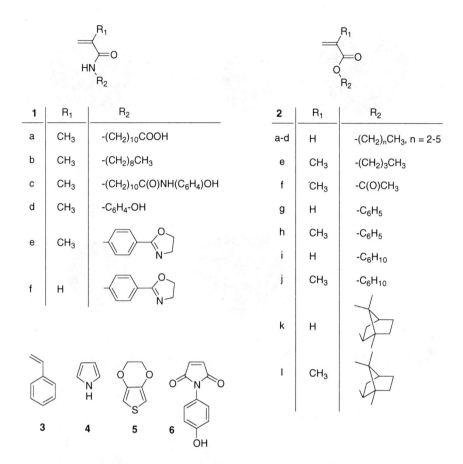

1	R$_1$	R$_2$
a	CH$_3$	-(CH$_2$)$_{10}$COOH
b	CH$_3$	-(CH$_2$)$_8$CH$_3$
c	CH$_3$	-(CH$_2$)$_{10}$C(O)NH(C$_6$H$_4$)OH
d	CH$_3$	-C$_6$H$_4$-OH
e	CH$_3$	
f	H	

2	R$_1$	R$_2$
a-d	H	-(CH$_2$)$_n$CH$_3$, n = 2-5
e	CH$_3$	-(CH$_2$)$_3$CH$_3$
f	CH$_3$	-C(O)CH$_3$
g	H	-C$_6$H$_5$
h	CH$_3$	-C$_6$H$_5$
i	H	-C$_6$H$_{10}$
j	CH$_3$	-C$_6$H$_{10}$
k	H	
l	CH$_3$	

3 **4** **5** **6**

Scheme 2. Suitable guest monomers for the complexation with methylated β-cyclodextrin.

Cyclodextrin complexes were polymerized radically in water at 60°C in the presence of 5 mol% of 2,2'-azobis(2-amidinopropane) dihydrochloride as initiator. The polymers precipitated at about 30 min after the initiator was added. ^{1}H NMR- and IR-spectra clearly showed that the oxazoline ring is still present in the polymer. The remaining oxazoline functions offer possibilities of further modification, e.g. ionic grafting of alkyloxazolin.[17]

However, the preparation and polymerization of equimolar cyclodextrin complexes of commercial standard monomers like styrene (**3**) and (meth)acrylic derivatives (**2**) opens an alternative route to polymer synthesis simply from the water phases without using surfactants. Water-soluble **2f**/methylated β-CD or **3**/methylated β-CD complexes can be easily obtained from an aqueous solution of methylated β-CD and equimolar amounts of methyl methacrylate (**2f**) or styrene (**3**), respectively, to give clear homogenous solutions of the complexed monomers.

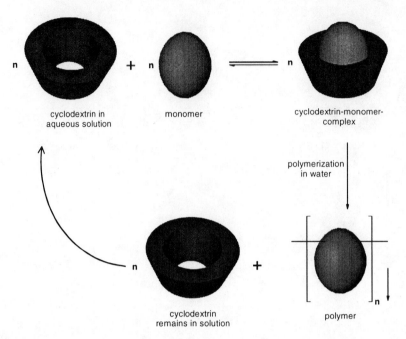

Scheme 3. Schematic presentation of complexation and polymerization of β-cyclodextrin-complexed monomers.

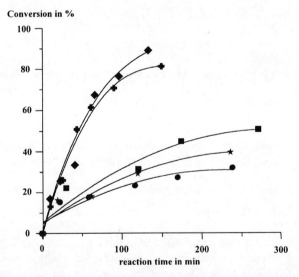

Figure 3. The influence of methylated β-cyclodextrin on the polymerization rates: ✚ N-methacryl-oyl-11-aminoundecanoic acid (**1a**)/CD complex in water; ◆ N-methacryloyl-11-aminononane (**1b**)/CD complex in water; ■ N-methacryloyl-11-aminoundecanoic acid (**1a**)/CD complex in water/DMSO; φ N-methacryloyl-11-aminoundecanoic acid (**1a**) in water/DMSO; ● N-methacryloyl-11-aminononane (**1b**) in water/DMSO.

A free radical redox initiator system, e.g., $K_2S_2O_8$/NaHSO$_3$, was used to initiate the polymerization of methylated β-CD complexed monomers in aqueous solution at 60°C. It was shown in both cases, that polymerization from homogenous, aqueous CD solution is faster and ends up with higher yields and molecular weights than polymerization in an organic solvent under similar conditions.[18]

4. FEATURES OF THE POLYMERIZATION METHOD

This new cyclodextrin-mediated polymerization method differs strongly from classical methods like emulsion polymerization of hydrophilic monomers. For example, due to the inclusion of the monomers into the CD-cavities the copolymerization parameters may be influenced significantly. As an interesting observation, it was found that the reactivity ratios of the methylated β-CD complexed monomers isobornyl acrylate (**2k**) and n-butyl acrylate (**2b**) ($r_{2k/\beta\text{-CD}} = 0.3 \pm 0.1$), ($r_{2b/\beta\text{-CD}} = 1.7 \pm 0.1$) differ significantly from the r-values of the corresponding uncomplexed monomers in organic solution **2b** ($r_{2k} = 1.3 \pm 0.1$), ($r_{2b} = 1,0 \pm 0.1$).[19]

Additionally, copolymerization of methylated β-CD complexed monomers n-butyl acrylate (**2b**), n-hexyl acrylate (**2d**) and cyclohexyl acrylate (**2i**) lead to the following reactivity ratios: copolymerization of **2b**/methylated β-CD and **2d**/methylated β-CD: ($r_{2b/\beta\text{-CD}} = 1.01 \pm 0.01$, $r_{2d/\beta\text{-CD}} = 1.04 \pm 0.01$); copolymerization of **2i**/methylated β-CD and **2d**/methylated β-CD: ($r_{2i/\beta\text{-CD}} = 0.74 \pm 0.01$, $r_{2d/\beta\text{-CD}} = 1.28 \pm 0.01$); copolymerization of **2i**/methylated β-CD and **2b**/methylated β-CD: ($r_{2i/\beta\text{-CD}} = 0.75 \pm 0.04$, $r_{2b/\beta\text{-CD}} = 1.13 \pm 0.01$). Only, the copolymerization of water-soluble **2b** complex with **2d** complex leads to a nearly ideal statistical copolymer. In contrast to that, the copolymerization in absence of cyclodextrin in DMF/H$_2$O-solution lead to nearly ideal statistical copolymers in all cases: **2b** and **2d**: ($r_{2b} = 1.01 \pm 0.01$, $r_{2d} = 0.91 \pm 0.03$); **2i** and **2d**: ($r_{2i} = 1.04 \pm 0.04$, $r_{2d} = 1.04 \pm 0.08$); **2i** and **2b**: ($r_{2i} = 0.95 \pm 0.02$, $r_{2b} = 1.24 \pm 0.04$).[20] It can be concluded, that the interaction between monomer and cyclodextrin are liable substantially for the control of copolymerization reactivity ratio. According to these experiments, it can be postulated, the stronger the interaction between host and guest the lower the r-value.

Finally, the influence of the hydrophilic character of acrylate guest monomers on the initial polymerization rate v_0 has also been investigated.[21] For kinetic studies, the methylated β-CD complexed monomers **2a-d, 2i** were polymerized in water by free radical mechanism using the water-soluble initiator 2,2'-azobis(2-amidinopropane) dihydrochloride (AAP). The concentration of non reacted monomer in the reaction mixture was deteminated using UV-spectroscopy or high performance liquid chromatography (HPLC). The intial reaction rate v_0 increases from methylated β-CD complexed monomer **2a** (12.5 ± 1.18) via **2b** (27.5 ± 0.83) to **2c** (44.2 ± 3.54) to **2i** (49.4 ± 1.18) up to **2d** (75.8 ± 2.55) x 10^{-6} mol·L^{-1}·s^{-1}.

However, the water-solubility of uncomplexed monomer increases in the order n-hexyl acrylate > n-pentyl acrylate > cyclohexy acrylate > n-butyl acrylate > n-propyl acrylate. The effect of the reduced water-solubility of the monomers **2a-d** is caused by the increasing hydrophobic character of the side group. Moreover, the stability of host-guest complex generally increases with geometrical fit of the guest in the CD cavity.

Also, the complex stabilities and dynamics, which are influenced by the sterical demands of the guest monomers, have an influence on the overall rate of polymerization. But, the increase of the initial rate v_0 with the increasing hydrophobic character of the complexed monomers seems to be the most important factor for the involved monomers **2a-d**. It was postulated that, the more hydrophobic the monomer, the higher the local concentration of complexes close to the active radical chain end of the phase separated polymer, which leads to higher values of the initial rate v_0.

Also, for the determination of the chain-transfer constants the complexed monomers **2f** and **3** were polymerized for 4 h at 80°C using 0.3% (w/w) of the water-soluble free radical initiator AAP in the presence of different amounts of the water-soluble chain-transfer agent N-acetyl-L-cystein from 0 to 3.0 mol%. Relatively high chain-transfer constants of N-acetyl-L-cysteine were found in the case of the complexed monomers **2f** and **3** in water ($Cs_{2f} = 1.7 \pm 0.3$ and $Cs_3 = 2.6 \pm 0.3$). In contrast, the chain-transfer constants in DMF/water mixture are significantly lower for uncomplexed monomers **2f** and **3**, respectively ($Cs_{2f} = 0.7 \pm 0.1$ and $Cs_3 = 0.7 \pm 0.1$).

From these results it can be concluded, that the molecular weight of polymer obtained from complexed monomers can be controlled more effectively using hydrophilic thiol containing chain-transfer agents than in the case of polymerization of the corresponding uncomplexed monomers in an organic medium.[22] Furthermore, the chain transfer constants of methylated β-CD complexed dodecanethiol $Cs_{2e} = 0.5 \pm 0.2$ for **2f**/methylated β-CD and $Cs_3 = 2.2 \pm 0.3$ for **3**/ methylated β-CD were observed in water, which also differ significantly from the corresponding values obtained from the polymerization experiments with uncomplexed monomers in the presence of uncomplexed dodecanethiol in an organic solution ($Cs_{2f} = 3.1 \pm 0.3$ and $Cs_3 = 3.4 \pm 0.3$).[23] That means that in this case, the molecular weight of polymers obtained from organic solution can be controlled more effectively by the uncomplexed hydrophobic thiol as the chain transfer agent.

Host-guest complexes of methyl methacrylate (**2f**) and methylated β-cyclodextrin were polymerized in aqueous medium using atom-transfer radical polymerization ATRP.[24] 4,4'-di-(5-nonyl)-2,2'-bipyridyl (dNbpy) was used as a very effective ligand in the ATRP of **2f**, which was complexed with large (approximately 4-fold molar) excess of methylated β-cyclodextrin in water. The polymerization of **2f**/methylated β-CD complex was initiated by ethyl 2-bromoisobuturate (EBIB) in the presence of copper(I)bromide as the catalyst.

It was found that the polymerization of **2f** under these condition has a *"living character"*. The obtained polymer has a much lower polydispersity (PD = 1.3 - 1.8) than those obtained from conventional free-radical polymerization. Additionally, it was observed that higher conversion of **2f** could be achieved using dNbpy/β-CD as the ligand compared to bipyridine (bipy). Furthermore, the block copolymerization of poly-methyl methacrylate bearing a bromoester endgroup with **3**/β-CD was also carried out under ATRP conditions in aqueous medium.

5. N-ISOPROPYL ACRYLAMIDE AS A MONOMER

Based on the above described observation, it was thus interesting to investigate the possibility to copolymerize a suitable CD-complexed hydrophobic monomer with a more hydrophilic comonomer. In this connection, the copolymerization of complexed monomers n-butyl methacrylate (**2e**), cyclohexyl methacrylate (**2j**), isobornyl acrylate (**2k**), isobornyl methacrylate (**2l**) and styrene (**3**) with the water-soluble N-isopropyl acrylamide (NIPAAm) was evaluated.[25]

Homopolymers from NIPAAm are well known and have been of great interest because of their lower critical solution temperature (LCST) behavior in water-solution. This means that the polymer is soluble at lower temperatures in water and precipitates above the LCST upon heating. Those kinds of polymers are potentially useful in the pharmaceutical or medical area.

Some papers described the polymerizations and characterizations of homo-PNIPAAM and copolymers with water-soluble comonomers such as 2-hydroxyethyl methacrylate, or itanoic acid. In the case of hydrophobic comonomers it is necessary to copolymerize in an organic solvent or to use eventually the emulsion polymerization process. However, the different solubilities of the monomers in water make the copolymerization of water-soluble and water-insoluble monomers by classical emulsion polymerization often difficult.

The methylated β-CD complexes of **2e**, **2j-l** and **3** were copolymerized with NIPAAm in water in molar ratios of 1:9 (90 mol% of NIPAAm) and 1:4 (80 mol% of NIPAAm), respectively for 12 h at 80°C using 5.5 mol% of the water-soluble free radical initiator 2,2'-azobis(2-aminopropane) dihydrochloride (AAP). The obtained copolymers were characterized using ^{1}H NMR spectroscopy and size exclusion chromatography (SEC). The composition of the copolymers were determined by ^{1}H NMR spectroscopy. In all cases, the content of hydrophilic comonomer-sequences (NIPAAm) in the copolymers is slightly higher than the initial monomer concentration below 80 mol% of initial concentration of NIPAAm.

For comparison the copolymerization of styrene (**3**) with NIPAAm at the same initial concentrations at molar ratios of 1:2 (67 mol% of NIPAAm) and 1:1 (50 mol% of NIPAAm), respectively, were also carried out under similar conditions in a mixture of N,N'-dimethylformamide (DMF) and water (9:1) without methylated β-CD. Surprisingly, it was found by means of the SEC data that, for all molar ratios of the incorporated monomer units, the weight averages M_w of the copolymers obtained from complexed styrene (**3**) and NIPAAm are about 3 to 7 times higher than in the case of the corresponding copolymers obtained from uncomplexed styrene (**3**) and NIPAAm in DMF/water-solution. In all cases, UV-signal and RI-signal of SEC-elution of the monomodalic SEC curves deviated only slightly by about 10% of the maximum. Regarding to SEC diagrams it is obvious that the polymers do not contain free residual methylated β-CD.

Furthermore, the **3**/methylated β-CD complex was copolymerized with various molar ratios of sodium 4-(acrylamido)-phenyldiazosulfonate (**7**) at 40°C using 1.5 mol% of the water-soluble free radical initiator 2,2'-azobis-(N,N'-dimethyleneisobutyramidine) dihydrochlorid (VA 044).[26] In all cases, the azo content of the polymer was below that of the monomer feed, determined by elemental analysis and UV spectroscopy. These results indicate that the diazosulfonate monomer (**7**) is less reactive than the complexed styrene (**3**). Even after extensive washing procedures and dialysis over several days, a small proportion of cyclodextrin was still left in the copolymer. The strong fixation of β-CD to the polymer may be explained by the fact that sulfonates tend to form complexes with CDs favored by H-bonding between the sulfonate groups and the hydroxy groups of the cyclodextrin.

The polymers obtained containing the diazosulfonate monomer (**7**) were soluble in DMF or dimethyl sulfoxide (DMSO). The molecular weight of the polymers were in the range M_n = 14,000-24,000 g mol^{-1}, with an unexpected low polydispersity PD = 2.4-2.5. Due to the high absorption coefficient (above 15,000 L mol^{-1} cm^{-1}) of the diazosulfonates, rapid photochemical decomposition was observed upon irradiation with UV light of suitable wavelength. The irradiation of the styrene-based copolymers led to decomposition of the azo chromophores. This loss of the polar sulfonate group causes a change in solubility of the polymer. As an example, the irradiation of the copolymers (10-30 mol% diazosulfonate) as films led to a DMF insoluble crosslinked material.

6. PHENOL DERIVATIVES

The use of cyclodextrin to polymerize phenol derivatives in water as solvent via an oxidative enzyme catalyzed mechanism was also investigated.[14] The enzyme-catalyzed process is expected to be an alternative route for the preparation of phenol polymers without the use of toxic formaldehyde, which is a coupling comonomer for the synthesis of conventional phenolic resins.

Because of the poor solubility of most phenols, the enzyme-catalyzed oligomerization was often carried out in a mixture of a water-miscible organic solvent and an aqueous buffer. Furthermore, the conversion to oligomers of higher molecular mass is often prevented. The phenolic dimers and trimers formed in the early stages of the reaction are insoluble in water and precipitate immediately.

To overcome these problems, N-(4-hydroxyphenyl)-maleimide (**6**), N-methacryoyl-11-aminoundecanoyl-4-hydroxyanilide (**1c**) and 4'-hydroxymethacrylanilide (**1d**) were simply complexed in water by stirring an equimolar amount of methylated β-CD and phenol. Oxidative oligomerizations of the cyclodextrin-complexed monomers were carried out in water at pH = 7 (controlled by a phosphate buffer) using horse radish peroxidase (HRP) as catalyst in presence of H_2O_2 (Scheme 4).

The oligomers **8-10**, which were obtained from complexed monomers showed an high-molecular mass fractions up to 3,000 g mol^{-1}, verified e.g., by MALDI-TOF-MS. FT-IR-spectra of oligomers **8-10** show no phenoxy ether linkages, since no additional peaks at 1150-1070 cm^{-1} and 1275-1200 cm^{-1} were observed (Figure 4). However, the oligomerization rate of complexed **6** in aqueous 1,4-dioxane was about 10% faster than that of the uncomplexed monomer. This indicates that the reactive sites of the phenoxy radical were not blocked by the cyclodextrin molecules.

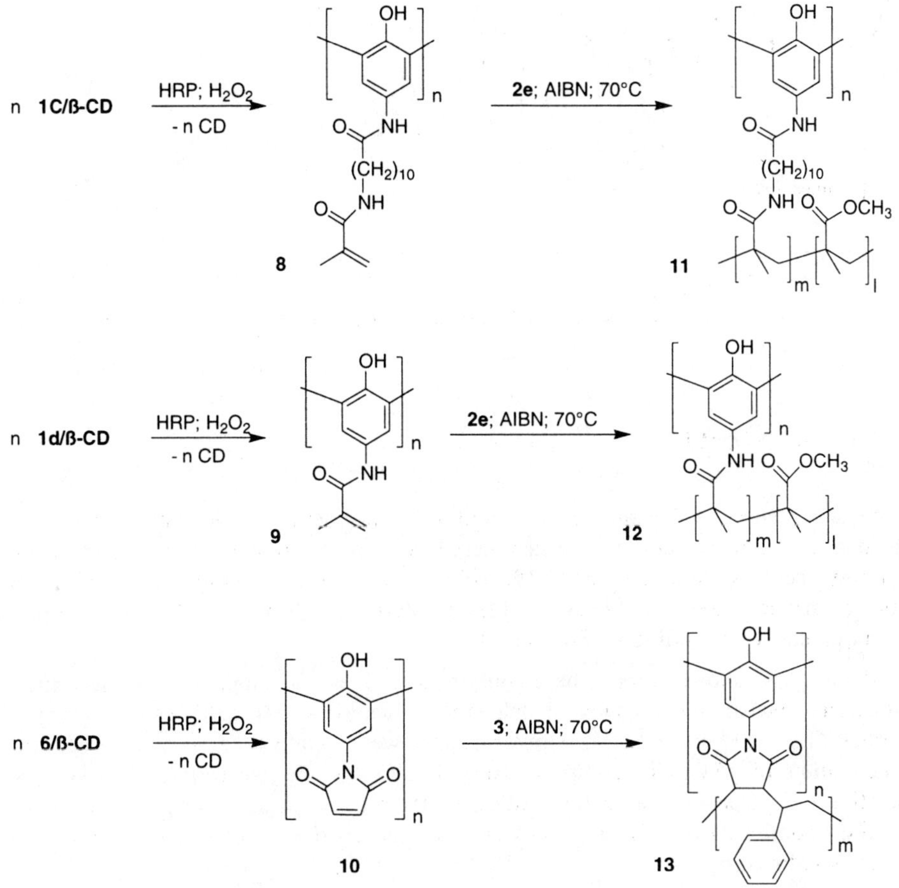

Scheme 4. Oxidative oligomerization of complexed phenol derivatives **1c, d,** and **6.**

The free radical copolymerization of oligomer **10** with styrene (**3**) using AIBN as catalyst in THF led to the formation of three dimensional networks **13**, and the AIBN initiated copolymerization of oligomers **8** or **9** with methyl methacrylate (**2f**) in THF resulted in powdery material **11** or **12**, which were insoluble in common organic solvents.

In contrast to the enzyme catalyzed oligomerization, free radical polymerization of complexed monomer **1d** in water at 70°C, initiated by the redox system $K_2S_2O_8/NaHSO_3$, resulted in the formation of poly(4'-hydroxymethacrylanilide). The conversion of the methacryl groups by free radical polymerization was confirmed by ^{1}H-NMR spectroscopy. It is interesting to note that the formation of phenoxy radicals by the enzyme catalyzed oligomerization does not initiate a free radical polymerization of the methacryl groups.

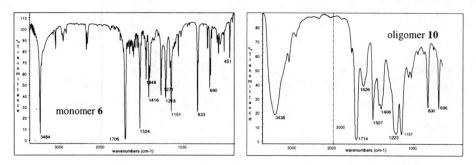

Figure 4. FT-IR spectra of N-(4-hydroxyphenyl)-maleimide (**6**) and corresponding oligomer **10**.

7. OTHER MONOMERS

A further successful example of cyclodextrin mediated polymerization method is the oxidative oligomerization of five members heterocyclic aromatic guests as pyrrol (**4**) or 3,4-ethylene-dioxythiophene (**5**).[13] The colorless crystalline of (**4**)/α-CD or (**5**)/α-CD complexes that are barely soluble at room temperature were polymerized in water at 60 - 70°C under oxidative conditions (Scheme 5).

IR analysis showed that these polymers did not contain any cyclodextrin. Conductivity measurements showed that these materials have the same electrical properties (10 - 100 S cm^{-1}).[27,28] Similar results were obtained upon electrochemical polymerization of (**5**)/α-CD complex using LiClO$_4$ as the electrolyte. Due to the substantial π-electron delocalization analog to their backbones, these polymers show interesting (nonlinear) optical properties and become good electronic conductors when oxidized or reduced.

Such polymers open up a variety of potential practical applications such as information storage, optical signal processing, electromagnetic interference (EMI) shielding, and solar energy conversion, as well as rechargeable batteries, light emitting diodes, field effect transistors, printed circuit boards, sensors and antistatic materials.[29,30]

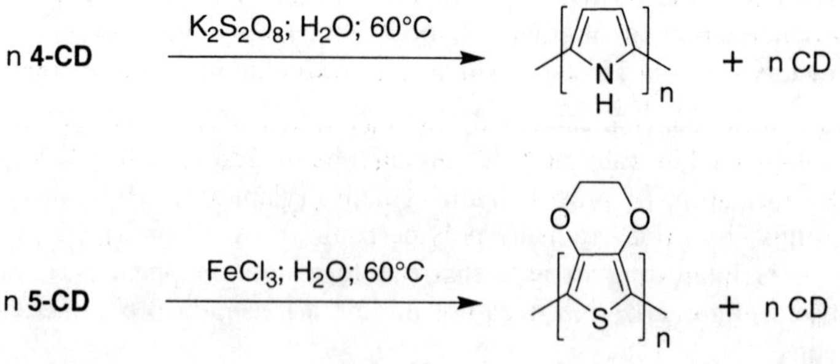

n **4-CD** $\xrightarrow{\text{K}_2\text{S}_2\text{O}_8;\ \text{H}_2\text{O};\ 60°\text{C}}$ [structure] + n CD

n **5-CD** $\xrightarrow{\text{FeCl}_3;\ \text{H}_2\text{O};\ 60°\text{C}}$ [structure] + n CD

Scheme 5. Oxidative polymerization of monomers **4**/β-cyclodextrin or **5**/β-cyclodextrin complexes in water.

In conclusion these investigations demonstrate the successful application of cyclodextrins in polymer synthesis in aqueous solution *via* free radical polymerization or via oxidative recombination mechanism. Some special aspects of cyclodextrins were found concerning the kinetics, chain transfer reaction and copolymerization-parameters.

8. REFERENCES

1. J. Szejtli and T. Osa, "Cyclodextrins", in *Comprehensive Supramolecular Chemistry*, Vol. 3, (Pergamon Press, Oxford, 1996).
2. J. Szejtli, *"Cyclodextrin Technology"* (Kluwer Academic Publisher, Dordrecht, 1998).
3. A. Harada, *Acta Polym.* **49**, 3 (1998).
4. S. Nepogodiev and J. F. Stoddart, *Cyclodextrin Based Catenanes and Rotaxanes*, *Chem. Rev.* **98**, 1959 (1998).
5. G. Wenz, *Angew. Chem.* **106**, 851 (1994).
6. W. Hermann, B. Keller, and G. Wenz, *Macromolecules* **28**, 4966 (1997).
7. M. Born and H. Ritter, *Angew. Chem.* **107**, 342 (1995).
8. M. Born, T. Koch, and H. Ritter, *Acta Polym.* **45**, 68 (1994).
9. M. Born and H. Ritter, *Macromol. Rapid Commun.* **12**, 471 (1991).
10. O. Noll and H. Ritter, *Macromol. Rapid Commun.* **18**, 53 (1997).
11. J. Jeromin and H. Ritter, *Macromol. Rapid Commun.* **19**, 377 (1998).
12. P. Glöckner and H. Ritter, *Designed Monomers and Polymers,* (in press).
13. J. Storsberg and H. Ritter, *Adv. Mater.* **12**, 567 (2000).
14. M. H. Reihmann and H. Ritter, *Macromol. Chem. Phys.* **201**, 798 (2000).
15. J. Jeromin and H. Ritter, *Macromolecules* **32**, 5236 (1999).
16. J. Jeromin, O. Noll, and H. Ritter, *Macromol. Chem. Phys.* **199**, 2641 (1998).
17. M. Fischer and H. Ritter, *Macromol. Rapid Commun.* **21**, 142 (2000).
18. J. Storsberg and H. Ritter, *Macromol. Rapid Commun.* **21**, 236 (2000).
19. P. Glöckner and H. Ritter, *Macromol. Rapid Commun.* **20**, 602 (1999).
20. S. Bernhardt, P. Glöckner, and H. Ritter, *Polymer Bulletin* **46**, 153 (2001).
21. S. Bernhardt, P. Glöckner, A. Theis, and H. Ritter, *Macromolecules* **34**, 1647 (2001).
22. P. Glöckner, N. Metz, and H. Ritter, *Macromolecules* **33**, 4288 (2000).
23. P. Glöckner and H. Ritter, *Macromol. Chem. Phys.* **201**, 2455 (2000).
24. J. Storsberg, M. Hartenstein, A. H. E. Müller, and H. Ritter, *Macromol. Rapid Commun.* **21**, 1342 (2000).
25. P. Casper, P Glöckner, and H. Ritter, *Macromolecules* **33**, 4361 (2000).
26. J. Storsberg, P. Glöckner, M. Eigner, U. Schnöller, H. Ritter, B. Voit, and O. Nuyken, *Designed Monomers and Polymers*, **4**, 9 (2001).
27. J. Rodriguez, H.-J. Grande, and T. F. Otero, in: *Handbook of Organic Conductive Molecules and polymers*, edited by H. S. Nalwa (Wiley, Chichester, UK, 1997), Vol. 2, p. 417.
28. G. Heywang and F. Jonas, *Adv. Mater.* **4**, 116 (1992).
29. *Handbook of Conducting Polymers*, 2nd ed., edited by T. A. Skotheim, R. L. Elsenbaumer and J. R. Reynolds (Marcel Dekker, New York 1998).
30. J. D. Stenger-Smith, *Prog. Polym. Sci.* **23**, 57 (1998).

SUPRAMOLECULAR COMPOUNDS OF CYCLODEXTRINS WITH [60]FULLERENE

Chivikula N. Murthy and Kurt E. Geckeler[*]

1. INTRODUCTION

The concept of the 'Chemistry beyond the molecule' has been gaining importance for the last few years and has resulted in several reviews.[1-3] The formation of supramolecular structures, assemblies, and arrays, held together by weak intermolecular interactions and non-covalent binding, has been used in applications being anticipated in both materials science research and biological systems. Among the supramolecular structures with noncovalent binding, the host-guest chemistry of fullerenes, specifically [60]fullerene, has been a topic of interest for almost a decade.

The interest in [60]fullerene chemistry is due to its unique physical and chemical properties. These include (a) its optical limiting property due to its electronic absorption in the entire UV-Vis range, (b) efficient singlet oxygen generating ability, (c) radical scavenging character, and (d) superconductivity on doping with alkali metals. While the properties (a) and (d) are useful in materials research, the properties (b) and (c) have applications in the biomedical area.

The covalent functionalization of [60]fullerene has been studied and new methods have been developed of incorporating the hydrophobic molecule with a low solubility in most organic solvents.[4,5] Novel supramolecular structures with enhanced solubility in both polar and nonpolar solvents have been prepared.[6,7] A number of host molecules have been found to form supramolecular structures with [60]fullerene including calix-[3,5,6,8]arenes,[8-11] calix[4]naphthalene,[12] cyclotri-veratrylene,[13] triptycenes,[14] porphyrin,[15] γ-cyclodextrin,[16] and β-cyclodextrin.[17]

* Laboratory of Applied Macromolecular Chemistry, Department of Materials Science and Engineering, Kwangju Institute of Science and Technology, 1 Oryong-dong, Buk-gu, Kwangju 500-712, South Korea.

Advanced Macromolecular and Supramolecular Materials and Processes
Edited by K. Geckeler, Kluwer Academic/Plenum Publishers, 2003

2. CYCLODEXTRIN INCLUSION COMPLEXATION

2.1. Fundamentals

The cyclodextrins are cyclic sugar molecules and the rich chemistry of these molecules has been extensively studied. These sugars are designated as α, β, or γ- cyclodextrins depending on whether the ring contains 6, 7, or 8 glucopyranose units, respectively. These molecules have a hydrophobic cavity and a hydrophilic exterior thus making them suitable candidates for generating supramolecular architectures based on non-covalent interactions with other molecules. This non-covalent interaction between cyclodextrins and other host molecules has been effectively used to solubilize water-insoluble molecules, keeping in view a number of applications including drug delivery systems.[18]

The solubility of β-cyclodextrin in water is the least among the cyclodextrins. However, it is the cheapest commercially available cyclodextrin, being more than sixty times cheaper than γ-cyclodextrin and also the most widely used in pharmaceutical applications.

[60]Fullerene discovered in 1985 has been found to have a rich chemistry and recent research has shown the biological activity of water-soluble [60]fullerene derivatives such as the DNA-cleaving ability and anti-HIV activity.[19,20] However, as [60]fullerene is a nonpolar molecule, it is insoluble in polar solvents. Therefore, attempts have been made to make [60]fullerene water-soluble by forming inclusion complexes with cyclodextrins.

2.2. Molecular Dimensions

Much of the research on inclusion complexation of cyclodextrins and [60]fullerene has centered on γ-cyclodextrin.[21,22] No reports are available in the literature on the inclusion complexes of the [60]fullerene molecule and β-cyclodextrin, and there are only a few reports on the inclusion complexation of [60]fullerene with cyclodextrin derivatives. This has resulted in the derivatization of β-cyclodextrin to enhance its water solubility and its subsequent complexation with [60]fullerene.[23,24]

The first reports on the inclusion complexation of [60]fullerene were based on heterogeneous reaction media and all have stated that only γ-cyclodextrin forms inclusion complexes, whereas β-cyclodextrin and α-cyclodextrin do not form complexes.[16,21,22] The reason given was the comparable size of [60]fullerene and the cavity diameter of the γ-cyclodextrin molecule, and that the [60]fullerene molecule does not intrude deeply into the γ-cyclodextrin cavity. In fact, there are several reports that describe the experimental failure to prepare inclusion complexes of β-cyclodextrin with [60]fullerene.

A comparison of the cavity diameters of α–, β–, and γ– cyclodextrin is given in Figure 1.

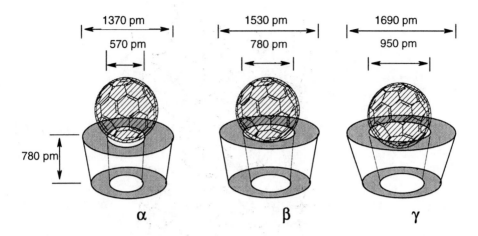

Figure 1. Comparison of the molecular dimensions of the different cyclodextrins with respect to [60]fullerene.

From the dimensions given in Figure 1 it becomes evident that the dimensions of the cyclodextrins do primarily not rule out the formation of an inclusion complex between β-cyclodextrin and [60]fullerene. The cavity diameter of β-cyclodextrin is 780 pm and the outer rim diameter is 1530 pm on the polar side of the molecule, as compared to 950 pm and 1690 pm, respectively, for the γ-cyclodextrin. On the other hand, it has been reported from crystallographic studies that the β-cyclodextrin cavity is suitable for spherically shaped guests such adamantane derivatives,[25] and the space enclosed by the head-to-head dimer is ~2.5 times wider than the cavity of the single molecule.[26]

It is also known that chemical modification of β-cyclodextrin changes the size and the shape of the cavity. For example, by methylation, the depth of the cavity is extended from 780 pm to ~ 1100 pm and the hydrophilic nature of both ends of the cyclodextrin cavity is changed due to the methyl groups attached to the O-2, O-3, or O-6 of the cyclodextrin. Thus, the cavity can fully accommodate adamantanol and naphthalene derivatives.[27,28]

2.3. Molecular Modeling

Molecular modeling studies are efficient tools to study supramolecular structures and have the advantage of offering the possibility of predicting structures even before carrying out an experiment. Energy-minimization calculations on the inclusion complex of β-cyclodextrin and [60]fullerene showed that it is possible to form a 2:1 complex. This is shown in the wire-frame model in Figure 2.

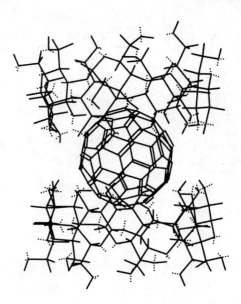

Figure 2. Energy-minimized model of a 2:1 complex between β-cyclodextrin and [60]fullerene.

Several dimensions can be obtained from the molecular modeling studies and from the energy-minimization calculations (Table 1).

Table 1. Dimensions of the energy-minimized models of the 2:1 inclusion complexes between cyclodextrins and [60]fullerene

	Dimension	
	β	γ
Nearest C-C distance (pm)	400 ~ 450	400 ~ 450
Nearest O-H distance (pm)	700 ~ 900	500 ~ 600
Accessible C_{60} surface (%)	~ 45	~ 15

In addition, the surface of [60]fullerene accessible to interact with the molecular environment was found to be higher in the case of the complex with β-cyclodextrin than for that with γ-cyclodextrin. It is evident that in a 2:1 complex between β-cyclodextrin and [60]fullerene, about 45% of the surface of the [60]fullerene is exposed when compared to about 15% for the complex between γ-cyclodextrin and [60]fullerene.

It is interesting to note that previous studies on the inclusion complex of γ-cyclodextrin and [60]fullerene have reported both 1:1 and 2:1 complexes between the two, whereas only a 2:1 complex was detected between β-cyclodextrin and [60]fullerene.

This result has important implications in the application of this supramolecular complex in biomedical research.

2.4. Solubility and Stability

Water-solubility is one of the major requirements for the emerging applications of supramolecular molecules and complexes in biomedical research. Since the discovery of the DNA-cleaving ability of [60]fullerene and its radical scavenging activity, efforts have been directed towards making the essentially nonpolar and hydrophobic [60]fullerene water-soluble. Two main trends have emerged in this direction: (a) to derivatize the [60]fullerene core, or (b) to form an inclusion complex with a water-soluble host. One of the drawbacks of the derivatized [60]fullerene is the concerns on the toxicity of the resulting products that is not known.

The other approach has been to form inclusion complexes where the essential characteristic properties of the [60]fullerene moiety are retained. Though a number of host molecules have been found to include [60]fullerene in their cavity, their application has been hindered due to their insolubility in water. Consequently, efforts have been directed towards the formation of inclusion complexes between cyclodextrins and [60]fullerene.

Based on the molecular modeling studies and analysis it has been seen that, apart from γ-cyclodextrin, it is possible to form an inclusion complex between β-cyclodextrin and [60]fullerene, given the optimal conditions for the complex formation. Table 2 summarizes the details of the different approaches reported for the complex formation of cyclodextrins and [60]fullerene.

Table 2. Comparison of the synthetic procedures adopted for the inclusion complex formation between cyclodextrins and [60]fullerene

Cyclodextrin	Solvent	Temperature	Year	Ref.
gamma	Water	Reflux	1992	16
gamma	Water/methanol	Reflux	1994	21
beta	Toluene/DMF	25 °C	2001	17

Several authors have reported the unsuccessful attempts to form inclusion complexes between β-cyclodextrin and [60]fullerene. However, all these reactions reported were attempted in heterogeneous reaction media, in which [60]fullerene is insoluble. In contrast, a change of the reaction medium to a homogeneous one resulted in a successful complex formation (60%) with a good yield.[17]

It has also been reported that the complex is stable and that concentrations of [60]fullerene as high as 1×10^{-4} (mol dm^{-3}) could be attained in water.

This is significant when compared to the solubility of the cyclodextrins and their inclusion complexes (see Table 3).

Table 3. Solubility of cyclodextrins and their inclusion complexes with [60]fullerene. CD = cyclodextrin.

Sample	Solvent	Solubility (mg ml^{-1})	Ref.
α-CD	Water	145	18
β-CD	Water	18.5	18
γ-CD	Water	232	18
C_{60}	Toluene	2.8	30
β-CD/C_{60} (2:1)	Water	0.4	17
γ-CD/C_{60} (2:1)	Water	0.33	21

Apart from the solubility of the supramolecular structures in aqueous media, the other essential requirement for such compounds is their stability relative to that of the separated species in solution under the actual conditions of medium and temperature. This stability determines their potential applications in the biomedical and pharmaceutical areas.

Stability constants have been measured using different techniques, and also absorption spectroscopy has been used as a technique to measure the stability constants of such complexes.[16] It was found that the β-cyclodextrin/[60]fullerene complex is quite stable in aqueous solution and its stability is comparable to that of the γ-cyclodextrin/[60]fullerene complex. The values determined by using the UV-Vis spectroscopy (Benesi-Hildebrand method) are given in Table 4.

Table 4. Comparison of data for the inclusion complexation of cyclodextrins (CDs) with [60]fullerene.

Constant	Unit	β–CD	Ref.	γ–CD	Ref.
Equilibrium constant	dm^6 mol^{-2}	1.69 x 10^5	17	4.0 x 10^4 2.6 x 10^7	29 21
Extinction coefficient	dm^3 mol^{-1} cm^{-1}	2,159	17	12,200	21

2.5. Radical Scavenging Properties

One of the outstanding features of C_{60} is its radical scavenging property. In a model study using a stable free-radical diphenyl picryl hydrazyl (DPPH) it was found that the β-CD/[60]fullerene complex showed a good radical scavenging activity with time. The gradual reduction of a typical absorption maximum is shown in Figure 3.

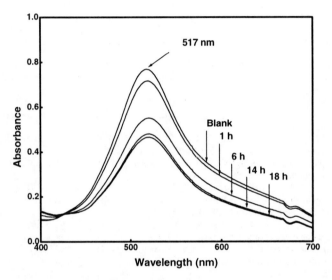

Figure 3. Radical scavenging behavior of the β–cyclodextrin/[60]fullerene inclusion complex with time studied by absorption spectroscopy.

3. CONCLUSIONS

Several inclusion complexes of cyclodextrins with [60]fullerene have been studied featuring quite different properties. Essential is their water-solubility with respect to biomedical applications. Compared to the γ-cyclodextrin complex with C_{60}, the β-cyclodextrin complex has a three times higher accessible surface and is about 60 times cheaper. In conjunction with the radical scavenging property of [60]fullerene, the resulting supramolecular compounds have a high potential for applications in the photodynamic therapy.

4. ACKNOWLEDGEMENTS

The authors gratefully acknowledge the generous financial support from the Ministry of Science and Technology and the Ministry of Education, Korea, as well as from the Kwangju Institute of Science and Technology.

5. REFERENCES

1. J.–M. Lehn, *Supramolecular Chemistry: Concepts and Perspectives* (VCH, Weinheim, 1995).
2. F. Diedrich and G. M. Lopez, Supramolecular fullerene chemistry, *Chem. Soc. Rev.* **28**, 263-277 (1999).
3   G. R. Desiraju, Chemistry beyond the molecule, *Nature* **412**, 397-400 (2001).
4. R. Taylor and D. R. M. Walton, The Chemistry of Fullerenes, *Nature* **363**, 685-693 (1993).
5. K. E. Geckeler, Macromolecular Fullerene Chemistry, *Trends Polym. Sci.* **2**, 355-360 (1994).
6. S. Samal, B.-J. Choi, and K. E. Geckeler, The first water-soluble main- chain polyfullerene, *Chem. Commun.* **15**, 1373-1374 (2000).
7. S. Samal, and K. E. Geckeler, Cyclodextrin-fullerenes: A new class of water-soluble fullerenes, *Chem. Commun.* **13**, 1102-1103 (2000).
8. N. S. Issacs, P. J. Nicols, C. L. Raston, C. A. Sandoval, and D. J. Young, Solution volume studies of deep cavity inclusion complex of C_{60}: p-benzylcalix[5]arene, *J. Chem. Soc., Chem. Commun.* **19**, 1839-1840 (1997).
9. J. L. Bourdelande, J. Font, R. Gonzalez-Moreno, and S. Nonell, Inclusion complex of calix[8]arene-C_{60} : photophysical properties and its behavior as singlet molecular oxygen sensitizer in solid state, *J. Photochem. Photobiol.* A, **115**, 69-71 (1998).
10. A. Ikeda, M. Yoshimura, H. Udzu, C. Fukuhara, and S. Shinkai, Inclusion of [60]fullerene in a homo-oxacalix[3]arene-based dimeric capsule crosslinked by a Pd(II)-pyridine interaction, *J. Am. Chem. Soc.* **121**, 4296-4297 (1999).
11. T. Liu, M. X. Li, N. Q. Li, Z. J. Shi, Z. N. Gu, and X. H. Zhou, Electrochemical behavior of the $(C_{60})_2$-calix[6]arene inclusion complex films, *Electroanalysis* **11**, 1227-1232 (1999).
12. P. E. Georghiou, S. Mizyed, and S. Chowdhury, Complexes formed from [60]fullerene and calix[4]-naphthalenes, *Tetrahed. Lett.* **40**, 611-614 (1999).
13. H. Matsubara, A. Hasegawa, K. Shiwaku, K. Asano, M. Uno, S.Takahashi, and K. Yamamoto, Supra-molecular inclusion complexes of fullerenes using cyclotriveratrylene derivatives with aromatic pendants, *Chem. Lett.* **9**, 923-924 (1998).
14. E. M. Veen, B. L. Feringa, P. M. Postma, H. T. Jonkman, and A. L. Speck, Solid state organization of C_{60} by inclusion crystallization with tryptycenes, *J. Chem. Soc., Chem. Commun.* **17**, 1709-1710 (1999).
15. K. Tashiro, T. Aida, J-Yu. Zheng, K. Kinbara, K. Saigo, S. Sakamoto, and K. Yamaguchi, A cyclic dimer of metalloporphyrin forms a highly stable inclusion complex with C_{60}, *J. Am. Chem. Soc.* **121**, 9477-9478 (1999).
16. T. Andersson, K. Nilsson, M. Sundhal, G. Westman, and O. Wennerstrom, C_{60} embedded in γ-cyclodextrin: a water-soluble fullerene, *J. Chem. Soc., Chem. Commun.* 604-606 (1992).
17. C. N. Murthy and K. E. Geckeler, The water-soluble β-cyclodextrin-[60]fullerene complex, *Chem. Commun.* 1194-1195 (2001).
18. J. L. Atwood, J. E. D. Davies, D. D. MacNicol, and F. Vogtle, in: *Comprehensive Supramolecular Chemistry*, Vol.3, Cyclodextrins, Eds. J. Szejtli and T. Osa (Elsevier, New York, 1996).
19. R. Sijbesma, G. Srdanov, F. Wudl, J. A. Castoro, C. Wilkins, S. H. Friedman, D. L. DeCamp, and G. L. Kenyon, Synthesis of a Fullerene Derivative for the Inhibition of HIV Enzymes, *J. Am. Chem. Soc.* **115**, 6510-6514 (1993).
20. S. H. Friedman, D. L. DeCamp, R. Sijbesma, G. Srdanov, F. Wudl, and G. L. Kenyon, Inhibition of the HIV-1 Protease by Fullerene Derivatives: Model Building Studies and Experimental Verification, *J. Am. Chem. Soc.* **15**, 6506-6509 (1993).
21. K. I. Priyadarsini, H. Mohan, A. K. Tyagi, and J. P. Mittal, Inclusion complex of γ-cyclodextrin-C_{60}: Formation, characterization, and photophysical properties in aqueous solutions, *J. Phys. Chem.* **98**, 4756-4759 (1994).
22. K. Kanazawa, H. Nakanishi, Y. Ishizuka, T. Nakamura, and M. Matsumoto, An NMR study of the buck-minsterfullerene complex with cyclodextrin in aqueous solution, *Fullerene Sci. Technol.* **2**, 189-194 (1994).
23. D. D. Zhang, J. W. Chen, Y. Yang, R. F. Chai, and X. L. Shen, Studies of methylated β-cyclodextrin and C_{60} inclusion complexes, *J. Incl. Phenom. Molec. Recogn. Chem.* **16**, 245-253 (1993).
24. H. Hu, Y. Liu, D. D. Zhang, and L. F. Wang, Studies on water-soluble α-, β-, or γ-cyclodextrin pre-polymer inclusion complexes with C_{60}, *J. Incl. Phenom.* **33**, 295-305 (1999).
25. J. A. Hamilton, and M. N. Sabesan, Structure of a complex of cycloheptaamylose with 1-adamantane carboxylic acid, *Acta Cryst., Sec. B* **38**, 3063-3069 (1982).
26. K. H. Jogun, E. Eckle, and K. Bartels, Dimeric β-cyclodextrin complexes may mimic membrane diffusion transport, *Nature* **274**, 617 (1978).

27. M. Czugler, E. Eckle, and J. J. Stezowski, The crystal and molecular structure of a 2,6-tetradeca-o-methyl-ß-cyclodextrin adamantanol complex, *J. Chem. Soc., Chem. Commun.* 1291-1293 (1981).
28. K. Harata, The x-ray structure of an inclusion complex of heptakis(2,6-di-O-methyl)-β-cyclodextrin with 2-naphthoic acid, *J. Chem. Soc., Chem. Commun.* 546-548 (1993).
29. A. Buvari-Barcza, T. Braun, and L. Barcza, On the formation of water-soluble buckminsterfullerene-γ-cyclodextrin complexes, *Supramol. Chem.* **4**, 131-133 (1999).
30. R. S. Rouff, D. S. Tse, R. Malhotra, and D. C. Lorents, Solubility of C_{60} in a variety of solvents, *J. Phys. Chem.* **97**, 3379-3383 (1993).

STRUCTURE AND FUNCTION OF POLYMERIC INCLUSION COMPLEX OF MOLECULAR NANOTUBES AND POLYMER CHAINS

Kohzo Ito, Takeshi Shimomura, and Yasushi Okumura*

1. INTRODUCTION

Various organic supramolecules with unique structures have attracted great interests of many scientists.[1-3] Harada et al. prepared a polyrotaxane supramolecule in which cyclodextrin (CD) molecules of cyclic form were threaded on a polymer chain with bulky ends[4,5] and then synthesized molecular nanotubes with the diameter smaller than carbon nanotubes by crosslinking the adjacent CD units in the polyrotaxane.[6] This molecular nanotube, highly soluble in several kinds of solvents such as water, has a constant inside diameter (0.45nm) and a longitudinal length of submicron order controllable by varying the length of the polyrotaxane.

Owing to the infinitesimal inside diameter of the molecular nanotube, a polymer chain included in the nanotube has an extended conformation, such as a planar zigzag one, with no degrees of freedom other than a translational motion along its longitudinal axis. Therefore, the inclusion of a polymer chain into a molecular nanotube is entropically unfavorable and promoted by attractive interaction such as hydrophobic one between the chain and the nanotube. In other words, heating results in the dissociation of the polymer chain from the nanotube with recovery of the intrinsic entropy of random conformation as shown in Fig.1. Moreover the conformational entropy and inclusion interaction are roughly estimated to be proportional to the length of the nanotube and polymer chain. Consequently, the free energy changes drastically with inclusion or dissociation between a long molecular nanotube and polymer chain, which leads to a transitional behavior.

We have recently investigated the inclusion complex formation between molecular nanotubes and polymer chains theoretically[7,8,12] and experimentally[9-11,13-16] As mentioned above, this polymeric supramolecular system has a feature that the inclusion-dissociation behavior becomes cooperative and sharp compared with the inclusion

* Graduate School of Frontier Sciences, University of Tokyo, 7-3-1 Hongo, Bunkyo-ku, Tokyo, 113-8656, Japan (*kohzo@k.u-tokyo.ac.jp*).

Advanced Macromolecular and Supramolecular Materials and Processes
Edited by K. Geckeler, Kluwer Academic/Plenum Publishers, 2003

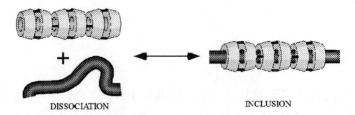

DISSOCIATION   INCLUSION

Figure 1. Schematic diagram of the inclusion complex formation of molecular nanotube and linear polymer chain. When the polymer chain is included into the nanotube, the chain conformation is confined to rodlike one, all trans or planar zigzag configuration, because of the extremely small inside diameter of the nanotube. Then the polymer chain loses the conformational entropy. On the other hand, heating promotes the dissociation of the linear chain from the nanotube.

complex formation between cyclic molecule and small molecular compounds.[7] And we have proposed some functional supramolecules such as insulated molecular wire[14] and topological gel[16]. In this chapter, we will introduce these theoretical and experimental works on the polymeric supramolecular system of molecular nanotubes and polymer chains.

2. THEORY ON INCLUSION AND DISSOCIATION BEHAVIOR OF POLYMERIC INCLUSION COMPLEX

First of all, we have theoretically treated the inclusion-dissociation behavior of the nanotubes and linear polymer chain of the same length in solutions with the Flory-Huggins lattice model.[7] This model allows the theory to incorporate the interaction among polymer chains. We obtained the total free energy of a system containing nanotubes, polymer chains and solvents to determine the fraction of the nanotube occupied by polymer chains in the cavity. The results show us that the free energy profile as a function of the inclusion fraction is quite different between good and poor solvents. In good solvent, the free energy profile is always concave upwards and hence the polymer chains are gradually dissociated from or included into the nanotube. This behavior becomes sharp with increasing length of the nanotube and polymer chain, and eventually reaches a transitional one without hysteresis in the infinite limit.

On the other hand, the free energy profile can have two local minima at nearly full inclusion and dissociation states in poor solvent. Then inclusion-dissociation transition occurs with hysteresis loop as shown in Fig.2. Furthermore, the theory predicts that as polymer chains are added to a nanotube solution with poor solvent, the chains are first included into nanotubes, the inclusion saturates at proper concentration and then the chains are drastically dissociated from the nanotube in reverse. This is because polymer chains in poor solvent have attractive interaction with each other. It is worthwhile noting that these transitional behaviors require long nanotubes and polymer chains in high concentration. When a small molecular compound forms inclusion complex with a small cyclic molecule, the free energy gap between the inclusion and dissociation states is slight compared with the polymeric inclusion complex. Therefore the inclusion complex formation of small molecular compounds would hardly show the inclusion-dissociation transition. The transitional behavior is an important feature of the polymeric inclusion complex.

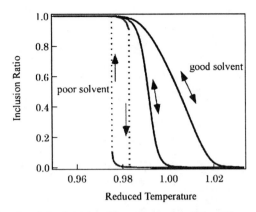

Figure 2. The inclusion-dissociation behavior of the molecular nanotube and linear polymer chain. In good solvent, the polymer chain is included into or dissociated from the nanotube gradually. This behavior becomes sharp with increasing length of the nanotube and chain. On the other hand, the inclusion-dissociation transition is expected to occur with hysteresis loop in poor solvent.

Next we theoretically treated the inclusion complex formation of molecular necklace in which CDs were threaded on a polymer chain without bulky ends.[8] It was reported that many hydrogen bonds between adjacent CDs made important contributions to the inclusion complex formation. Accordingly we calculated the total free energy of the system, taking account of the interaction between adjacent CDs in the theory. The results indicated that if the interaction between adjacent CDs is small, CD molecules gradually included or dissociated the polymer chain as the amount of CDs or temperature is varied. The inclusion-dissociation behavior became cooperative and sharp with increasing adjacent interaction similar to the inclusion complex formation with the molecular nanotube.

3. EXPERIMENTAL RESULTS OF INCLUSION COMPLEX FORMATION

3.1 Optical Absorption Spectra

It was reported that the molecular nanotube included iodine molecules in the cavity.[6] However, it had not been confirmed whether the linear chain was actually included into the small cavity. We experimentally investigated the inclusion behavior of polyethylene glycol (PEG) with different contour lengths and some diblock copolymers of PEG and alkyl chains with larger hydrophobicity into the molecular nanotube by measuring the optical absorption spectra of solution with iodine as a probe.[9,10]

When the nanotube includes iodine in the cavity, the solution color changes from yellow to red since the alignment of iodine in the cavity increases π conjugation. This is

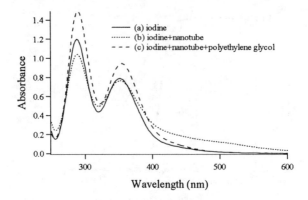

Figure 3. The optical absorption spectra of aqueous solutions of (a) iodine only, (b) iodine and molecular nanotube, and (c) iodine, nanotube and polyethylene glycol (PEG). The absorbance at longer wevelength increases with addition of nanotubes to iodine solution, and reversely decreases with further addition of PEG.

similar to the well-known iodine-starch reaction. If polymer chains added to the solution push out iodine and instead are included into the nanotube, the solution color should revert from red to yellow. Figure 3 shows this process. When the nanotube is added to the iodine solution, the absorbance at wavelength longer than ca. 400nm increases, which corresponds to the color change from yellow to red. And a further addition of PEG to the solution reduces the absorbance as shown in Fig.3. This indicates that PEG replaces iodine in the cavity of the nanotube. Incidentally, addition of polypropylene glycol (PPG) thicker than PEG exerted no change in solution color. This is because PPG is too thick to be included into the nanotube and replace iodine.

When PEG is added to solution of iodine and nanotube, the absorbance decreases with increasing concentration of PEG. Correspondingly we can determine the inclusion fraction of PEG in the nanotube from the absorbance at longer wavelength, e.g., 550nm. We measured the fraction of the nanotube including PEG as the concentration or contour length of PEG is varied. The results indicated that the inclusion fraction increases sharply with contour length of PEG. This is in qualitative agreement with our theory.

Furthermore, we calculated the total free energy of a system containing nanotubes, linear chains, iodine and solvent to evaluate the dependence of the inclusion fraction on the polymer concentration. By fitting the theoretical curves to the experimental data, we eventually obtained the inclusion energy of PEG in the molecular nanotube. The value $2.7kT$ per monomer unit of PEG, obtained independently of the polymer length, is equal to the inclusion energy between α-CD and propanol, where kT is the thermal energy at room temperature (T=300K).

The experimental results indicate that replacement of iodine in the cavity requires a large amount of PEG compared with iodine. This is because iodine is more hydrophobic than PEG and the cavity of the molecular nanotube has high hydrophobicity. Then

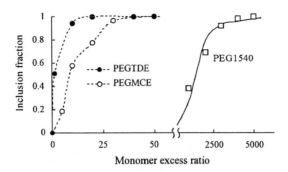

Figure 4. The dependence of the inclusion fraction of polyethylene glycol PEG1540 (Molecular Weight: 1540), polyethylene glycol tridecyl ether (PEG-TDE) and polyethylene glycol monocetyl ether (PEG-MCE) on the monomer excess ratio of polymer chains to the molecular nanotube. Of three polymer chains, PEG-TDE is the shortest and most hydrophobic one while PEG1540 is the longest and most hydrophilic. In spite of the shortest chain, the inclusion fraction of PEG-TDE increases most sharply with polymer concentration. This indicates that the hydrophobicity of a linear chain greatly promotes the inclusion complex formation with the molecular nanotube since the cavity is hydrophobic.

hydrophobic polymer chains are expected to replace iodine in the nanotube by smaller amount than PEG. We compared the inclusion behavior of polyethylene glycol tridecyl ether (PEG- TDE) and polyethylene glycol monocetyl ether (PEG-MCE) with PEG1540 (Molecular Weight: 1540).[10] Of three polymer chains, PEG-TDE is the shortest and most hydrophobic one while PEG1540 is the longest and most hydrophilic. Figure 4 shows the experimental results of the inclusion behavior. In spite of the shortest chain, the inclusion fraction of PEG-TDE increases most sharply with polymer concentration. This indicates that the hydrophobicity of a linear chain greatly promotes the inclusion complex formation with the molecular nanotube since the cavity is hydrophobic. This means that the nanotube has a molecular recognition of hydrophobicity.

3.2 Induced Circular Dichroism

Induced circular dichroism is widely used for confirmation of inclusion complex formation between cyclodextrin and chromophore such as azobenzene. We applied this technique to direct investigation of the polymeric inclusion complex formation of the molecular nanotube and PEG modified with azobenzene at both ends (PEG-Az).[11]

We prepared some PEG-Az chains of different length. The solution of PEG-Az or the nanotube showed no circular dichroism at optical absorption band of azobenzene, typically 350nm. However, almost the same spectral profile of circular dichroism as the optical absorption spectrum of azobenzene emerged in the mixture of them. The same profile between the circular dichroism and absorbance clearly indicates the induced circular dichroism, namely, that PEG-Az forms the inclusion complex with the molecular nanotube. We confirmed the inclusion complex formation of the nanotube and linear chain by the induced circular dichroism spectroscopy.

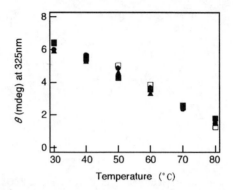

Figure 5. The temperature dependence of ellipticity of the nanotube and PEG-Az at 325nm. The symbols show the difference in molecular weight (Mw) of PEG: ● Mw=600, □ Mw=1000, ▲ Mw=2000, ■ Mw=4000. The polymer chains are dissociated from nanotube with increasing temperature.

Next we measured the dependence of ellipticity for PEG-Az of different length on the polymer concentration and found that the increase of ellipticity saturated at different concentration ratios of PEG-Az to the nanotube, which are almost proportional to the polymer length. This means that longer PEG-Az chains can fill nanotubes by smaller amount. If azobenzene groups alone of PEG-Az are included in both ends of the nanotube like caps of the tube, the saturation ratio should be independent of the polymer length. Therefore, it was concluded that PEG-Az chains were deeply included into the capillary of nanotube.

Furthermore, we measured the temperature dependence of ellipticity as shown in Fig.5. It is seen that the ellipticity decreases with increasing temperature. This indicates that heating promotes dissociation of PEG-Az from nanotubes since the induced circular dichroism is proportional to the number of the inclusion complex. This tendency is in qualitative agreement with our theoretical prediction.

4. APPLICATION OF POLYMERIC INCLUSION COMPLEX

4.1 Molecular Linear Motor of Nanotube on Polymer Rail

Now let us introduce some applications of the polymeric inclusion complex to novel kinds of functional supramolecules. First of all, we have theoretically proposed switching inclusion complex of the molecular nanotube and block copolymer.[12]

Let us consider a diblock copolymer consisting of A and B chains of the same length as the nanotube. Both chains can be included into the nanotube; the A chain has inclusion energy and conformational entropy larger than the B chain. Then it is possible to form a polyrotaxane of the molecular nanotube and diblock coplymer with bulky ends.

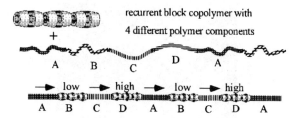

Figure 6. Nanotube and a recurrent block copolymer comprising linear polymer chains of four different kinds. If we properly choose the inclusion energies and conformational entropies of four chains, the nanotube travels on the block copolymer rail stepwise in one direction by temperature undulation.

The difference between two chains in the inclusion energy and conformational entropy yields a switching behavior that the nanotube includes the A chain at low temperature and the B chain at high temperature. We calculated the total free energy of a system containing many switching inclusion complexes, taking account of the miscibility between two chains. The theoretical results indicated that the nanotube shifts gradually with varing temperature on a miscible block copolymer while switching transition occurs with a hysteresis loop in immiscible block copolymers.

Next let us consider a recurrent block copolymer comprising linear chains of four different kinds as shown in Fig.6. If we properly choose the inclusion energies and conformational entropies of four chains, a nanotube travels on the block copolymer rail stepwise in one direction with undulating temperature. This means a molecular linear motor operated by temperature undulation. It is worth noting that if a cyclic molecule is substituted for a nanotube, it will move in a random direction by temperature undulation. This is because the thermal agitation dominates the movement of the small molecule. In contrast, the nanotube, more generally the polymeric system, overcomes the thermal agitation. This is another feature of the polymeric inclusion complex as well as the transitional behavior.

4.2 Dendritic Supramolecular Structure

We have recently formed self-assembling dendritic supramolecule of the molecular nanotube and a three-armed starpolymer.[13] Here the starpolymer joints three nanotubes, which leads to a network or dendritic supramolecular structure.

The starpolymer was synthesized from polyethylene glycol monocetyl ether (PEG-MCE) of Mw=1250 and 1,3,5-benzenetricarbonyl trichloride. The contour length of PEG-MCE is about half as long as the molecular nanotube. The inclusion complex formation between the molecular nanotube and the starpolymer was confirmed with iodine by the optical absorption spectra as mentioned before.[10] And we measured the quasi-elastic light scattering of the solution of the nanotube and starpolymer.[10] Then the hydrodynamic radius in the solution increased from 56nm to 300nm after the starpolymer was added to the nanotube solution. These results indicate that the nanotube and starpolymer form a large inclusion complex.

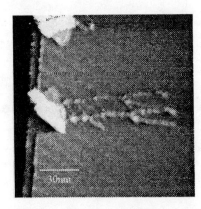

Figure 7. STM image of a trifurcated dendritic inclusion complex formed by molecular nanotubes and three-armed starpolymers. The longitudinal straight line on the left is the step-edge of HOPG. The dendritic structure is fixed to the step-edge. The double image of the structure is due to a rough tip of the STM probe.

To confirm the inclusion complex formation, we performed scanning tunneling microscopy (STM) observation. We first covalently fixed PEG-MCE along the step-edge of highly oriented pyrolytic graphite (HOPG) and next immersed the HOPG substrate in solution of the molecular nanotube, followed by dry. When we observed the substrate by STM, it was seen that some rodlike structures lay perpendicular to the step-edge. The length (20-30nm) corresponded to those of the nanotube. This indicates that the nanotube included PEG-MCE chain fixed to the step-edge on HOPG. Next we immersed HOPG modified with PEG-MCE into solution of the nanotube and starpolymer, and then observed the substrate by STM. Figure 7 shows the STM image, where a trifurcated dendritic structure is seen. Since this structure has three branches of ca. 30nm in length, we concluded that the starpolymer jointed three nanotubes and formed the trifurcated dendritic inclusion complex with the nanotube.

4.3 Insulated Molecular Wire

Much attention has recently been paid to molecular devices. Conjugated conducting polymers were regarded as a promising candidate for molecular wire connecting among electrodes and functional organic molecules such as molecular diode. However, it is quite difficult to actually use conducting polymers for molecular wire and even the conductivity of a long single chain has not been measured yet. This is mainly because conducting polymer chains generally form fibrils entangled complicatedly. Certainly, some conducting polymer chains with alkyl side chains are soluble in organic solvents and can be isolated in very dilute solution. However, the polymer chain forms coiled conformation then because of the large conformational entropy. The coiled conformation consists of randomly distributed trans and gauche configurations. The trans configuration, namely, the planar zigzag or coplanar one, delocalised electron due to π conjugation while the gauche configuration breaks the conjugation to form defect. This indicates that high conductivity is not expected in the coiled conformation of the conducting polymer but in the all trans configuration, rodlike conformation. Consequently, to use a conducting polymer chain for molecular wire, we have to isolate a single chain and extend it to rodlike conformation. Incidentally, we can somewhat stretch a conducting polymer chain by adding poor solvent or doping, but then the extended polymer chain rapidly aggregates to become insoluble.

To resolve the difficulty in using the conducting polymer for molecular wire, we applied the inclusion complex formation with CD to a conducting polymer chain.[14] What is the advantage of covering the conducting polymer chain with CDs? The coverage should reduce the attractive interaction between polymer chains considerably because CD has hydrophilic outside and highly soluble in various organic solvents. This means that one can easily isolate a single conducting polymer chain by covering with CDs. Next it is also expected that the conducting polymer chain forming the inclusion complex with CDs would be confined to rodlike conformation because the inside diameter of CD is extremely small. This means that the polymer chain has all trans, planar zigzag or coplanar configuration. Therefore, the π conjugation system should spread over a whole chain of the conducting polymer, which results in high conductivity. This inclusion complex has a molecular wire as an axis, which is covered with insulating cyclic molecules. Therefore it can be regarded as insulated molecular wire.

A conducting polymer used in the inclusion complex formation is emeraldine base polyaniline (PANI), which is highly soluble in n-methyl-2-pyrrolidone (MP) and has the average contour length of 300nm. Other soluble conducting polymers have too bulky side chains such as hexyl one to be included into the fine nanotube. We mixed MP solution of PANI with aqueous solution of α-, β- and γ-CDs at various temperatures. Then blue precipitation appeared in only the solution with β-CD at low temperature below ca. 275K. It has been reported that very high concentration of rodlike inclusion complexes yields precipitation, which is therefore used as an evidence of the inclusion complex formation. Accordingly, the experimental results suggested that only β-CD formed the inclusion complex with PANI because of a close fit. This is consistent with a report that aniline molecule, monomer of PANI, forms the inclusion complex with β-CD only. Furthermore, it is another important point of the experimental results that the precipitation was observed at low temperature. This is in qualitative agreement with the theoretical prediction that the inclusion behavior should be promoted at low temperature as mentioned before.

Next we investigated the structure of the inclusion complex by the electric birefringence spectroscopy and STM. The electric birefringence appears in solution containing rodlike molecules. The experimental results of the electric birefringence showed that large electric birefringence was observed in the solution of β-CD and PANI at low temperature below ca. 275K and drastically decreased down to zero with increasing temperature. This means that β-CD and PANI forms rodlike inclusion complex at lower temperature while β-CD is dissociated from PANI at higher temperature. On the other hand, Fig. 8 shows a STM image of HOPG substrate, on which low-temperature solution of β-CD and PANI was dropped and spincoated. As

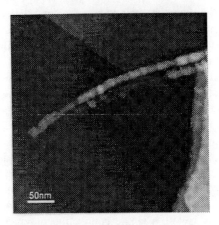

Figure 8. STM image of an insulated molecular wire formed by β-CD and PANI. The length is almost equal to the average contour length 300nm of PANI and the height is close to the outside diameter of β-CD. The double image of the structure is due to a rough tip of the STM probe.

shown in Fig.8, a rodlike structure is caught by an exfoliation of step on HOPG. The length is almost equal to the average contour length 300nm of PANI and the height is close to the outside diameter of β-CD. Consequently, we concluded that the rodlike structure was identified as the inclusion complex of the β-CD and PANI, namely, the insulated molecular wire.

Moreover, we investigated the insulation effect of β-CD on oxidization of PANI by iodine.[15] It was reported that when iodine was added to MP solution of emeraldine base PANI, the solution color changed from blue to violet owing to the oxidization of PANI by iodine. Accordingly, the optical absorption spectroscopy determines whether PANI is oxidized or not. When we added iodine to solution of β-CD and PANI at 275K, the solution color did not change although the solution color of PANI at 275K changed to violet by addition of iodine. This indicates that β-CD fully covers PANI and prevents oxidization of PANI by iodine at low temperature. Next we heated up the solution of the insulated molecular wire and iodine to 288K. Then the solution color shifted to violet ca. 4 hours after heating. This suggests that β-CD was slowly dissociated from PANI at 288K and then PANI was oxidized by iodine. Namely, PANI is not oxidized by iodine as long as β-CD covers PANI perfectly.

4.4 Topological Gel

Polymeric system shows self-assembled higher-order structures formed by a variety of intramolecular and intermolecular interactions. The structures and physical properties in nanoscale exert a great influence on the macroscopic properties. Very recently, we have reported a novel kind of gel other than the conventional physical and chemical gels.[16] The gel has high modulus, transparency and swellability arising from the nanoscale peculiar structure.

We first synthesized a polyrotaxane in which a PEG chain with large molecular weight is sparsely included by α-CD. By chemically cross-linking α-cyclodextrins contained in the polyrotaxanes in solutions, we got transparent gels with good tensibility, low viscosity and large swellability in water. In this gel, the polymer chains with bulky end groups are neither covalently cross-linked like chemical gels nor attractively interacted like physical gels, but are topologically interlocked by figure-of-eight cross-links. It is expected that the figure-of-eight cross-links can pass the polymer chains freely to equalize the 'tension' of the threading polymer chains just like pulleys. Therefore, the nanoscopic heterogeneity in structure and stress may be automatically relaxed in the gel. Then we call this topological gel by figure-of-eight cross-links a 'polyrotaxane gel' or 'topological gel'.

On tensile deformation, the polymer chains in the chemical gel are broken gradually due to the heterogeneous polymer length between fixed cross-links. On the other hand, the polymer chain in the polyrotaxane gel can pass through the figure-of-eight cross-links acting like pulleys to equalize the tension of the polymer chains cooperatively. Note that the equalization of tensions can occur not only in a single polymer chain, but also among adjacent polymers interlocked by the figure-of-eight cross-links. We call this 'pulley effect'. The physical properties of the topological gel are supposed to mainly result from the pulley effect.

The topological gel is a real example of the sliding gel theoretically investigated so far and can be regarded as the third gel other than the chemical and physical gels, where the polymer network is interlocked by topological restrictions. The concept of the topological gel is important not only in the creation of high-performance gels or rubbers, but also as a new framework of artificial molecular motors based on the sliding motion just like the actin and the myosin.

5. CONCLUDING REMARKS

We have so far introduced our recent theoretical and experimental works on the polymeric supramolecules of the inclusion complex between the molecular nanotubes and polymer chains. The fundamental feature of the polymeric inclusion complex is a transitional behavior different from the inclusion complex between small molecules. We have proposed some applications of the polymeric inclusion complex such as the insulated molecular wire and the topological gel. They are expected to develop into nanoscale electronic devices and nanomachines.

It may be stressed that there are many tools consisting of a string and ring or tube. This means that they are excellent combination and produce novel functions which are not given by each of them. Although we regularly use the combination in macroscopic scale, we have not sufficiently applied it to functional materials in nanoscopic scale yet. Since polymer chains have nanoscopic diameter and macro- or mesoscopic contour length, we should employ polymers for interconnection between nano- and macroscopic regions effectively. The nanoscopic structure composed of a combination between a polymer chain and ring or tube exerts large influence on macroscopic properties in the polymeric supramolecular system. The topological gel is a good example. The interconnection between nano- and macroscopic regions is another important feature of the polymeric inclusion complex.

ACKNOLEDGEMENTS

We are indebted to Prof. Akira Harada for introducing us to this new field and helpful advice about synthesis of the molecular nanotube. This study was partly performed through Special Coordination Funds of the Ministry of Education, Culture, Sports, Science and Technology of the Japanese Government and partly supported by CREST of JST (Japan Science and Technology).

REFERENCES

1. G. Wenz, Cyclodextrins as building blocks for supramolecular structures and functional units, *Angew Chem., Int. Ed. Engl.* **33**, 803-822(1994).
2. D. Philp and J. F. Stoddart, Self-assembly in natural and unnatural systems, *Angew Chem., Int. Ed. Engl.* **35**, 1154-1196(1996).
3. A. Harada, Preparation and structures of supramolecules between cyclodextrins and polymers, *Coordination Chemistry Reviews* **148**, 115-113(1996).
4. A. Harada and M. Kamachi, Complex formation between poly(ethylene glycol) and α-cyclodextrin, *Macromolecules* **23**, 2821-2823(1990).
5. A. Harada, J. Li and M. Kamachi, The molecular necklace: a rotaxane containing many threaded α-cyclodextrin , *Nature* **356**, 325-327 (1992).
6. A. Harada, J. Li and M. Kamachi, Synthesis of a tubular polymer from threaded cyclodextrins, *Nature* **364**, 516-518 (1993).
7. Y. Okumura, K. Ito and R. Hayakawa, Inclusion-dissociation transition between molecular nanotubes and linear polymer chains, *Phys. Rev. Lett.* **80(22)**, 5003-5006(1998).
8. Y. Okumura, K. Ito and R. Hayakawa, Theory on inclusion behavior between cyclodextrin molecules and linear polymer chains in solutions, *Polym. Adv. Technol.* **11**, 815-819(2000).
9. E. Ikeda, Y. Okumura, T. Shimomura, K. Ito and R. Hayakawa, Inclusion behavior between molecular nanotubes and linear polymer chains in aqueous solutions, *J. Chem. Phys.* **112(9)**, 4321-4325(2000).
10. Y. Okumura, E. Ikeda, T. Shimomura, K. Ito and R. Hayakawa, Inclusion complex formation between polyethylene glycol chains and molecular nanotubes in aqueous solutions, *Rep. Prog. Polym. Phys. Jpn.* **40**, 95-98(1997).
11. M. Saito, T. Shimomura, Y. Okumura, K. Ito and R. Hayakawa, Temperature dependence of inclusion-dissociation behavior between molecular nanotubes and linear polymers, *J. Chem. Phys.* **114(1)**, 1-3 (2001).
12. Y. Okumura, K. Ito and R. Hayakawa, Switching transition of molecular nanotubes forming an inclusion complex with block copolymers in solutions, *Phys. Rev. E* **59(4)**, 3823-3826(1999).
13. Y. Okumura, K. Ito, R. Hayakawa and T. Nishi, Self-assembling dendritic supramolecule of molecular nanotubes and starpolymers, *Langmuir* **26**, 10278-10280(2000).
14. K. Yoshida, T. Shimomura, K. Ito and R. Hayakawa, Inclusion complex formation of cyclodextrin and polyaniline, *Langmuir* **15(4)**, 910-913(1999).
15. T. Shimomura, K. Yoshida, K. Ito and R. Hayakawa, Insulation effect of an inclusion complex formed by polyaniline and β-cyclodextrin in solution, *Polym. Adv. Technol.* **11**, 837-839(2000).
16. Y. Okumura and K. Ito, The polyrotaxane gel: a topological gel by figure-of-eight cross-links, *Adv. Mater.* **13(7)**, 485-487(2001).

RECEPTIVE AND RESPONSIVE MOLECULES: DESIGN, SYNTEHSIS, AND EVALUATION

Byeang Hyean Kim[*], Gil Tae Hwang, and Su Jeong Kim

1. INTRODUCTION

Receptive macrocycles constitute a large spectrum of compounds involving both artificial as well as naturally occurring substances such as crown ethers, cryptands, cyclophanes, porphyrins, and macrolides.[1] These compounds play an important role in the exciting area of host-guest supramolecular chemistry. Our interest in the field was initiated by some of our salient observations during the total synthesis of a natural and potassium ion-receptive macrocycle, namely, nonactin.[2] After the accomplishment of the total synthesis, we soon recognized the importance of novel, efficient macrocyclization methods and focus our efforts on the design and synthesis of various artificial receptive macrocyclic systems. In this chapter, we present our synthetic results towards the natural 32-membered macrocyclic nonactin and other artificial macrocycles based on our cycloadditive macrocyclization and condensation methods. We also describe herein the molecular recognition study of our host systems with many guest molecules and our ongoing project on the design and synthesis of various photoresponsive and self-responsive molecular systems.

2. SYNTHESIS OF NATURAL RECEPTIVE MACROCYCLES, NONACTIN[3]

Nonactin (1) is an ionophoric macrotetrolide isolated from a variety of *Streptomyces cultures*.[2] The special feature of nonactin is its ability to bind alkali metal cations, particularly potassium.[4] The antibiotic activity of nonactin can be correlated to its ionophoric properties. The 32-membered macrocycle, nonactin is composed of two subunits of (-)-nonactic acid (2a) and two subunits of (+)-nonactic acid (2b), arranged in an alternating order. Our synthetic method of total synthesis of nonactin is based on the efficient generation of nonactic acid subunits and the highly effective macrocyclization method.

---

[*] Byeang Hyean Kim, Gil Tae Hwang, and Su Jeong Kim, Department of Chemistry, Center for Integrated Molecular Systems, Pohang University of Science and Technology, Pohang, 790-784, Korea.

Advanced Macromolecular and Supramolecular Materials and Processes
Edited by K. Geckeler, Kluwer Academic/Plenum Publishers, 2003

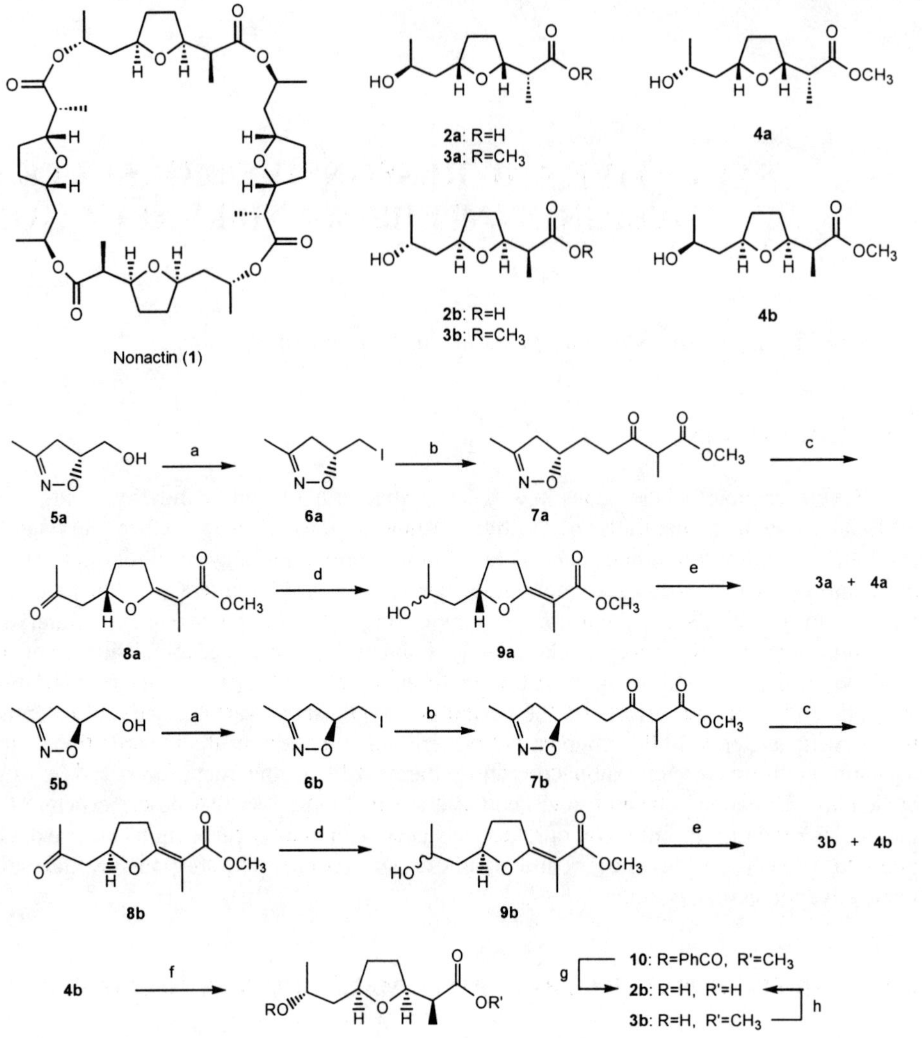

Nonactin (1)

2a: R=H
3a: R=CH₃

4a

2b: R=H
3b: R=CH₃

4b

Scheme 1. *Reagent and conditions*: (a) I₂, PPh₃, imidazole, Et₂O/CH₃CN (3/1), 90%(**6a**), 95%(**6b**) (b) Methyl 2-methylacetoacetate, NaH, *n*-BuLi, 10% HMPA/THF, 0℃, 81% (**7a**), 81% (**7b**); (c) (1) Ra-Ni, H₂, B(OH)₃, MeOH/H₂O (7/1), (2) oxalic acid, CH₂Cl₂, reflux, 74% overall (**8a**), 77% overall (**8b**) (d) L-selectride, THF, -78℃, 97% (**9a**), 95% (**9b**); (e) 5% Rh/Al₂O₃, H₂, MeOH, 11% (**3a**)+68% (**4a**) 15% (**3b**)+62% (**4b**); (f) PhCO₂H, DEAD, PPh₃, THF; (g) 2N NaOH, MeOH, 91% overall (f and steps); (h) 2N NaOH, 100%.

The synthetic pathways towards (-)-methyl 8-*epi*-nonactate (**4a**) and (+)-nonactic acid (**2b**) from **5a** and its antipode **5b** are shown in Scheme 1. The synthetic scheme was born of the thorough retrosynthetic analysis,[5] which render the required synthesis much shorter. This synthesis represents one of the most effective syntheses of (+)-nonactic acid so far explored.

For the macrotetramerization of nonactin subunits, we first concentrated on the cyclodimerization method with linear dimer of nonactin subunits under various conditions, but suffered from low reaction yields. Therefore we switched our synthetic approach to the macrocyclization method of the linear tetramer of nonactin subunits for the total synthesis. The synthetic scheme toward the final target nonactin is summarized in Scheme 2.

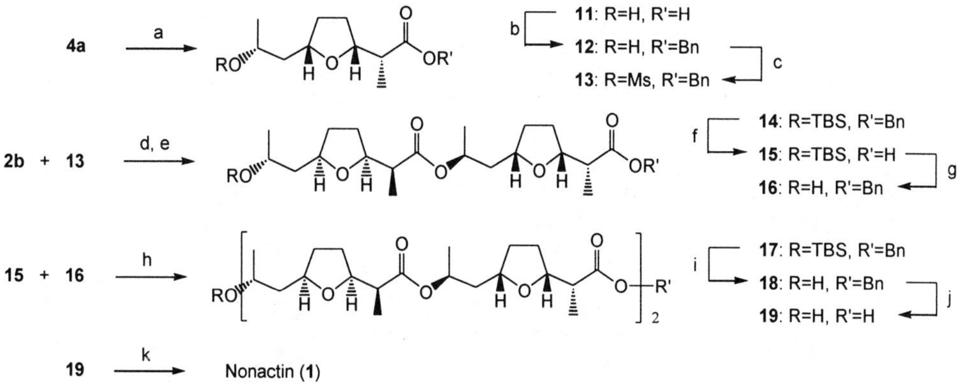

Scheme 2. *Reagent and conditions*: (a) 2N NaOH, 100%; (b) KOBu^t, BnBr, DMF, 60℃, 95%; (c) MsCl, Et$_3$N, CH$_2$Cl$_2$, 0℃, 96%; (d) KOBu^t, DMF, 60-70℃; (e) TBSCl, imidazole, DMAP, DMF, 43% overall; (f) 5% Pd/C, H$_2$, THF, 98%; (g) 40% HF/CH$_3$CN (5/95), 96%; (h) (1) 2,4,6-trichlorobenzoyl chloride, Et$_3$N, THF, (2) DMAP, C$_6$H$_6$, RT, 87%; (i) 40% HF/CH$_3$CN (5/95), 97%; (j) 5% Pd/C, H$_2$, THF, 99%; (k) (1) 2,4,6-trichlorobenzoyl chloride, Et$_3$N, THF, (2) DMAP, C$_6$H$_6$, reflux, 54%.

3. ARTIFICIAL RECEPTIVE MACROCYCLIC SYSTEMS

3.1. Design of Novel Macrocycles

Development of efficient synthetic pathways for macrocycles is very important in supramolecular chemistry. Several synthetic routes such as (a) simple cyclization, (b) cyclization in conjugation with another molecule (capping), (c) condensation of two or four identical or different units have been envisaged in one of the famous textbooks on macrocyclic chemistry written by B. Dietrich, P. Viout, J.-M. Lehn.[6] We have envisioned an additional novel pathway for macrocycles by using quadruple or double cycloadditions[7] as cornerstones (Figure 1). We call this method *cycloadditive macrocyclization* reaction. Another important feature of this method is presence of a well-defined transition state, due to which this novel cycloadditive macrocyclization can afford various macrocycles in a stereoselective manner.

Our synthetic strategy for crown ether-type cyclophanes is based on the multiple cycloadditions between bifunctional dipoles and bifunctional dipolarophiles (Scheme 3). The bifunctional dipoles (bis-nitrile oxides) were generated from the corresponding dialdehyde (isophthaldehyde, terephthaldehyde, and 2,6-pyridinedicarboxaldehyde) by Huisgen method,[8] and for the bifunctional dipolarophiles, divinyl ethers and diacrylates were selected. With appropriate combination of bifunctional dipoles and bifunctional dipolarophiles, the ring size of crown ether-type cyclophanes could be controlled through

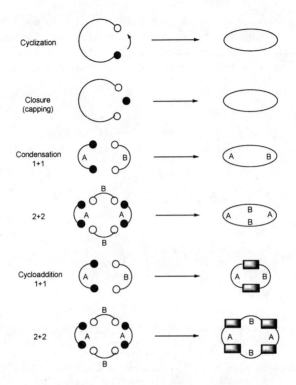

Figure 1. Possible methods of cyclization.

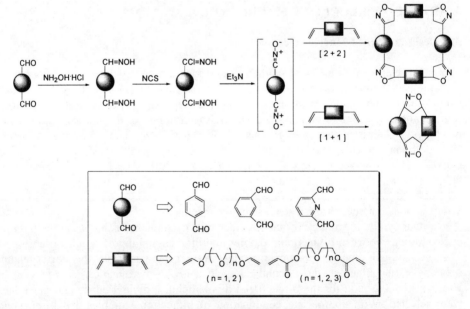

Scheme 3. Cycloadditive macrocyclization method.

either [2+2] quadruple cycloadditions or [1+1] double cycloadditions.

Three-dimensional macrocycles (3D-macrocycles) play an essential role in supramolecular chemistry and it is very important to develop efficient synthetic pathways toward them. We have envisioned a new pathway for 3D-macrocycles by extending our novel quadruple cycloadditive macrocyclization (QCM)[7] as a key step (Scheme 4).[9,10] Compound **20** is a bis-dipolarophile with a calixarene, cyclodextrin, or porphyrin moiety and compound **21** is a bis-dipole with nitrile oxide functionality. Reaction between **20** and **21** may give 3D-macrocycle **22** in a one-pot procedure if the required reaction geometry and transition states for QCM are allowed. Compared to other macrocyclization methods, this modular QCM approach has several advantages: (1) good overall yield in relatively fewer number of steps, (2) excellent stereoselectivity due to the presence of a well-defined transition state in [3+2] dipolar cycloaddition, (3) accessibility to structural diversity by changing bis-dipole and bis-dipolarophile components.

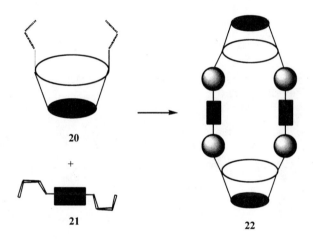

Scheme 4. 3D-macrocycles based on QCM.

3.2. Synthesis of Artificial Macrocycles

3.2.1. Crown Ether-Type Cyclophanes[7]

Divinyl ether (diethylene glycol and triethylene glycol) dipolarophiles were cycloadded with different bifunctional dipoles. In all cases, macrocycles formed as major products were via [2+2] quadruple cycloadditions. Synthesis of the 40-membered macrocycle **23** is one such representative reaction (eq 1). The isolated yield of the final quadruple cycloadducts **23** was 27%, which corresponds to 72% yield per cycloaddition.

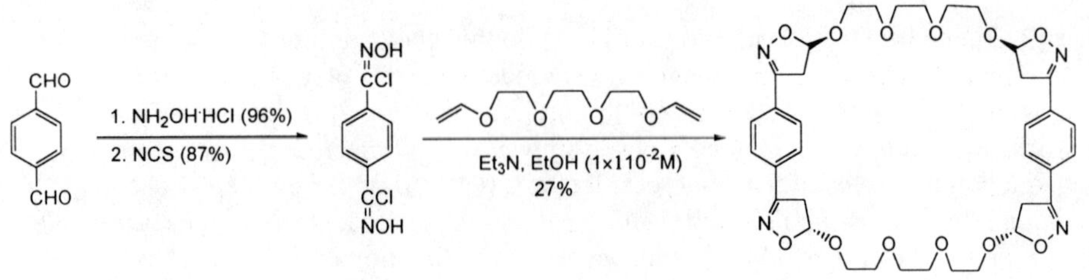

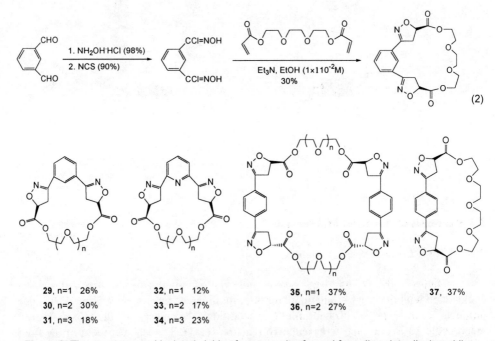

Figure 2. The structures and isolated yields of [2+2] quadruple cycloadditive macrocycles fro divinyl ether dipolarophiles.

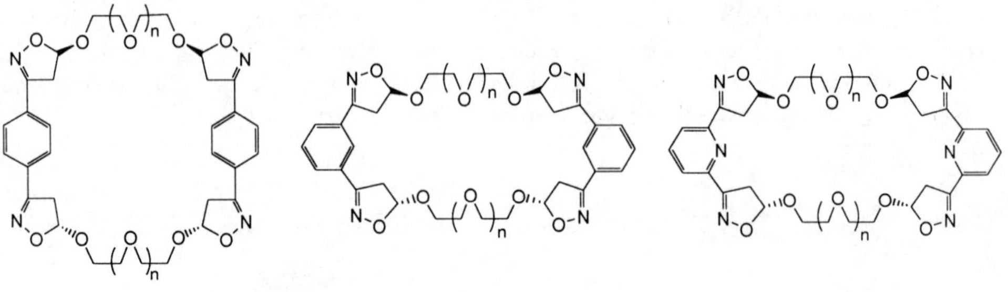

23, n=2   27%
24, n=1   14%

25, n=2   22%
26, n=1   23%

27, n=2   10%
28, n=1   14%

29, n=1   26%
30, n=2   30%
31, n=3   18%

32, n=1   12%
33, n=2   17%
34, n=3   23%

35, n=1   37%
36, n=2   27%

37, 37%

Figure 3. The structures and isolated yields of macrocycles formed from diacrylate dipolarophiles.

Figure 2 summarizes the structures and isolated yields (final quadruple cycloadditions) of major macrocyclic products.

Diacrylate dipolarophiles provided either [2+2] quadruple cycloadducts or [1+1] double cycloadducts as major products depending on the geometry of the bifunctional dipoles and the chain length of the dipolarophiles. With meta-related bifunctional dipoles (isophthaldinitrile oxide and 2,6-pyridinedinitrile oxide), [1+1] double cycloadducts were formed in good yields. A typical synthesis of 21-membered macrocycle **30** is shown in eq. 2. In reactions with a para-related bifunctional dipole (terephthaldinitrile oxide), both diethylene glycol diacrylates afforded [2+2] quadruple cycloadducts. However, in the case of tetraethylene glycol diacrylate, the chain length was long enough to react with terephthaldinitrile oxide in [1+1] fashion and so the [1+1] double cycloadduct **37** was formed as the major product. Figure 3 shows the structures and isolated yields of macrocycles formed from diacrylate dipolarophiles.

3.2.2. Silicon-Bridged Macrocycles[9]

Silicon-bridged macrocycles are of interest due to their additional coordination sites compared to carbomacrocycles and good chemophysical properties including solubility. We report here the facile synthesis of novel silacyclophanes by using the quadruple cycloadditive macrocyclization (QCM) methodology[4] and intramolecular nitrile oxide dimerization,[5] and their X-ray crystal structures.

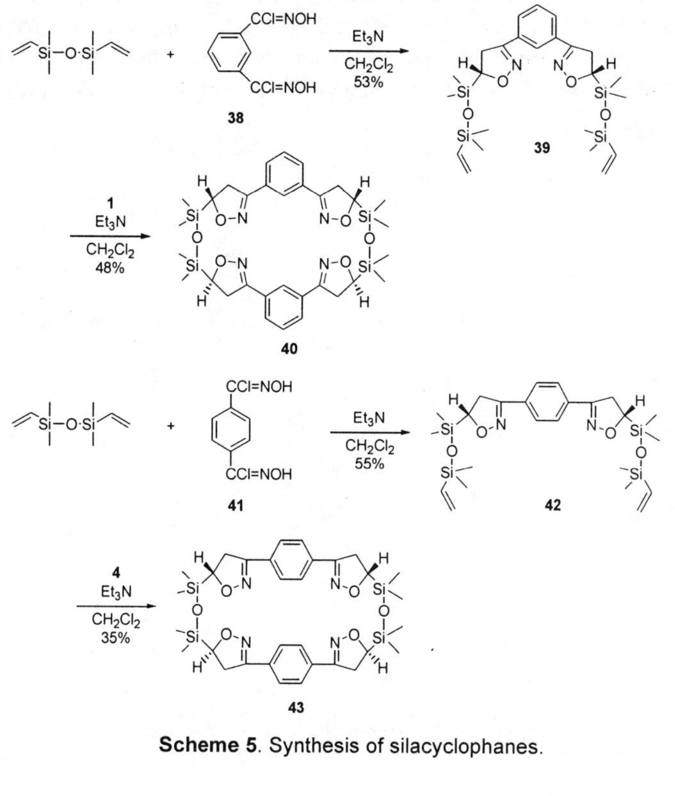

Scheme 5. Synthesis of silacyclophanes.

Silamacrocycle **40** was synthesized in two step sequence by using QCM methodology (Scheme 5).[7] The structure of silacyclophane **40** was confirmed by X-ray crystallography (Figure 4). In a similar fashion, silacyclophanes **43** was prepared by QCM methodology between terephthaldinitrile oxide and 1,3-divinyltetramethyldisiloxane.

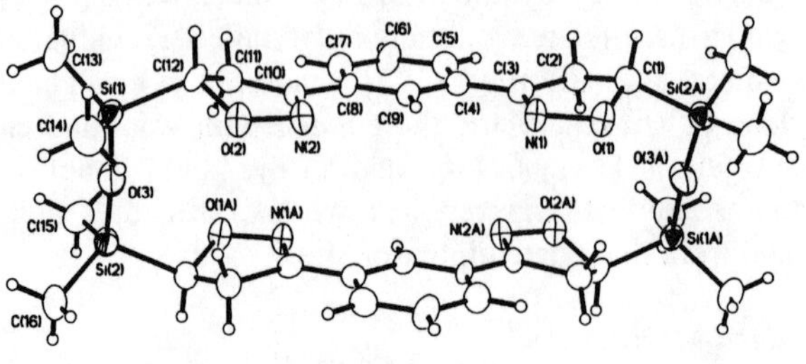

Figure 4. X-Ray crystal structure of **40**.

In the case of QCM with 2,6-pyridinedintrile oxide, two macrocyclic cycloadducts were isolated. The first one is the regular 2+2 cycloadduct **45** (8%) and the other product is a [2+1] triple cycloadduct **46** (25%) (Scheme 6). Silamacrocycle **45** is the pyridine version of compound **40** and is formed through QCM in one pot. Formation of compound **46** proceeds *via* double cycloadditions followed by intramolecular nitrile oxide dimerization. Generation of rather unusual product **46** may be attributable to the stability of 2,6-pyrininedinitrile oxide and the proximity between two' nitrile oxide moieties. The chemical structure of **46** was confirmed by X-ray crystallography (Figure 5).

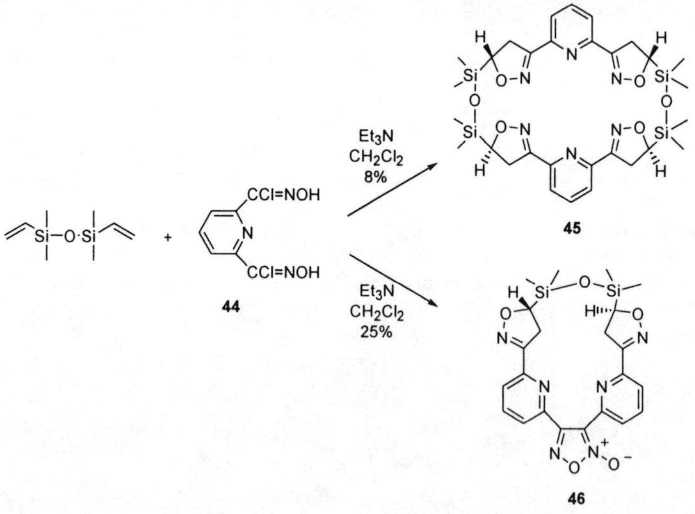

Scheme 6. Synthesis of a triple cycloadduct **46**.

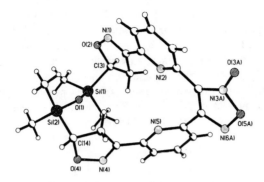

Figure 5. X-Ray crystal structure of **46**.

3.2.3. Cyclophane-Type Bis-Calix[4]arenes[10]

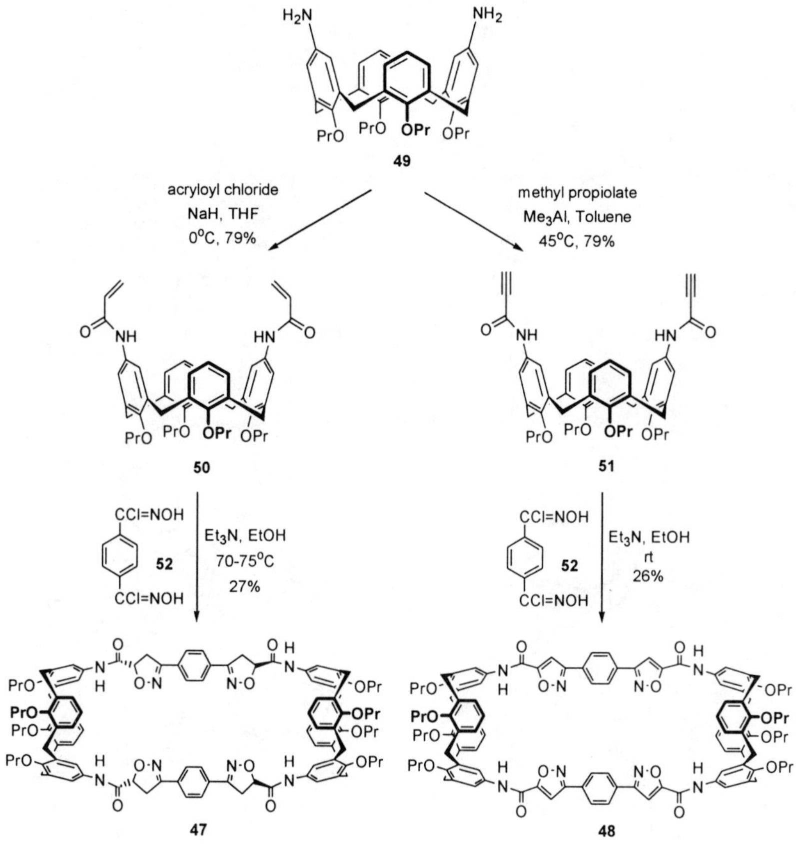

Scheme 7. Synthesis of bis-calix[4]arenes based on QCM.

Synthetic routes for the bis-calix[4]arenes **47** and **48** are shown in Scheme 7. *N*-acryloylation and *N*-propiolation of the diamine **49**[11] afforded the bifunctional dipolarophiles **50** and **51**, respectively. Cycloadditions between the dihydroximoyl chloride **52** and compound **50** in the presence of triethylamine in ethanol then provided the bis-calix[4]arene **47** in 27% isolated yield, which corresponds to a 72% yield per cycloaddition. Similarly, the bis-calix[4]arene **48** was prepared by cycloadditions between compounds **51** and **52** in 26% yield. Both bis-calix[4]arenes **47** and **48** were prepared in only two steps from diamine **49** in 21% overall yield.

3.2.4. Ultramacrocycle with Four Calix[4]arenes[12]

We describe here a single step synthesis of novel quadruple calix[4]arene **53** and full characterization of the compound including the determination of the X-ray crystal structure. To our knowledge, the X-ray structure of **53** is the first one reported in the area of quadruple calix[4]arenes.

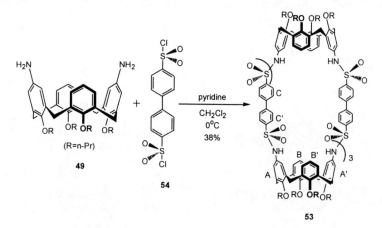

Scheme 8. Synthesis of a quadruple calix[4]arene **53**.

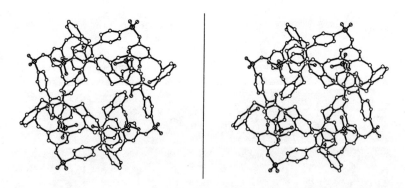

Figure 6. The stereoscopic view of the crystal structure of the quadruple calix[4]arene **53**. Hydrogen atoms, propyl units, and co-crystallized solvent molecules are left out for clarity.

Synthesis of compound **53** is shown in Scheme 8. The formation of the quadruple calix[4]arene **53** results from a high degree of molecular assembly between diaminocalix[4]arenes[11] and the biphenyl linkers through eight sulfonamide bonds.

Figure 6 shows the stereoscopic view of the crystal structure of the quadruple calix[4]arene **53**. Colorless crystals of compound **53** suitable for X-ray structural investigations were obtained from a chloroform–methyl alcohol mixture. The crystal structure has an inversion center (*i*) and all four calix[4]arenes point outward. In this compound there are four divergent cavities for binding guests.

3.2.5. Bis-Calix[4]arenes with Imine Linkages[13]

Synthetic routes for the bis-calix[4]arenes **55-59** are shown in Scheme 9. The condensation reactions of dialdehydes with 1.1 equiv. of diaminocalix[4]arene **49**[11] in

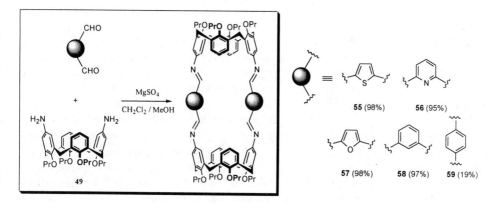

Scheme 9. Synthesis of bis-calix[4]arenes with imino linkages.

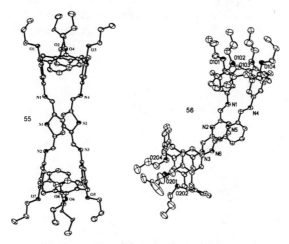

Figure 7. X-ray crystal structures of bis-calix[4]arene **55-56**. Two chloroform solvent molecule for **55**, and one methanol and a half of chloroform solvent molecules for **56** are omitted for clarity.

refluxing CH₂Cl₂/MeOH (1:1 v/v) in the presence of MgSO₄ for 24h provided the corresponding bis-calix[4]arenes **55-58** in excellent yields (95% ~ 98%, Scheme 9). However, in the case of bis-calix[4]arene **59**, the isolated yield was only 19%. This result suggests that the efficiency of macrocyclization *via* imine condensation depends strongly on the geometry of the aldehyde or amine.

Figure 7 shows the X-ray crystal structures of the bis-calix[4]arenes **55-56**. These bis-calix[4]arenes are all nanometer sized macrocycles as shown in **55** (1.7nm long), and **56** (1.8nm long), and have good-sized cavities for host-guest complexation. The X-ray crystal structures clearly show that these calix[4]arenes adopt pinched cone conformations[8] in the solid state.

3.3. Molecular Recognition Study

To show the utility of the bis-calix[4]arene hosts, we have carried out the binding studies between the bis-calix[4]arenes and viologen-type guest molecules.[14] Viologens are well known as oxidation-reduction indicators and have also interesting biological activities. Diverse ranges of properties in the viologens are best understood and categorized in terms of their redox chemistry and the electron-deficient nature of the viologen dication, and the viologens occupy a pivotal place in the field of electro-active organic molecules.[15] The structures of the guests for the molecular recognition with the host **55** are shown in Figure 8. The titration experiments were carried out by ¹H NMR spectroscopy in CDCl₃/CD₃OD (2:1 v/v, 300K). The signals of imine and aromatic protons in the bis-calix[4]arene **55** shifted down-field when the viologens **60-64** were added to the host solution (the measured complexation-induced $\Delta\delta$ value of imine proton of **55** was 0.30 ppm from δ 7.27 to δ 7.57 when 1.0 equiv. of viologen **60** was added), whereas no change in the chemical shift values of **55** are detected when the compounds **65-67** were added. This indicates that suitable size of *N*-alkyl groups and the presence of the bipyridinium dication in viologens are essential to the inclusion process. The 1:1 complex was confirmed by Job plot between bis-calix[4]arene **55** and the viologens (**62** and **64**). Nonlinear least-squares fitting analysis[16] of the complexation with **55** provided the association constant 727M⁻¹ for **60**, 515M⁻¹ for **61**, 320M⁻¹ for **62**, 154M⁻¹ for **63**, and 124M⁻¹ for **64**. The signals of protons in the bis-calix[4]arene **56** did not shift upon addition of the viologen **60**, **62**, **63**, and **64** and other bis-calix[4]arenes **57-59** showed

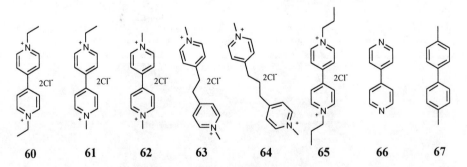

Figure 8. The structures of the guest molecules.

very weak affinities toward the viologen guests. These results indicate that overall shape of host molecules and the electron density of aromatic linkers (thiophene, benzene, furan, pyridine) are the important factors in the binding affinity with the viologen guests.

4. RESPONSIVE MOLECULAR SYSTEMS

One of our research focus in the recent past have been on the design and development of various photoresponsive molecular systems which can act as light harvesting molecular antenna, tunable photo fluorophores, etc. So far we have designed and synthesized various compounds (Figure 9) via acetylenic couplings. Some of them show good photophysical properties including better quantum yields. Another area of our research interest is on the synthesis of self-responsive molecules which can self assemble leading to the formation of macroscopic structures like organogels, vesicles, liposomes and so on. Full details of these ongoing projects will be reported in due course.

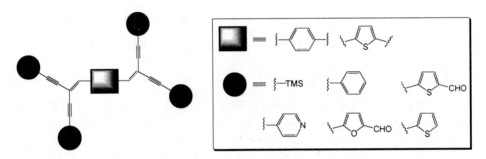

Figure 9. Synthesized fluorophores.

5. ACKNOWLEDGEMENTS

We sincerely thank our co-workers whose names are listed in the cited literatures. We are grateful to CIMS and BK21 for financial support.

6. REFERENCES

1. (a) T. A. Robbins, C. B. Knobler, D. R. Bellew, and D. J. Cram, *J. Am. Chem. Soc.* **1994**, *116*, 111. (b) D. J. Cram. *Nature* **1992**, *356*, 29. (c) T. K. Park, and J. Rebek, Jr., *Angew. Chem., Int. Ed. Engl.* **1990**, *29*, 245. (d) Q. Feng, T. K. Park, and J. Rebek, Jr., *Science* **1992**, *256*, 1179. (e) E. Fan, S. A. Van Arman, S. Kincaid, and A. D. Hamilton, *J. Am. Chem. Soc.* **1993**, *115*, 369. (f) J. Yang, E. Fan, S. J. Geib, and A. D. Hamilton, *ibid* **1993**, *115*, 5314. (g) F. Vögtle, *Supramolecular Chemistry* (Wiely, New York, 1991). (h) H.-J. Schneider, and H. Dürr, *Frontiers in Supramolecular*

Organic Chemistry and Photochemistry (VCH, Weinheim, 1991). (i) J. L. Atwood, *Inclusion Phenomena and Molecular Recognition* (Plenum Press, New York, 1990).

2. (a) R. Corbaz, L. Ettlinger, E. Gaumann, W. Keller-Schien, F. Kradolfer, L. Neipp, V. Prelog, and H. Zahner, *Helc. Chim. Acta* **1995**, *38*, 1445. (b) J. Dominquez, J. D. Dunitz, H. Gerlach, and V. Prelog, *ibid* **1962**, *45*, 129. (c) H. Gerlach, and V. Prelog, *Liebigs Ann. Chem.* **1963**, *669*, 121. (d) M. Dobler, *Ionophores and their structure* (Wiley, New York, 1981). (e) B. T. Kilbourn, J. D. Dunitz, L. A. R. Pioda, and W. Simon, *J. Mol. Bio.* **1967**, *30*, 559.

3. (a) J. Y. Lee, and B. H. Kim. *Tetrahedron Lett.* **1995**, *41*, 4177. (b) J. Y. Lee, and B. H. Kim. *Tetrahedron* **1996**, *52*, 571.

4. J. H. Prestegard, and S. I. Chan, *J. Am. Chem. Soc.* **1970**, *92*, 4440.

5. (a) P. A. Bartlett, and K. K. Jemstedt, *Tetrahedron Lett.* **1980**, *21*, 1607. (b) B. H. Kim, and J. Y. Lee, *ibid* **1992**, *33*, 2557. (c) B. H. Kim, and J. Y. Lee, *ibid* **1993**, *34*, 1609. (d) I. Fleming, and S. K. Ghosh, *J. Chem. Soc. Chem. Commun.* **1994**, 2285, (e) M. Ahmar, C. Duyck, and I. Fleming, *Pure & Appl. Chem.* **1994**, *66*, 2049. (f) G. Solladié, and C. Dominguez, *J. Org. Chem.* **1994**, *59*, 3898 and references cited therein.

6. B. Dietrich, P. Viout, and J.-M. Lehn, *Macrocyclic Chemistry* (VCH, Weinheim, 1993).

7. B. H. Kim, E. J. Jeong, and W. H. Jung. *J. Am. Chem. Soc.* **1995**, *117*, 6390.

8. R. Huisgen, *Angew. Chem., Int. Ed. Engl.* **1963**, *2*, 565.

9. C. W. Lee, G. T. Hwang, and B. H. Kim, *Tetrahedron Lett.* **2000**, *41*, 4177.

10. G. T. Hwang, and B. H. Kim, *ibid.* **2000**, *41*, 10055.

11. D. M. Rudkevich, W. Verboom, and D. N. Reinhoudt, *J. Org. Chem.* **1994**, *59*, 3683.

12. G. T. Hwang, and B. H. Kim, *Synth. Commun.* **2000**, *30*, 4205.

13. G. T. Hwang, and B. H. Kim, *Tetrahedron Lett.* **2000**, *41*, 5917.

14. Z. Zhong, A. Ikeda, and S. Shinkai, *J. Am. Chem. Soc.* **1999**, *121*, 11906.

15. a) P. M. S. Monk, *The Viologens* (John Wiley & Sons Ltd, Chichester, 1998). b) M. Asakawa, C. L. Brown, S. Menzer, F. M. Raymo, J. F. Stoddart, and D. J. Williams, *J. Am. Chem. Soc.* **1997**, *119*, 2614. c) D. B. Amabilino, P. R. Ashton, S. E. Boyd, J. Y. Lee, S. Menzer, J. F. Stoddart, and D. J. Williams, *Angew. Chem., Int. Ed. Engl.* **1997**, *36*, 2070.

16. R. S. Macomber, *J. Chem. Educ.* **1992**, *69*, 375.

PREPARATION, MORPHOLOGY, AND APPLICATION OF MONODISPERSE POLYMERIC SPHERES

Audist I. Subekti, Evelyn M. Rodrigues, and Robert P. Burford[*]

1. INTRODUCTION

Hydrogels have become increasingly important for a broad range of applications include contact lenses,[1] and other medical devices,[2] as intelligent supports for the release of low levels of additives including perfumes,[3] for foodstuff and thickeners. In addition to providing an inert matrix for many applications, some hydrogel polymers respond to the environment and so are increasingly used in 'smart'materials and devices. For example, gels based on certain substituted acrylamides show a lower critical solution temperature[4], and so can exhibit substantial volume change as a function of temperature. Poly(N-isopropyl acrylamide) undergoes a major transition at about $32°C$[5] and is used in many domestic and medical applications.

Similarly, many gel forming polymers show an ability to complex with metal ions and biomolecules, and so can be used for complexation as ion chromatograph column packings and for electrophoresis plate media. Again, it is known that polyelectrolytes and polyacids such as those containing acrylic acid can bind with cations and this binding is pH dependent.[6]

Some copolymers may show a combination of both pH and temperature sensitivity. For example, Guohua and Hoffman[7] produced polymers which undergo marked solubility changes in water in response to changes in temperature or pH. They prepared hydrogels of N-isopropylacrylamide (NIPAAm) and acrylic acid (AAc) with monomer ratios from 100/0 to 20/80wt%. Their concept was to graft temperature sensitive-side chains from the NIPAAm onto a pH sensitive backbone of the AAc. These hybrid gels have a cloud point (CP) is which higher than that of the PNIPAAm homopolymer and rapidly rises as the AAc content increases.

---

[*] Centre of Polymer Science and Engineering, School of Chemical Engineering and Industrial Chemistry, University of New South Wales, Sydney, NSW-2052, Australia.

Advanced Macromolecular and Supramolecular Materials and Processes
Edited by K. Geckeler, Kluwer Academic/Plenum Publishers, 2003

As the pH of the solution is lowered at a constant temperature, the gels shrink but still remain in solution. Potential applications of these copolymer-biomolecule systems were stated to include devices for drug delivery, diagnostics, separations, cell culture, and bio-reactions.

However, for some applications polymers are conveniently disposed in devices as particles, for example, in controlled release pharmaceuticals and column packings. The active ingredients of agricultural chemicals may be embedded within a polymeric host, or an active surface may be developed on the porous spheres, for example as ion exchange supports. Many such products are employed with a diameter ranging from 0.5 to about 2 or 3 mm, but their shape and size distribution are often irregular or broad, depending on the method of manufacturing.[8]

When it is desirable to ensure a very uniform performance, the use of particles of highly regular shape and size may be valuable. Some of the prior methods for making hydrophilic micro-spheres, such as inverse micro-emulsions and suspensions, need surfactant and can cause product fines. There are some techniques that lead to highly uniform but smaller sized spheres based on seed emulsions, such as the Shirasu porous glass method[9] and there are also several methods being developed in Norway.[10] However, most are tedious, multi-step, or lead to finer particles than those we seek.

Ruckenstein and Liang[11] revealed a method for making 1-2 mm diameter particles, which we have developed further, as briefly described below. The concept involves the surface gelation of droplets descending a column-heated oil, leading to a skin that prevents coalescence at the base of the column. We have studied the effects of several variables including flow rate, droplet size, the use of glycerol/water mixture as solvent for acrylamide monomers (to allow higher temperatures and so wider range of initiators), as well as several oil types. We currently use silicone oil, an initial temperature of 96°C and a multi-needle delivery system fed by a syringe pump.

In this paper, we wish to consider four aspects of hydrogel particle production and application. These are:

- Morphology of homo and copolymeric spheres
- Surface modification
- Swelling properties
- Cation binding

2. PREPARATION AND STRUCTURE OF MONODISPERSE SPHERICAL HYDROGELS

2.1. Description of the General Procedure

The Ruckenstein and Liang paper[11] was invaluable because it disclosed the basis of the method. In essence, it is a refined example of inverse suspension polymerisation, but the delivery of monomer, cross-linker and thermal initiator is controlled in such a way that uniform droplets are produced which are then preserved by surface gelation ("skin formation"). However, the paper described a rather primitive process, in which the solution was delivered using a hand-operated syringe, with the droplets from a single needle entering a hot paraffin oil filled column.

Bushell et al.[12] examined several process variables including alcohol, diol, and glycerol replacements for the aqueous solvent, solution delivery rate, silicone fluids of ranging viscosity, and column temperature. Particles ranging from about 1.4 to 2 mm were formed depending on conditions, but in all cases of quite narrow (ρ = 1.1) dispersity.

The use of alcohols tended to make subsequent soxhlet extraction more demanding without offering any obvious advantages. The present work uses the reactor shown schematically in Figure 1. Feed rates, via a piston-driven syringe, vary from 19.8 to 34.9 mL/minute depending on viscosity, and the 200 cS silicone oil to kept at 96 °C, controlled using a thermo-couple located in the oil itself.

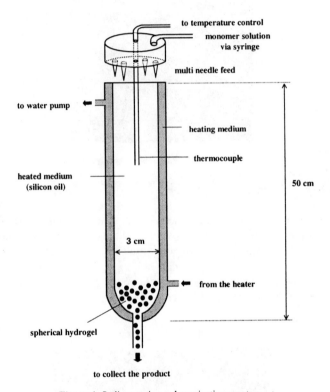

Figure 1. Sedimentation polymerisation reactor.

We observe in the column a thin skin around each droplet. This is typically either translucent or opaque, or has a different refractive index to the solution within. This forms after about 10 to 20 seconds, depending on feed rate and monomers. The 'caviar – like' particles reach the bottom of the column after about 40-60 seconds, but do not coalesce (Figure 2).

They are left on the base of the column for 2 hours. The temperature is lowered to 65°C over about 30 minutes, and then kept at that temperature for a further 1.5 hours prior to sphere removal. Product is formed using a multi-needle delivery device, allowing sufficient material for characterization.

Figure 2. Skin formation during the sedimentation process.

2.2. Morphology of Homopolymer Spheres

Whilst the Ruckenstein and Liang paper[11] foreshadowed that a range of hydrophilic monomers could potentially be converted in the spheres by their sedimentation method, in fact all published data was confined to acrylamide with N,N-methylene-bis-acrylamide crosslinker. We have homo- and copolymerized six different monomers, and find that their regularity and transparency vary from one to another. For example, the polyacrylamide has the appearance of 'macadamia nuts', as shown in Figure 3.

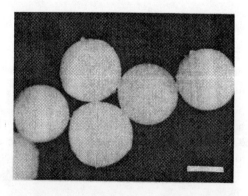

Figure 3. Morphology of polyacrylamide spheres (bar: ~ 1 mm) .

Anhydrous, extracted and dried polyacrylamide is a typically hard, crystalline material, so the opacity of the spheres is expected. The circumferential cracking may arise due to contraction when the water is removed during the acetone extraction step, or in the subsequent drying stages.

However, we find that many of the homo and particularly copolymer spheres are translucent or transparent, and are essentially defect free. Poly-dimethyl acrylamide product with high regularity and uniformity is shown by optical microscopy below (Figure 4).

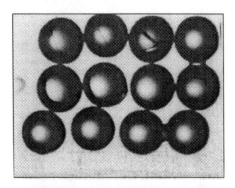

Figure 4. Morphology of poly(dimethyl acrylamide) spheres.

The smooth and regular surface of a DMAM sphere is given in the SEM photograph in Figure 5.

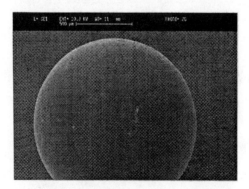

Figure 5. SEM photograph of poly(dimethyl acrylamide) spheres

2.3. Morphology of Copolymer Spheres

A basis upon which the sedimentation polymerisation procedure operates is that monomers have high propagation rate constants, Rp relative to Rt. This requirement is met by many acrylic monomers. We have performed screening experiments with the monomer combinations shown in Table 1.

Table 1. Copolymerisation results

Monomer		Success	Appearance	Uniformity
1	2			
N,N-dimethylacrylamide	Acrylic acid	High	Mostly clear, hard, smooth surface	High monodisperse, spherical, distinct
N-isopropylacrylamide	Acrylic acid	Moderate	White, opaque, hard	Reasonable monodispersity spherical
Acrylamide	Acrylic acid	High	Clear, hard, smooth surface	High monodisperse, spherical, distinct
2-hydroxyethyl methacrylate	Acrylamide	Moderate	Hazy, pale, slightly softer than others	Moderate monodisperse, spherical, distinct
N,N-dimethylacrylamide	Methacrylic acid	Requires further optimisation	White, opaque	-
N-isopropylacrylamide	Methacrylic acid	Requires further optimisation	White, opaque	-

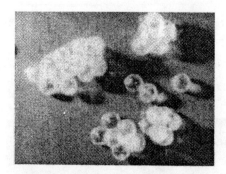

Figure 6. Morphology of poly(Aam-*co*-AAc 50) spheres.

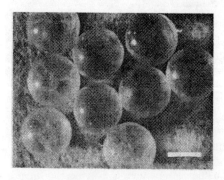

Figure 7. Morphology of poly(DMAM-co-AAc 33) spheres (bar: ~ 1 mm).

The copolymers are typically transparent and have regular structure without defects, as shown for the 1:1 acrylamide/acrylic acid and 2:1 N,N-dimethylacrylamide/acrylic acid copolymers shown in Figures 6 and 7.

It appears that for most copolymer ratios, the spheres are amorphous, but at higher ($\geq 66\%$) acrylic acid levels, some turbidity is noticed.

Magnetic resonance imaging (MRI) of hydrated acrylic acid/dimethyl acrylamide spheres gave evidence for a hydrophobic core-hydrophilic shell structure (Figure 8).

Note that the MRI tube is about five times the diameter of each sphere, so that about 15 individual spheres are shown in the image. Based on the higher water uptake shown by N,N'-dimethylacrylamide, we propose that the inner material is acrylic acid. Further MRI studies of other copolymers are in progress.

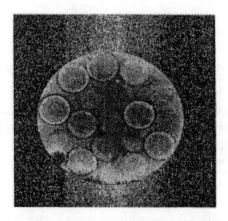

Figure 8. MRI of 1:1 PDMAM/AAc beads.

2.4. Cross-Sectional Structure

The original study by Ruckenstein and Liang included a medium magnification SEM photograph indicating that their polyacrylamide possessed low porosity. This is consistent with the polymerisation being essentially comparable with bulk conditions, with no phase inversion or separation involved.

The possibility of imparting porosity to these materials is, however, of very great interest, and so we have examined a sphere by making a section, which should be free of preparation artefacts. This was done using a focussed ion beam-milling machine, and the cross-sectioned of the interior of the polyacrylamide sphere is shown in Figure 9.

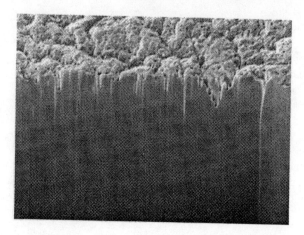

Figure 9. Cross-sectioned structure of polyacrylamide.

We have undertaken some screening experiments using sugar as a possible porosigen,[13] but although the presence, for example, of reducing sugars, was confirmed using Fehling's solution, SEM of fractured spheres showed little or no internal porosity. We are continuing to explore this aspect, however, using water-soluble polymers such as polyethylene glycol in the feed solution.

3. Surface Modification

For many applications, the surface properties of polymer particles control performance. It is evident from the range of monomers already investigated, and from the many other hydrophilic monomers which can be selected, that many functional groups are available for further reaction.

In addition, it would be possible to include both monomers and polymers in the feed solution which might allow spheres to become semi-interpenetrating polymer networks during spheres formation or by subsequent independent polymerisation using a non-radical route. Such methods might exploit biodegradable polymers, or have components, which subsequently break down, giving an alternative route to the porosigen method. In the present case, however, we shall focus on post-treatments and in particular describe the formation of polypyrroles and polyanilines.

When hydrated polyacrylamide spheres are immersed in neat pyrrole for 24 hours, the superficial pyrrole removed and then exposed to 0.6 M FeCl$_3$ solution for a further 24 hours, a black shell of the polypyrrole can be seen, as shown in the polished section in Figure 10.

If, however, the spheres are immersed in 0.3 M aqueous pyrrole solution and the same procedure followed, the polypyrrole forms throughout the particle, which at least at this magnification (Figure 11) suggests an IPN-like structure.

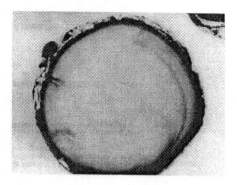

Figure 10. Polished-sectioned of polyacrylamide-polypyrrole IPN.

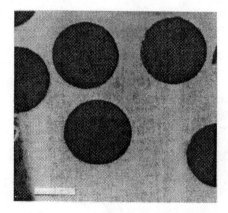

Figure 11. Optical microscopy picture of poly(acrylamide-polypyrrole) IPN (bar: ~ 1 mm)

We have performed similar experiments with aniline, but generally the thick coatings are rather crust-like and readily dislodge from the spherical substrate. Further refinements should allow good coatings. We have treated polyaniline encrusted spheres with acid and base and noted colour changes associated with the various forms such as emeraldine depending upon the level of protonation. We plan to extend this work with several of the newer copolymer spheres we have recently made.

4. EFFECT OF ENVIRONMENT ON VOLUME AND WATER UPTAKE

Previous studies of the N-isopropyl acrylamide/acrylic acid copolymer have shown responses to both pH and temperature,[14] but we have found that such materials are difficult to make in high quality using the sedimentation polymerisation method. We are continuing to pursue this combination, but present have made the poly(N,N'-dimethylacrylamide/acrylic acid) system, which to our knowledge has not previously been reported. In this work homo and copolymers of dimethyl acrylamide and acrylic acid, using mole ratios of 2:1, 1:1, and 1:2 were made, using in each case either 5 and 10% "Bis" cross-linker.

Duplicate samples of spheres were incubated at 40°C at pH's of 1, 4, 7, 9, and 14 for 24 h and sizes were measured using a Wild M3C stereomicroscope. For the temperature-dependent swelling studies, the spheres were swollen in a solution buffered at pH 7 at temperatures from 20 to 60°C with 5°C increments. Initial equilibration was performed at 20 °C for 24 hours, followed by weighing and photographic recording. The same spheres were then heated at 5°C of increments for 30 minutes, followed by rapid measurement and then returned to the controlled-temperature water bath.

To see the trends in sphere size as a function of pH, we have normalized volumes on the basis of the volumes measured at a pH of 1. This allows comparison between each copolymer, even when the initial size of the dry spheres varies. All dry spheres made, with the exception of the acrylic acid homopolymer (diameter: ~ 1.3 mm) were in fact between 1.7 and 2.1 mm. We recognise that this procedure is partially flawed, as the masses at pH = 1 vary, but the relative change in volume are demonstrated.

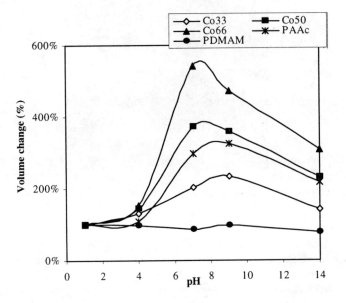

Figure 12. pH-dependent swelling properties.

The poly N,N'-dimethyl acrylamide shows little pH sensitivity, consistent with the inertness of the tertiary amide group. However, the acrylic acid homopolymer and the copolymers all show a volume maximum at either pH 7 or 9.

As shown in Figure 12 for copolymer spheres, shrinkage occurs at a pH below 4. As the pH increases to 7, the volume of the spheres increases significantly, and reaches their peak at pH 7 or 9. Based on the acid-base equilibrium for a weak acid, the degree of dissociation of the acid strongly depends on the pH of the surrounding environment.

The dissociation behaviour of the acid will temporarily disappear at the lower pH, while it is essentially completely ionized in a basic environment. As a consequence of repulsive interactions of the negative charges within the gels, the degree of swelling will be higher.[14,15]

Additionally, Lee and Shieh[16] also found the same phenomenon in their studies of NIPAAm, AAm and HEMA. They showed that the swelling ratios of the gels in the buffer solution pH 10.4 are higher than those in the buffered solutions of pH 4. This evidence showed that the amide groups on the copolymeric gels would be easily hydrolyzed to form carboxylate groups under basic conditions.

In a basic environment at elevated temperature (45°C and pH 10.4), the hydrolised reaction of the amide group on gel would be accelerated, and the repulsive effect of the carboxylate ions on the gels would increase. Both events would increase the swelling ratio higher and make the effect of de-swelling less significant.[16]

5. CATION BINDING

50 mg of dry copolymer beads were shaken in 25 mL $CuCl_2$ or $CrCl_3$ solutions, with concentrations from 10 μM to 1 M, for 24 hours at 25 °C. The beads were then removed and the concentration difference of the solution determined by AAS and/or ICP. The cation uptake in spheres shown in Figure 13 and indicates that positive transport has occurred.

More particularly, even at the lowest concentrations, permanent coloration of the spheres by the copper chloride solution could be seen visually, with the beads remaining a distinct green.

5.1. Trace Metal Analysis

50 mg of spheres made from the copolymer of N,N-dimethyl acrylamide and acrylic acid of varying composition were shaken in 25 mL of 10 ppm $NiCl_2$, $CuCl_2$ and $CrCl_3$ for 24 hours. The supernatant again was analysed by AAS and/or ICP, and the loss of cation assumed to be absorbed by the beads.

Figure 14 shows that copper is absorbed most, with nickel and chromium being taken up similar amounts at 0 and 33 mol% acrylic acid, but falling to about two third the amount as acrylic acid content further increases. It is noted that the concentration of the cation in the beads at higher acrylic acid content greater than in the adjacent solution.

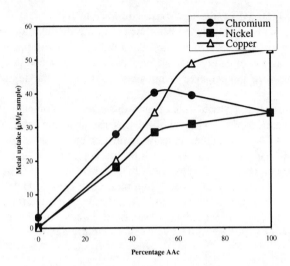

Figure 13. Trace metal uptake of the spheres at pH 6.

The poly dimethyl acrylamide spheres have only a tertiary amide group, which coordinates less with the cations, so that lower metal uptake is expected. The high complexation with the acrylic acid is expected, with such groups being employed in polymers widely used for ion exchange media.

5.2. Metal Uptake at Higher Concentration

Whilst the previous data for low cation concentration indicate that the spheres are potentially suitable for trace metal reduction in waster streams, the levels are too low to reveal detail of the mechanism of binding. Therefore, each of the copolymers of DMAM and AAc were immersed in 0.1 and 1 M solutions of copper and chromium chloride. The briefly washed and thoroughly dried spheres were then embedded in Araldite and polished to a diamond paste level of smoothness. The carbon coated specimens were examined in an EDX equipped Hitachi 4500 FESEM, using a voltage of 20kV to allow both Lα and Kα X rays to be generated. These are readily seen in the EDX spectrum for each element.

Line scanning selecting for the Kα of each element is readily undertaken. A low magnification SEM of 33 mol% PDMAM/AAc spheres, with the scan passing through the copper-free resin at the centre, is shown in Figure 14. The low background level, expected for the embedding resin, can be seen in the centre of the X ray scan. Whilst prologued analysis will improve the signal-to-noise characteristics of Figure 14b, our experience has shown that such traces are rarely smooth. This and other profiles suggest, but do not prove, that the metal concentration is uniform within a sphere. A second limitation with EDX is that, even with smooth samples of the type examined here, truly quantitative data is rarely possible to obtain.

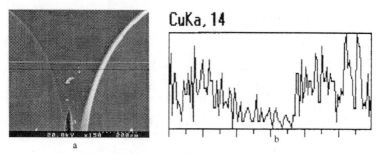

Figure 14. Line scanning through two different spheres after sorption, Cu 1 M (a = SEM, b = EDX).

The same samples were examined in a Cameca SX50 microprobe calibrated with pure (> 99.99%) copper and chromium. This was operated at 15 kV and 20 nA, using a LiF analysis crystal. Most samples were scanned over 500-900 µm in 20µm steps with a counting time of 20 seconds. Given the polymer density, this leads to X-ray collection from about 20 µm diameter regions.

An example of microprobe analysis for copper is shown in Figures 15 and 16. The first shows a low magnification image of a sphere, with white dots indicating the location of each analysis. Here, we begin the analysis in the embedding resin at the interface with the sphere surface.

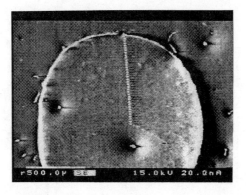

Figure 15. Line scanning through two different spheres after sorption, Cu 1 M.

The associated Figure 16 shows the data for copper in 33 mol% PDMAM/AAc spheres immersed in 0.1 and 1.0 M $CuCl_2$ solution. The low values at the left are for the Araldite resin and are essentially zero. The mean values of copper are 3.5 and 11 wt%, respectively, these again being higher than the original copper concentration in the solution. It should notes that such uptake in strongly dependent upon pH, but a detailed description is beyond the scope of present paper.

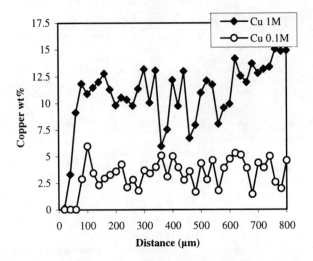

Figure 16. Point analysis graph for 33 mol% PDMAM/AAc spheres.

A further observation is that the traces are quite "noisy", as found for EDX. It is possible that at the 20 µM level, irregularities in the texture of the hydrogel exist, leading to compositional changes at the microscopic level. We have not found evidence for this by any of the present techniques available, but will pursue this aspect further. It is plausible that a heterogeneous morphology may be present which is not evident in the dry gel particles.

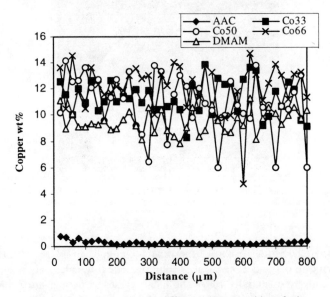

Figure 17. Copper sorption for different AAc composition of spheres.

We have examined the role of PDMAM/AAc ratio upon copper (II) ion uptake at the low "natural" pH of 1. The data have all been collected from the interior of single spheres, with uniform spacing as shown by the imprint of the micrograph in Figure 17.

At this low pH, the lowest uptake is actually for the poly acrylic acid homopolymer, which is highly protonated. The remaining copolymers and the PDMAM homopolymers all take up between 5 and 8wt% of Cu^{2+}. In molar terms, this equals to a metal ion for two pairs of repeat units in the 1:1 copolymer. At this concentration, "saturation" of the polymer is not complete by complexation to the metal ion.

We are proceeding to examine more concentrated salt solutions, and also to study ion competition. It is noted from the related studies of salt interaction with polyamides, that salt can coordinate directly with the amide carbonyl group, or via hydrogen bonding.[17,18] In principle, therefore, higher metal uptake is likely.

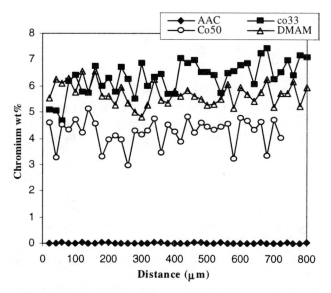

Figure 18. Chromium sorption for different AAc composition of spheres.

The uptake for several compositional spheres immersed in 1M $CrCl_3$ under the same condition is shown in Figure 18. The microprobe data were in this case all collected from each particle, as shown for one analysis in Figure 19.

It can be seen that the uptake in acrylic acid at pH 0.92 is negligible, whilst the other spheres have Cr concentration ranging from about 4 to 6.7 wt%. Such levels are somewhat lower than for the copper, suggesting that the complexation is less effective. The concentration profile across each sphere shows no significant trend, the level being essentially constant if "noise" is ignored. In this regard, the results are comparable with those obtained for the copper chloride.

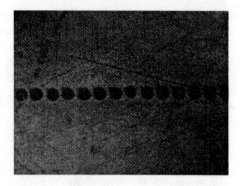

Figure 19. Individual elemental analysis for Cr using the Cameca SX50 microprobe.

6. ACKNOWLEDGEMENT

The authors wish to acknowledge financial support for Audist Subekti via an AusAID postgraduate scholarship. Assistance on microscopic analysis was provided by M. Dickson, V. Piegerova, and B. Searle. Assistance on magnetic resonance imaging (MRI) was by A. Whittaker. Resin embedding was performed by Mr. R. Flosman.

7. REFERENCES

1. J.C. Wheeler, et.al., Evolution of hydrogel polymers as contact lenses, surface coatings, dressings, and drug delivery systems [Review], *J. of Long-Term Effects of Medical Implants* **6**, 207-217 (1996).
2. Nedo, Japan, Intelligent Polymers, http//www.nedo.go.jp
3. C.J. Kim, *CHEMTECH* **24,** 36-39 (1994).
4. S.W. Shalaby, in *Water-soluble Polymer: Synthesis, Solution Properties and Applications, Thermo-responsive gels*, ACS Series, edited by Charles L. McCormick, George B. Butler, p. 502-505 (1991).
5. H.G. Schild, in *Water-soluble Polymer: Synthesis, Solution Properties and Applications, Probes of the LCST of Poly N-isopropylacrylamide*, edited by Charles L. McCormick, George B. Butler, ACS Series, p. 249-259 (1991).
6. C. Heitz, W. Binana-Limbele, J. François and C. Biver., Absorption and desorption of chromium ions by poly(acrylic acid) gels, *J. Appl. Polym. Sci.* **72** (4), 455-466 (1999).
7. C. Guohua, and A.S Hoffman, Graft copolymer that exhibit temperature-induced phase transitions over a wide range of pH, *Nature* **373** (6509), 49-52 (1995).
8. A. Lamprecht, H.R. Torres, U. Schäfer and C.M. Lehr, Biodegradable microparticles as a two-drug controlled release formulation: a potential treatment of inflammatory bowel disease, *J. Contr. Release* **69**, 445-454 (2000).
9. K. Kandori, K. Kishi, and Ishikawa, T., Preparation of uniform silica hydrogel particles by SPG filter emulsification method. *Colloids Surf.*, **62,** 259-262 (1992).
10. J. Ugelstad, P. Stenstad, L. Kilaas, W.S. Prestvik, A. Rian, K. Nustad, R. Herje and A. Berge, Biochemical and biomedical application of monodisperse polymer particles, *Macromol. Symposia* **101,** 491-500 (1996).
11. E. Ruckenstein and H. Liang, Sedimentation Polymerization, *Polymer* **36** (14), 2857-2860 (1995).
12. K. Bushell, *Sedimentation Polymerisation*, Honours Project, UNSW (1996).
13. B. Wechlber, *Porous Polymer Particle*, Honours Project, UNSW (1999).

14. M.K. Yoo, Y.K. Sung, Y.M. Lee and C.S. Cho., Effect of polyelectrolyte on the LCST of poly(N-isopropyl acrylamide) in the poly(NIPAM-co-Acrylic Acid), *Polymer* **41** (15), 5713-5719 (2000).
15. J. Grignon and A.M. Scallan, The effect of pH and neutral salts upon the swelling properties of cellulose gels, *J. Appl. Polym. Sci.* **25**, 2829-2843 (1980).
16. W.F. Lee and C.H. Shieh, pH-thermoreversible hydrogels I. Synthesis and swelling behaviours of the (N-isopropyl acrylamide –co acrylamide-co- 2-hydroxyethyl methacrylate), *J. Appl. Polym. Sci.* **71** (2), 221-231 (1999).
17. A.P. More and A.M. Donald, Mechanical-properties of Nylon-6 after treatment with metal halides, *Polymer* **34**, 5093-5098 (1993).
18. P. Dunn and G.F. Sansom, The stress cracking of polyamides by metal salts, part I. metal halides, *J. Appl. Polym. Sci.* **13**, 1641-1655 (1969).

SUPRAMOLECULAR ASSEMBLIES CONSISTING OF NAPHTHALENE-CONTAINING ANIONIC AMPHIPHILES AND ONE-DIMENSIONAL HALOGEN-BRIDGED PLATINUM COMPLEXES

Nobuo Kimizuka*, Yasuhiro Hatanaka, and Toyoki Kunitake[†]

1. INTRODUCTION

Molecular wires have been attracting much interest due to their indispensable roles in molecular-scale electronic devises, and conventional researches are largely focused on π-conjugated polymeric systems.[1] They suffer from limitations on the type of elements that can be incorporated into the chains. In contrast to the π-conjugated wires, one-dimensional inorganic complexes are composed of a rich variety of metal ions. A family of halogen-bridged one-dimensional M^{II}/M^{IV} mixed valence complexes $[M(en)_2][M'X_2(en)_2]Y_4$ (M, M' = Pt, Pd, Ni, X = Cl, Br, I, en : 1,2,-diaminoethane, Y : counterions such as ClO_4) has been attracting much interest due to their unique physicochemical properties such as intense intervalence charge transfer (CT) absorption, semiconductivity, and large third-order nonlinear optical susceptibilities.[2] They are not soluble in organic media and when dispersed in water, the one-dimensional structure is disrupted and dissociate into constituent molecular complexes. We have recently developed a new strategy to solubilize such one-dimensional structures in organic media, by the formation of polyioncomplexes consisting of anionic lipids and the mixed valent complexes.[3-5] In this study, we have newly synthesized naphthalene containing sulfonate amphiphiles and solution characteristics of the supramolecular complexes are discussed.

2. DESIGN AND SYNTHESIS OF NAPHTHALENE-CONTAINING AMPHIPHILES AND THEIR SUPRAMOLECULAR ASSEMBLIES

We have reported previously that naphthalene-containing ammonium amphiphiles form stable bilayer membranes in water and they display specific exciton interactions among the highly organized chromophores.[6] Efficient energy migration among chromophores organized in bilayer assemblies and their electron transfer characteristics have been also shown.[6,7] By the introduction of such organic chromophores in the lipid/one-dimensional mixed valent Pt^{II}/Pt^{IV} system, it is expected that photo-induced electron transfer between the chromophores and inorganic one-dimensional chain cause change in electronic states of the one-

*Nobuo Kimizuka, Yasuhiro Hatanaka, Department of Applied Chemistry, Faculty of Engineering, Kyushu University, Hakozaki 6-10-1, Fukuoka 812-8581, Japan. [†]Toyoki Kunitake, current addresss, Frontier Research System, RIKEN, Wako, Saitama, 351-0198, Japan.

Advanced Macromolecular and Supramolecular Materials and Processes
Edited by K. Geckeler, Kluwer Academic/Plenum Publishers, 2003

one-dimensional inorganic chains.   Such photophysical interplay of organic and inorganic components in mesoscopic supramolecular assemblies[8] is an issue of great importance in nanochemistry and is largely unexplored.

We newly prepared naphthalene-containing sulfonate amphiphiles **1**, **2** and their supramolecular assemblies $[Pt(en)_2][PtCl_2(en)_2](\mathbf{1, 2})_4$. Amphiphiles **1** and **2** are single-chained and double-chained compounds, respectively. In the alkyl chain moiety of **2**, ether linkages and multiple amide bonds with chiral, L-glutamate unit are introduced. Introduction of the ether units serve to increase the lipophilicity of the amphiphile[8,9] and the amide bonds enhance the intermolecular association especially in organic media.[10] These compounds are synthesized by the alkylation reaction of corresponding bromides with 6-hydroxynaphthalenesulfonate in ethanol. The structure and purity of the amphiphiles were confirmed by TLC-FID, IR, NMR spectroscopy and by elementary analysis. When anionic amphiphiles are polyioncomplexed with the cationic one-dimensional platinum chains, alkyl chains of the lipids serve as solvophilic parts and they impart solubility to the assemblies. On the other hand, polyion complex moieties are solvophobic and thus amphiphilic supramolecular structures are obtained in organic media. Such an amphiphilic design is a indispensable feature in the design of mesoscopic-scale supramolecular assemblies.[8] Ternary complexes of $[Pt(en)_2][PtCl_2(en)_2](\mathbf{1}$ or $\mathbf{2})_4$ was prepared by the reported procedure[3-5] and the composition was confirmed by elementary analysis.

$$CH_3\text{-}(CH_2)_{11}\text{-}O\diagup\text{naphthalene}\diagup SO_3^-Na^+ \qquad \mathbf{1}$$

$$CH_3\text{-}(CH_2)_{11}\text{-}O\text{-}(CH_2)_3\text{-}N\text{-}C\text{-}CH\text{-}N\text{-}C\text{-}(CH_2)_{10}\text{-}O\diagup\text{naphthalene}\diagup SO_3^-Na^+$$

$$CH_3\text{-}(CH_2)_{11}\text{-}O\text{-}(CH_2)_3\text{-}N\text{-}C\text{-}CH_2 \qquad \mathbf{2}$$

3. SOLUTION CHARACTERISTICS OF LIPID/ONE-DIMENSIONAL PLATINUM COMPLEXES IN ORGANIC MEDIA

3.1. Electron Microscopy

The polyioncomplex $[Pt(en)_2][PtCl_2(en)_2](\mathbf{1})_4$ gave a yellow colloidal (liquid crystalline) dispersion in chloroform, while $[Pt(en)_2][PtCl_2(en)_2](\mathbf{2})_4$ was homogeneously dispersed in chloroform with yellow color. Apparently, double-chained amphiphile provides enhanced solubility in organic media. The observed intense colors in organic media are ascribed to intervalence charge transfer absorption (CT ; $Pt^{II} \rightarrow Pt^{IV}$) of chloro-bridged platinum complexes,[2] and this clearly indicates that one-dimensional mixed-valence chains of $[Pt(en)_2][PtCl_2(en)_2]$ remain intact as lipophilic polyion complexes.[3-5]

Figure 1 displays electron micrographs of (a) $[Pt(en)_2][PtCl_2(en)_2](\mathbf{1})_4$ and (b) $[Pt(en)_2][PtCl_2(en)_2](\mathbf{2})_4$, respectively. The samples are not negatively stained and the dark portions indicate the presence of mixed-valent platinum chains. $[Pt(en)_2][PtCl_2(en)_2](\mathbf{1})_4$ provided rectangular microcrystals, while flexible nanofibers were found for $[Pt(en)_2][PtCl_2(en)_2](\mathbf{2})_4$. It is apparent that the polyion complexes are dispersed in organic media as mesoscopic superstructures, and the chemical structure of the lipids exerts a decisive role on the morphology of the assemblies. Recently, structures of polyion complexes formed from $[Pt(en)_2][PtI_2(en)_2]$ and n-alkylsulfonates are determined by X-ray diffraction

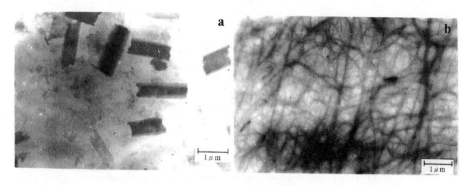

Figure 1. Electron micrographs of (a) [Pt(en)$_2$][PtCl$_2$(en)$_2$](1)$_4$ in chlorocyclohexane and (b) [Pt(en)$_2$][PtCl$_2$(en)$_2$](2)$_4$ in chloroform. Samples are not negatively stained.

experiments and the single-chained amphiphiles adopt interdigitated structure in the crystalline samples.[11] The lower solubility and crystalline microstructure observed for [Pt(en)$_2$][PtCl$_2$(en)$_2$](1)$_4$ would be also ascribed to the interdigitation of alkyl chains. On the other hand, such interdigitation does not take place for the double chained complex [Pt(en)$_2$][PtCl$_2$(en)$_2$](2)$_4$ and consequently, it showed enhanced solubility with nano-architectures.

3.2. Spectral Characteristics

UV-vis absorption spectra of [Pt(en)$_2$][PtCl$_2$(en)$_2$](1)$_4$ in chlorocyclohexane and [Pt(en)$_2$][PtCl$_2$(en)$_2$](2)$_4$ in chloroform are shown in Figure 2, together with their temperature dependence. They displayed charge transfer absorption bands at 370 nm ([Pt(en)$_2$][PtCl$_2$(en)$_2$](1)$_4$) and at 380 nm ([Pt(en)$_2$][PtCl$_2$(en)$_2$](2)$_4$), respectively. These absorption peaks are red shifted compared to the molecularly dispersed PtCl$_2$(en)$_2$ complex (absorption maximum, ca. 325 nm)[4] and therefore the chloro-bridged mixed valence chains are maintained in the dispersions. Interestingly, when the chlorochclohexane dispersion of ([Pt(en)$_2$][PtCl$_2$(en)$_2$](1)$_4$ was heated, its CT-absorption intensity at 370 nm started to decrease and disappeared above 60 °C, as shown in the inset (a). Upon cooling, the absorption intensity recovered below 50 °C and large thermal hysteresis was observed. The observed reversible thermochromism indicates that the halogen-bridged one-dimensional chain of [Pt(en)$_2$][PtCl$_2$(en)$_2$](1)$_4$ is disrupted at higher temperatures and they dissociate into component molecular complexes [Pt(en)$_2$](1)$_2$ and [PtCl$_2$(en)$_2$](1)$_2$, whereas they are re-assembled to the original supramolecular assemblies by cooling. Such self-assembling properties have been also reported for the other [Pt(en)$_2$][PtCl$_2$(en)$_2$](lipid)$_4$ complexes.[4] The large thermal hysteresis would be ascribed to the formation of interdigitated crystalline structure that required supercooling of the molecularly dispersed state. On the other hand, heating of the chloroform dispersion of [Pt(en)$_2$][PtCl$_2$(en)$_2$](2)$_4$ caused decrease of the CT-band intensity at lower temperatures (ca. 25 °C), and the intensity was lost at 45 °C (inset in Figure 2b). Cooling of the chloroform dispersion displayed smaller thermal hysteresis, probably due to the enhanced intermolecular association force by the hydrogen bonds between multiple amide groups. It is interesting that chemical structure of anionic amphiphiles determines not only solubility and morphology but also thermal characteristics of the supramolecular assemblies. The double-chained amphiphiles would be surrounding the periphery of one-dimensional Pt^{II}/Pt^{IV} chains, and this molecular orientation must impart enhanced lipophilicity to the supramolecular assembly.

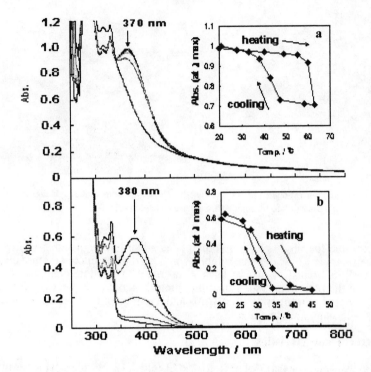

Figure 2. UV-vis absorption spectra (temperature dependence) of (a) $[Pt(en)_2][PtCl_2(en)_2](\mathbf{1})_4$ in chlorocyclohexane and (b) $[Pt(en)_2][PtCl_2(en)_2](\mathbf{2})_4$ in chloroform. Concentration, 0.05 unit mM. Temperature dependence of the CT peak intensities are shown in the insets.

Figure 3 compares fluorescence spectra of $[Pt(en)_2][PtCl_2(en)_2](\mathbf{2})_4$ in chloroform and that of **2** in ethanol-water mixture (1/1 by vol). The fluorescence of naphthalene chromophores near at 360 nm is remarkably quenched in the supramolecular assembly $[Pt(en)_2][PtCl_2(en)_2](\mathbf{2})_4$, and this is ascribed to photo-induced electron transfer or energy transfer from the aligned naphthalene chromophores to the one-dimensional chain of $[Pt(en)_2][PtCl_2(en)_2]$, as schematically shown in Figure 4.

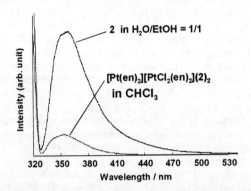

Figure 3. Fluorescence spectra of **2** in H_2O/EtOH = 1/1 and $[Pt(en)_2][PtCl_2(en)_2](\mathbf{2})_4$ in chloroform. Concentration, 0.02 unit mM. Excitation at 315 nm, at 20 °C

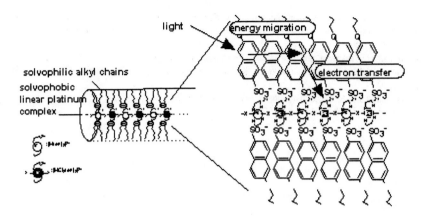

Figure 4. Schematic illustration of supramolecular assembly [Pt(en)$_2$][PtCl$_2$(en)$_2$](**2**)$_4$ in chloroform.

Apparently, electronic interaction between organic chromophores in the package molecules and inorganic molecular wire is achieved, and the present result opens a new design of photofunctional, inorganic molecular wires. Detailed study on the quenching process is now underway in these laboratories.

4. CONCLUSION

Together with the previous works,[3-5,12] the present results establish that the extended mixed-valence complex is maintained in organic media, by complexation with suitably designed anionic amphiphiles. Tuning of solubility, morphology and thermal characteristics of the supramolecular complex can be achieved by the use of suitably designed "packaging" amphiphiles. The double-chained amphiphile affords fine nano-fibrous structures and is conceived as a better molecular design as organic package molecules. It is also apparent that photoinduced electron transfer or energy transfer occurs between the aligned naphthalene chromophores and inorganic main chains, giving rise to photo-control of one-dimensional electronic states. The design of organic-inorganic electronic interactions would provide new functions in mesoscopic supramolecular assemblies.

5. REFERENCES

1. For example, G. Padmanaban and S. Ramakrishnan, *J. Am. Chem. Soc.* **122**, 2244 (2000).
2. H. Okamoto and M. Yamashita, *Bull. Chem. Soc. Jpn.* **71**, 2023 (1998) and references therein.
4. N. Kimizuka, N. Oda, and T. Kunitake. *Chem. Lett.* **1998**, 695.
5. N. Kimizuka, N. Oda, and T. Kunitake. *Inorg. Chem.* **39**, 2684 (2000).
6. N. Kimizuka, S. H. Lee, and T. Kunitake. *Angew. Chem. Int. Ed. Engl.* **39**, 389 (2000).
7. (a) N. Nakashima, N. Kimizuka, and T. Kunitake, *Chem. Lett.* 1817 (1985). (b) N. Kimizuka, T. Kunitake, *J. Am. Chem.Soc.* **111**, 3758 (1989).
8. (a) T. Kunitake, M. Shimomura, Y. Hashiguchi, and T. Kawanaka, *J. Chem. Soc., Chem. Commun*, 833 (1985). (b) H. Nakamura, H. Fujii, H. Sakaguchi, T. Matsuo, N. Nakashima, K. Yoshihara, T. Ikeda, and S. Tazuke, *J. Phys. Chem.* **92**, 6151 (1989).

9.   (a) N. Kimizuka, T. Kawasaki, K. Hirata, and T. Kunitake, *J. Am. Chem. Soc.* **120**, 4094 (1998) and references therein. (b) N. Kimizuka, T. Kawasaki, K. Hirata, and T. Kunitake, *J. Am. Chem. Soc.* **117**, 6360 (1995).

10. N. Kimizuka, T. Takasaki, and T. Kunitake, *Chem. Lett.* 1911 (1988).

11. N. Kimizuka, M. Shimizu, S. Fujikawa, K. Fujimura, M. Sano, and T. Kunitake, *Chem. Lett.* 967 (1998).

12. N. Matsushita and A. Taira, *Synthetic Metals*, **102**, 1787 (1999).

13. N. Kimizuka, *Adv. Mater.* **12**, 1461 (2000).

DIRECTLY LINKED AND FUSED OLIGOPORPHYRIN ARRAYS

Akihiko Tsuda, Naoya Yoshida, Aiko Nakano, Naoki Aratani, and Atsuhiro Osuka[*]

1. INTRODUCTION

The design and preparation of linear, rigid, rod-like, conjugated molecules with precise length and constitution have attracted much interest in light of their potential applications as liquid crystals, optical devices, sensors, molecular machines, and conductive molecular wires.[1] With these backgrounds, numerous attempts have been made to prepare the linear, rod-like, conjugated molecules.[2] In most cases, aromatic groups or chromophores are connected with rigid spacer molecules or directly without spacer, thereby to construct large π-electronic arrays in a repetitive manner.

Among these, porphyrins are one of the intriguing building units to construct rod-like conjugated molecules.[3] In spite of a large number of covalently linked porphyrin arrays, it still remains a great synthetic challenge to explore conjugated porphyrin arrays which are much larger in size, more rigid, and more extensively π-conjugated, being suitable as more realistic photonic or electronic molecular wires.

* Department of Chemistry, Graduate School of Science, Kyoto University, Sakyo-ku, Kyoto 606-8502, Japan.

Advanced Macromolecular and Supramolecular Materials and Processes
Edited by K. Geckeler, Kluwer Academic/Plenum Publishers, 2003

2. SINGLY-LINKED AND FUSED PORPHYRYN ARRAYS

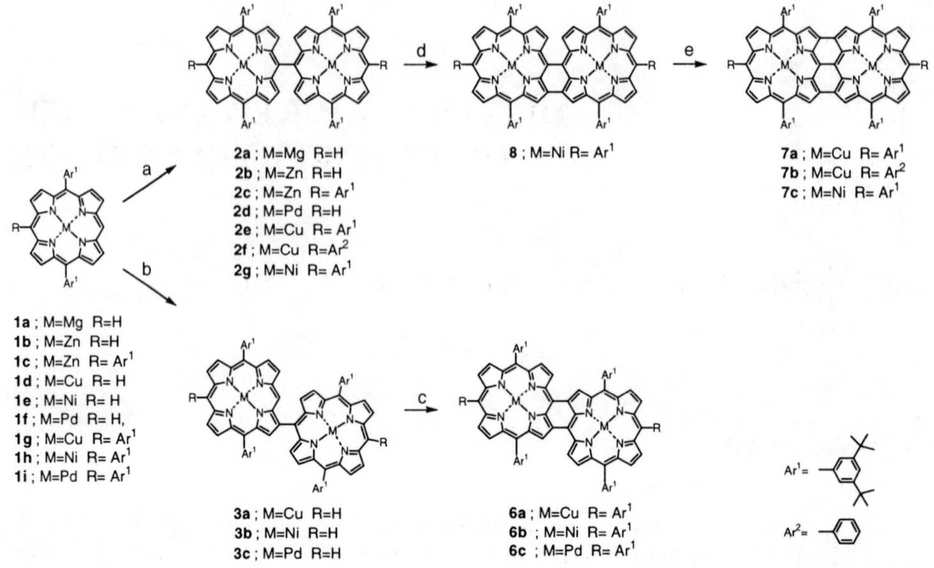

Scheme 1

2.1. *meso-meso-* and *meso-β*–Linked Diporphyrins

Recently, we have found that one-electron oxidation of porphyrins bearing sterically uncongested *meso*-position such as 5,15-diaryl-substituted metalloporphyrins and 5,10,15-triaryl-substituted metalloporphyrins is an effective route to prepare *meso-meso* and *meso-β* directly linked diporphyrins 2 and 3. Scheme 1 shows the synthetic routes to the singly-linked diporphyrins (paths a and b) staring from 5,15-diaryl- or 5,10,15-triaryl-substituted metalloporphyrins. Mg[II]- and Zn[II]-porphyrins **1a-c** gave *meso-meso*-linked diporphyrins **2a-c** by Ag[I] salt oxidation or electrochemical oxidation,[4] while the Cu[II]-, Ni[II],- and Pd[II]- porphyrins **1d-f** gave *meso-β*-linked diporphyrin **3a-c** by electrochemical oxidation.[5]

The molecular structure of **2** (Figure 1), which has been revealed by X-ray crystallography, shows that the two porphyrin rings in **2d** take a nearly orthogonal conformation (86.0°) and the bond connecting two porphyrin units is 1.47 Å long, being similar to C_2-C_3 bond length (1.48 Å), and thus tend to minimize the electronic interaction between the porphyrin components.

(a)   **(b)**

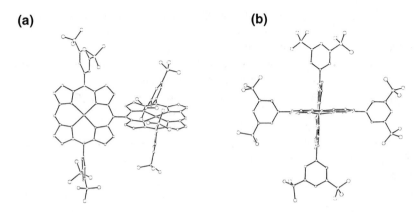

Figure 1. X-ray crystal structure of *meso-meso* singly-linked Pd^{II}-diporphyrin **2d**; (a) top view, (b) side view.

The molecular structure of **3** has been also confirmed by X-ray crystallography, and the similar orthogonal conformation was observed (not shown). The observed different regio-selectivity either giving **2** or **3** may be explained in terms of different HOMO orbital characters; namely A$_{2u}$–HOMO for Mg^{II}- and Zn^{II}- porphyrins, in which there is a large electron density at the *meso*-positions and A$_{1u}$-HOMO for Cu^{II}-, Ni^{II}- and Pd^{II}-porphyrins, in which there are nodes at the *meso*-positions and instead is a significant density at the *β*-positions.

2.2. Directly Linked Oligoporphyrin Arrays

As an extension, we prepared directly *meso-meso* singly-linked oligoporphyrins starting from 5,15-diaryl-Zn^{II}-porphyrin having two free *meso* positions (Scheme 2).[4] Linear polymeric porphyrin arrays were effectively obtained in the reaction with Ag^{I}-salt in CHCl$_3$ in the presence of 0.5% *N,N*-dimethylacetamide.[6] On the other hand, the discrete oligoporphyrin arrays having precise length were obtained by repeating Ag^{I}-salt promoted coupling reaction in CHCl$_3$ and subsequent repetitive GPC/HPLC (GPC = gel-preparation chromatography).[4]

Scheme 2

Recently we have achieved the synthesis of *meso-meso*-linked porphyrin arrays up to the 128-mer with ca. 108 nm molecular length.[7] To the best of our knowledge, the 128-mer is the longest discrete, rod-like organic molecule.

This coupling reaction is advantageous in light of its high regioselectivity occurring at the *meso*-position as well as its easy extension to large porphyrin arrays. The windmill porphyrin array **4a** in turn serves as an effective substrate for further coupling reactions, to give three-dimensionally arranged grid porphyrin arrays **5** (~48-mer) (Scheme 3).[8] In all Zn^{II}-metallated windmill porphyrin array **4b**, the S_1-energy level of the *meso-meso*-linked diporphyrin core is lower than that of the peripheral porphyrins, thereby allowing an energy flow from the peripheral porphyrins to the central diporphyrin core.

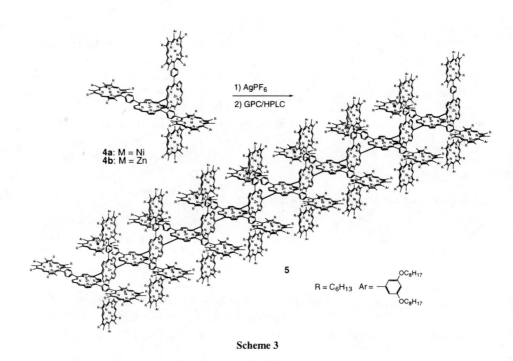

1) AgPF$_6$

2) GPC/HPLC

4a: M = Ni
4b: M = Zn

5

R = C$_6$H$_{13}$ Ar = —⟨ OC$_8$H$_{17}$ / OC$_8$H$_{17}$ ⟩

Scheme 3

At present, these giant molecules are still possessing a possibility to extend their one-dimensionally arranged structures, and therefore, the continuous synthetic challenges to produce further giant molecules are in progress in our laboratory.

2.3. Directly Fused Diporphyrins

Next' we found novel synthetic route to directly fused porphyrins through the oxidation of metallopoprhyins with tris(4-bromophenyl)aminium hexachloroantimonate (BAHA). The oxidation of Cu^{II}-, Ni^{II}- and Pd^{II}- porphyrins **1g-i** with A_{1u}-HOMO gave rise to formation of *meso-β* doubly-linked diporphyrins **6a-c** (path c in Scheme 1),[9] whereas the oxidation of *meso-meso* singly-linked diporphyrins **2e-g** with A_{1u}-HOMO gave rise to formation of *meso-meso,β-β,β-β*-triply-linked diporphyrins **7a-c** probably through *meso-meso β-β* doubly-linked diporphyrin **8** (path d and e in Scheme 1).[10]

In the oxidation reactions of **2e-g**, the peripheral chlorination at the *β*-positions was a serious side-reaction that was fairly suppressed by conducting the reaction in C_6F_6. All Zn^{II}- and Cu^{II}-complexes of directly fused diporphyrins can be easily converted to the free-base forms by treatment with acids such as HCl and H_2SO_4, and subsequent metal insertion led to various metal complexes as similar in the normal porphyrins.

The X-ray crystal structures have been obtained for all the compounds, *meso-β* doubly-linked Ni^{II}-diporphyrin **6b**, *meso-meso,β-β* doubly-linked Ni^{II}-diporphyrin **8** and *meso-meso,β-β,β-β* triply-linked Cu^{II}-diporphyrin **7b** (Figure 2(a) (c)).[11]

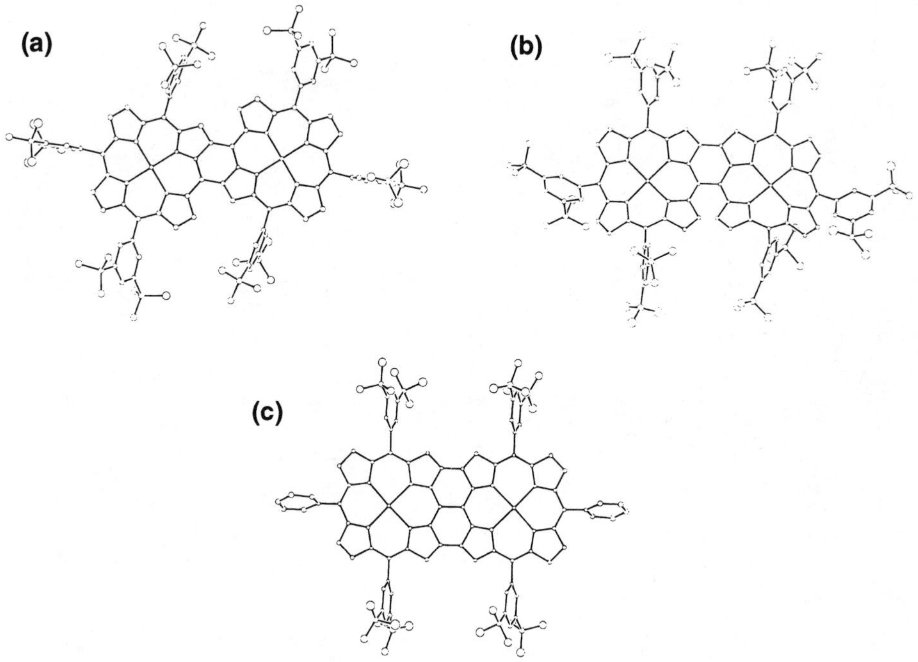

Figure 2. X-ray crystal structures of directly fused metallodiporphyrins: (a) *meso-β* doubly-linked Ni^{II}-diporphyrin **6b**, (b) *meso-meso β-β* doubly-linked Ni^{II}-diporphyrin **8**, and (c) *meso-meso,β-β,β-β–* triply-linked Cu^{II}-diporphyrins **7b**.

The two porphyrin rings in **6b** were coplanar but adopt a ruffled conformation in which the two newly *meso-β* bonds are 1.45 Å long, being shorter than that of **2d**. In *meso-meso β-β* doubly-linked Ni^{II}-diporphyrin **8**, the two porphyrin units have the opposite ruffling directions to circumvent the steric repulsion of the hydrogen atoms adjacent to the *meso-meso* connection, giving rise to large distance between the non-connected *β*-positions (3.20 Å). As to *meso-meso,β-β,β-β*–triply-linked Cu^{II}-diporphyrins **7b**, the two porphyrins take a completely coplanar conformation.

Figure 3 compares the absorption spectra of the Ni^{II}-porphyrin series. In contrast to the sharp Soret band of the monomer **1h**, the singly-linked diporphyrins **2g** and **3b** exhibit perturbed Soret bands but the spectral changes are both rather modest in the Q-bands in spite of the direct connection. This may be ascribed to their almost perpendicular conformations, which are unfavorable to the electronic conjugation. Therefore, the spectral changes in **2g** and **3b** can be qualitatively understood in terms of the exciton coupling between the transition dipole moments of the porphyrin, which depends upon the oscillator strength of the dipole transition and the diporphyrin geometry. In contrast, the fused diporphyrins **6b**, **7c**, and **8** display entirely altered absorption spectra, which are characterized by largely split Soret bands and red-shifted and intensified Q-bands. The altered electronic absorption spectra may be a direct consequence of electronic π-conjugation. Among the three, it is obvious that the triply

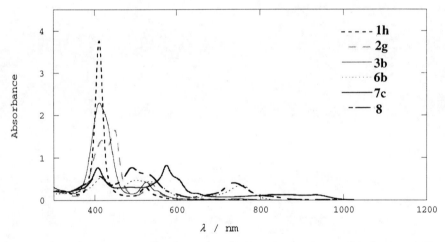

Figure 3. Absorption spectra of Ni^{II}-porphyrin **1h** (2×10^{-5} M) and diporphyrins **2g**, **3b**, **6b**, **7c**, and **8** (1×10^{-5} M) in CHCl$_3$ at room temperature.

linked diporphyrin **7c** displays the most altered absorption spectrum and its Q-band reaches into near-infrared region, reflecting a very small optical HOMO-LUMO gap.

From these spectral and above X-ray crystal structural studies of diporphyrins it can be found that the singly-linked diporphyrins, which adopt orthogonal conformation, mainly reflect Coulombic interporphyrin interaction to the absorption spectra. On the other hand, the double or triple connections between porphyrin components steadily extend the π-electronic communication over the arrays. The fused planar structures are very effective for the extension of electronic coupling between porphyrin components.

2.4. Fully Conjugated Porphyrin Tapes

Sc^{III}-catalyzed oxidation of *meso-meso* singly-linked Zn^{II}-diporphyrin with DDQ also led to an efficient formation of *meso-meso,β-β,β-β* triply-linked Zn^{II}-diporphyrin than in the case of the oxidation with BAHA.[12] This synthetic method was applicable to the higher *meso-meso* linked Zn^{II}-porphyrin arrays (up to the 12-mer) to give the corresponding fused tape-shaped porphyrin arrays in 62~91% yields (Scheme 4).

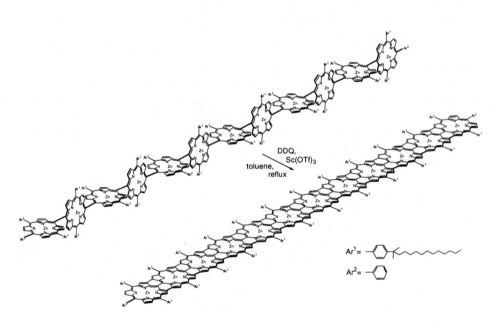

Scheme 4

The triply-linked oligoporphyrin arrays produced display the extremely red-shifted absorption bands, reflecting the extensively π-conjugated electronic systems over the array.

The lowest energy electronic absorption bands become more red-shifted and intensified upon the increase of the number of the porphyrins, and eventually reach into the infrared frequency. The extremely small HOMO-LUMO gaps, the exceedingly low oxidation potentials, and the linearly long rigid molecular shapes of the higher triply-linked porphyrin arrays encourage their potent use as a molecular wire.

3. CONCLUSION

Chemical or electrochemical oxidation of metalloporphyrins **1** bearing sterically uncongested *meso*-position led to the formations of directly *meso-meso-* and *meso-β* singly-linked diporphyrins **2** and **3**. One-dimensionally arranged *meso--meso*-linked oligo- or polymeric porphyrin arrays were also effectively formed by the oxidation of 5,15-diaryl-Zn^{II}-porphyrin with Ag^{I}-salt in $CHCl_3$. The discrete rod-like, *meso-meso*linked 128-mer and three-dimensionally arranged grid porphyrin 48-mer were obtained by repeating the oxidative coupling reactions.

Fused planar *meso-β* doubly-linked diporphyrins **6** were obtained by the strong oxidation of metalloporphyrins **1** with BAHA. In addition, the reaction of singly-linked diporphyrins **2** with BAHA led to the formation of *meso-meso β-β* doubly-linked and *meso-meso,β-β,β-β* triply-linked diporphyrins **8** and **7**.

Their X-ray crystal structures and systematic red-shifts of absorption spectra with increasing connection between porphyrin components indicated that electronic interactions over the arrays are increased in the order of $1 < 2 \approx 3 < 6 \approx 8 < 7$. The *meso-meso*-linked oligoporphyrins were also effectively converted to the tape-shaped, completely fused forms by Sc^{III}-catalyzed oxidation with DDQ.

4. REFERENCES

1. (a) E. Clar, *Ber.* **69**, 607 (1936). (b) R. E. Martin, F. Diederich, *Angew. Chem. Int. Ed.* **38**, 1350 (1999). (c) P. F. Schwab, M. D. Levin, and J. Michl, *Chem. Rev.* **99**, 1863 (1999). (d) H. L. Anderson, *Chem. Commun.* 2323 (1999). (e) C. Joachim, J. K. Gimzewski, and A. Aviram, *Nature*, **408**, 541 (2000). (f) V. Balzani, A. C. F. M. Raymo, and J. F. Stoddard, *Angew. Chem. Int. Ed.* **39**, 3348 (2000).

2. (a) J. Kao and A. C. Lilly, Jr, *J. Am. Chem. Soc.* **109**, 4149 (1987). (b) M. Pomerantz, R. Cardona, and P. Rooney, *Macromolecules*, **22**, 304 (1989). (c) J. Roncali, *Chem. Rev.* **97**, 173 (1997).

3. (a) N. Kobayashi, M. Numao, R. Kondo, S. Nakajima, and T. Osa, *Inorg. Chem.* **30**, 2241 (1991). (b) M. J. Crossley, P. L. Burn, S. J. Langford, and K. J. Prashar, *J. Chem. Soc., Chem. Commun.* 1921 (1995). (c) M. Graça, H. Vicente, L. Jaquinod, and K. M. Smith, *Chem. Commun.* 1771 (1999).

4. (a) A. Osuka and H. Shimidzu, *Angew. Chem., Int. Ed. Engl.* **36**, 135 (1997). (b) N. Yoshida, H. Shimidzu, and A. Osuka, *Chem. Lett.* 55 (1998). (c) T. Ogawa, Y. Nishimoto, N. Yoshida, N. Ono, and A. Osuka, *Chem. Commun.* 337 (1998).

5. T. Ogawa, Y. Nishimoto, N. Yoshida, N. Ono, and A. Osuka, *Angew. Chem. Int. Ed.* **38**, 176 (1999).

6. N. Yoshida, N. Aratani, and A. Osuka, *Chem. Commun.* 197 (2000).

7. (a) N. Aratani, A. Osuka, Y. H. Kim, D. H. Jeong, and D. Kim, *Angew. Chem., Int. Ed.* **39**, 1458 (2000). (b) Y. H. Kim, D. H. Jeong, D. Kim, S. C. Jeoung, H. S. Cho, S. H. Kim, N. Aratani, and A. Osuka, *J. Am. Chem. Soc.* **123**, 76 (2001).

8. (a) A. Nakano, A. Osuka, I. Yamazaki, T. Yamazaki, and Y. Nishimura, *Angew. Chem. Int. Ed.* **37**, 3023 (1998). (b) A. Nakano, T. Yamazaki, Y. Nishimura, I. Yamazaki, and A. Osuka, *Chem. Eur. J.* **6**, 3254 (2000).

9. A. Tsuda, A. Nakano, H. Furuta, H. Yamochi, and A. Osuka, *Angew. Chem., Int. Ed.* **39**, 558 (2000).

10. A. Tsuda, H. Furuta, and A. Osuka, *Angew. Chem., Int. Ed.* **39**, 2549 (2000).

11. A. Tsuda, H. Furuta, and A. Osuka, submitted for publication.

12. A. Tsuda and A. Osuka, submitted for publication.

ELECTRIC FIELD EFFECTS ON ABSORPTION AND EMISSION SPECTRA OF END-PHENYLETHYNYLATED *MESO-MESO* LINKED PORPHYRIN ARRAYS IN A POLYMER FILM

Yuji Iwaki, Nobuhiro Ohta, Aiko Nakano,and Atsuhiro Osuka[*]

1. INTRODUCTION

Supramolecular chemistry based on a porphyrin building block has attracted much attention in light of a biomimetic construction of biologically functional networks as well as a scientific investigation of new materials applicable as a molecular-scale devise.[1-6]

Just recently, mono-disperse, rodlike giant molecules composed of *meso-meso* directly linked porphyrin array were synthesized by one of the present authors.[7,8] These extremely long, discrete porphyrin arrays show an interesting spectral feature, which is very different from the porphyrin monomer. As reported in our previous paper,[9] the field effects on absorption spectra of these compounds show that the change in electric dipole moment following excitation ($\Delta\mu$) is non-zero. The magnitude of $\Delta\mu$ becomes larger with increasing the number of linked molecules especially with excitation into the splitting component of the Soret band located at the longer wavelength region. Then, it was suggested that photoinduced charge transfer occurs along the linked porphyrin array, though the porphyrin arrays have a symmetric electronic structure. Significant changes in electric dipole moment that accompany strongly allowed optical transitions are prerequisite to effective molecular first hyperpolarization.[10] Therefore, porpyrin arrays may show efficient behavior of the incident light frequency doubling.

[*] Yuji Iwaki and Nobuhiro Ohta, Research Institute for Electronic Science (RIES), Hokkaido University, Sapporo 060-0812, Japan, and Graduate School of Environmental Earth Science, Hokkaido University, Sapporo 060-0810, Japan. Aiko Nakano and Atsuhiro Osuka, Department of Chemistry, Graduate School of Science, Kyoto University, Kyoto 606-8502, Japan.
Corresponding authors: N. Ohta (E-mail, nohta@es.hokudai.ac.jp); A. Osuka (E-mail, osuka@kuchem.kyoto-u.ac.jp).

Advanced Macromolecular and Supramolecular Materials and Processes
Edited by K. Geckeler, Kluwer Academic/Plenum Publishers, 2003

125

In the present study, electric field effect on absorption spectra as well as on fluorescence spectra of phenylethynylated *meso-meso* linked porphyrin arrays have been examined in a PMMA polymer film with electric field modulation spectroscopy. By analyzing the Stark shifts observed in absorption spectra and in fluorescence spectra, a difference both in electric dipole moment and in molecular polarizability between the ground state and each of the excited states is evaluated, and the results are compared with the ones of the compounds having no phenylethynyl substitute. Based on the results of the field effects on fluorescence spectra, excitation dynamics and its field dependence of phenylethynylated *meso-meso* linked porphyrin arrays are also discussed.

2. EXPERIMENTAL

Molecular structures of phenylethynylated *meso-meso* linked zinc porphyrin arrays employed in the present study are shown in Figure 1. Hereafter, these oligomers are denoted by ZAc(N), according to the number of the connected porphyrins, i.e., N. Structures of *meso-meso* linked zinc porphyrin arrays having no phenylethynyl substitute, which are denoted by Z(N), employed in our previous study are also shown in Figure 1. Electric

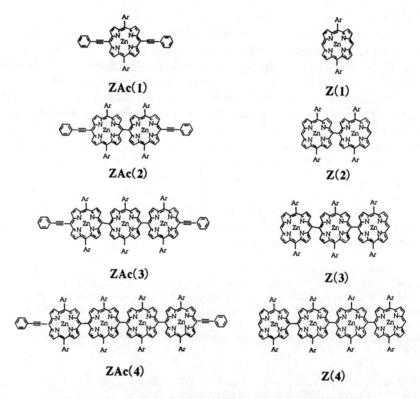

Figure 1. Molecular structures of Z(N) and ZAc(N). Ar = 3,5-ditertbutylphenyl.

field effects were examined for ZAc(N) doped in a PMMA polymer film for N from 1 to 4. The synthesis and characterization of ZAc(N) will be described elsewhere.[11]

A certain amount of benzene solution of PMMA containing porphyrin or porphyrin oligomer was poured onto the ITO-coated substrate by a spin coating technique. Then, a semitransparent aluminum (Al) film was deposited on the sample containing polymer film. The ITO and Al films were used as electrodes. Hereafter, applied electric field is denoted by F, and its strength is represented in rms.

Electric field effects on absorption and fluorescence spectra were examined using electric field modulation spectroscopy with the same apparatus as the one described elsewhere.[12] A sinusoidal ac voltage was applied, and the field-induced change in absorption intensity or emission intensity was detected with a lock-in amplifier (SR830, SRS) at the second harmonic of the modulation frequency. All the optical spectra were measured at room temperature under vacuum conditions. Hereafter, plots of the field-induced change in absorption intensity and in fluorescence intensity as a function of wavelength (or wavenumber) are called as electroabsorption (E-A) spectra and electrofluorescence (E-F) spectra, respectively.

3. THEORETICAL BACKGROUND

The level shift induced by F depends on the electric dipole moment and molecular polarizability at the states concerned. As a result, absorption spectra as well as emission spectra change in the presence of F. An expression for such a field-induced change in absorption intensity as well as in emission intensity was derived by Liptay and co-workers.[13] By assuming that the original isotropic distribution in rigid matrices such as a PMMA polymer film is maintained even in the presence of F, the spectral change in absorption intensity given in units of wavenumber, v, i.e., $\Delta A(v)$, in the presence of F may be given by the following equation: [13,14]

$$\Delta A(v) = (fF)^2 \{ AA(v) + Bvd[A(v)/v]/dv + Cvd^2[A(v)/v]/dv^2 \qquad (1)$$

where f is the internal field factor, A depends on the change in polarizability and hyperpolarizability of the transition moment, and B and C are given as follows:

$$B = [\Delta\bar{\alpha}/2 + (\Delta\alpha_m - \Delta\bar{\alpha})(3\cos^2\chi - 1)/10]/(hc) \qquad (2)$$

$$C = (\Delta\mu)^2 [5 + (3\cos^2\xi - 1)(3\cos^2\chi - 1)]/(30h^2c^2) \qquad (3)$$

where h is the Planck's constant and c is the light speed. Here, $\Delta\mu$ is the difference in electric dipole moment between the ground state and the excited state, i.e., $\Delta\mu = \mu_e - \mu_g$, and $\Delta\alpha$ is related to the difference in polarizability tensor, $\Delta\alpha = \alpha_e - \alpha_g$:

$$\Delta\mu = |\Delta\mu|; \Delta\bar{\alpha} = (1/3)\mathrm{Tr}(\Delta\alpha) \qquad (4)$$

$\Delta\alpha_m$ denotes the diagonal component of $\Delta\alpha$ with respect to the direction of the transition dipole moment, χ is the angle between the direction of F and the electric vector of the excitation light, and ξ is the angle between the direction of $\Delta\mu$ and the transition dipole moment.

Electric field-induced change in fluorescence intensity, i.e., $\Delta I_F(\nu)$, observed at the second harmonic of the modulation frequency in a polymer film is also given by a similar equation to Eq. (1), i.e., by a linear combination of fluorescence spectrum, its first and the second derivative spectra. The first and the second derivative components correspond to the spectral shift and the spectral broadening resulting from the difference in molecular polarizability and electric dipole moment between the fluorescent state and the ground state, respectively.

When a nonradiative process competes with emissive processes, a field-induced change in transition moment as well as a field-induced change in nonradiative process will induce a change in fluorescence intensity. The E-F spectra which give the same shape as the emission spectrum corresponds to the field-induced change in emission quantum yield. By evaluating the magnitude of the change in emission intensity, therefore, electric field effects on excitation dynamics can be discussed.

4. RESULTS AND DISCUSSION

E-A spectra of ZAc(N) doped in a PMMA polymer film, where N ranges from 1 to 4, were obtained in the region from 300 to 700 nm with a field strength of 0.75 MVcm^{-1}. The results are shown in Figure 2, together with the corresponding absorption spectra. The magnitude of ΔA is proportional to the square of the applied field strength in every case. Before the E-A spectra across the Soret bands of ZAc(N) are discussed, the E-A spectra of Z(1) - Z(4) are briefly reviewed. Note that ZAc(N) is a phenylethynylated compound of Z(N) (see Figure 1).

As reported in a previous paper,[9] the Soret band of Z(1) shows a sharp peak at 413.7 nm in PMMA, whereas this band splits and shows two strong peaks in porphyrin arrays. The splitting component of the Soret band located at the shorter wavelength side, referred as band (A), gives nearly the same position, i.e., at 417.7, 415.9 and 416.0 nm for Z(2), Z(3) and Z(4), respectively, while another strong splitting component is located in the longer wavelength region, referred as band (B), gives a red-shift of the absorption peak with an increase of the number of the linked porphyrins, i.e., at 452.4, 480.1 and 492.6 nm for Z(2), Z(3) and Z(4), respectively. The transition moment for bands (A) and (B) is considered to be perpendicular and parallel to the linked porphyrin array, respectively. The interaction between two transition dipoles whose directions are the same induces a larger energy separation, resulting in the red-shift of band (B). On the other hand, two dipoles whose directions are perpendicular with each other give no interaction, resulting in a fixed peak position of band (A). In fact, the suggested perpendicular arrangement of the neighboring porphyrins implies no dipole-dipole interaction for band (A) of Z(N). The E-A spectra of Z(N) with N ≥ 2 across the band (B) are close in shape to the second derivative of the absorption spectrum, suggesting that permanent electric dipole moment is induced following excitation into the band (B). The magnitude of the change in electric dipole moment ($\Delta\mu$) for excitation at the band (B) of Z(N) was shown to increase with increasing N, i.e., with increasing the number of the linked porphyrins. Then, it has been proposed that photoinduced charge transfer occurs along the linked porphyrin array following excitation into the band (B), even for electronically symmetric *meso-meso* linked porphyrin complexes.

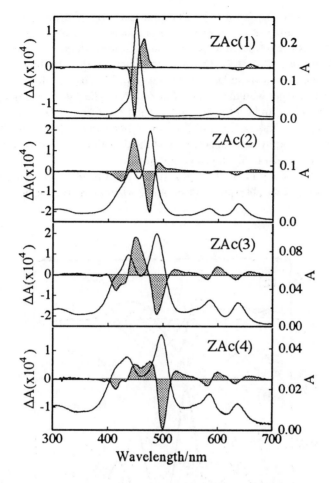

Figure 2. E-A spectra of ZAc(N) with N from 1 to 4 in a PMMA film (from top to bottom), together with the absorption spectra (solid line). Applied field strength was 0.75 MVcm⁻¹.

The Soret band of ZAc(1) shows a sharp peak at 450 nm in a PMMA polymer film, as in the case of Z(1). In ZAc(2) - ZAc(4), on the other hand, Soret bands give an exciton splitting, as in the case of porphyrin oligomers of Z(N). A strong splitting component of ZAc(N) located at the longer wavelength side gives a marked red-shift with increasing the number of the linked porphyrins. The peak position is at 474.8, 489.0 and 497.3 nm for ZAc(2), ZAc(3) and ZAc(4), respectively. These bands, denoted by band (β), which correspond to band (B) of Z(N), are considered to give the transition dipole whose direction is along the linked porphyrin array. An exciton splitting component of ZAc(N) located at the shorter wavelength side, which may correspond to band (A) of Z(N), is located at 432.9, 436.5 and 435.7 nm for ZAc(2), ZAc(3) and ZAc(4), respectively.

These bands are denoted by band (α). The peak position of these bands are nearly independent of the number of the linked porphyrins, and the direction of the transition dipole of these bands is considered to be perpendicular to the linked porphyrin array, as in the case of the bands (A) of Z(N). In ZAc(N), further, another splitting component of the Soret band appears as a shoulder of the band (α). This band, which is referred as band (γ), becomes stronger with increasing N in ZAc(N). The peak position of the band (γ) is roughly determined to be 415, 413 and 414 nm for ZAc(2), ZAc(3) and ZAc(4), respectively. Thus, the peak position of the band (γ) is also nearly independent of the number of the linked porphyrins, suggesting that the direction of the transition moment of this band is perpendicular to the linked porphyrin array.

The E-A spectrum of ZAc(1) in the region of the Soret band is very similar in shape to the first derivative of the absorption spectrum,[15] whereas the E-A spectra of ZAc(N)

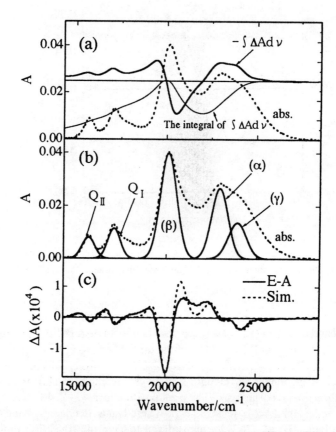

Figure 3. A series of optical spectra of ZAc(4). (a) Absorption spectrum (dotted line), - $\int\Delta A d\nu$ (thick solid line) and the integral of $\int\Delta A d\nu$ (thin solid line); (b) absorption spectrum (dotted line) and bands (α), (β) and (γ) and Q_I and Q_{II} bands (solid line); (c) E-A spectrum observed with a field strength of 0.75 MVcm^{-1} (solid line) and the simulated spectrum (dotted line). A continuum of the background is subtracted in the absorption spectra shown in (a) and (b).

with $N \geq 2$ are very different from simple first derivatives of the absorption spectra. As will be mentioned below, the E-A spectra across the band (β) of ZAc(N) are very similar in shape to the second derivative of the absorption spectra.

The analysis of the E-A spectra is described in detail by taking ZAc(4), as an example. The integral of ΔA, i.e., $\int \Delta A dv$, and the integral of $\int \Delta A dv$ are calculated as a function of v. The shape of $-\int \Delta A dv$ across the band (α) is roughly the same as the absorption band, implying that the E-A spectrum of band (α) is close in shape to the first derivative of the absorption spectrum. Actually, the E-A spectrum of band (α) includes the shape of the second derivative of the absorption spectrum, as is known from a slight blue-shift of the peak in $-\int \Delta A dv$ in comparison with the peak of the absorption band (α). The shape of $\int \Delta A dv$ across the band (β) is close in shape to the first derivative of the absorption spectrum, in contrast with the shape across the band (α). The further integral of $\int \Delta A dv$ gives a shape similar to the absorption spectrum (see Figure 3), indicating that

Table 1. The magnitude of $\Delta \mu$ and $\Delta \alpha$ for excitation into each of the Soret bands and Q bands of ZAc(N) with N from 1 to 4

		peak position (nm)	$\Delta \mu$ (D)	$\Delta \alpha$ $(4\pi\varepsilon_0 \text{Å}^3)$
ZAc(1)	Soret band	450.0	-	60
	Q band	646.0	-	25 (20)[a]
ZAc(2)	band (α)	432.9	-	210
	(β)	474.8	7.8 (12.0)[b]	-20 (-60)[b]
	(γ)	416.7	-	280
	Q_I	583.0	3.5	20
	Q_{II}	637.8	2.9 (2.1)[a]	60 (200)[a]
ZAc(3)	band (α)	436.5	-	230
	(β)	489.3	11.7 (16.6)[c]	-70 (-320)[c]
	(γ)	413.0	7.9	140
	Q_I	586.0	3.8	120
	Q_{II}	637.0	3.5 (2.6)[a]	70 (230)[a]
ZAc(4)	band (α)	435.7	2.9	280
	(β)	497.3	13.0 (21.5)[d]	-280 (-390)[d]
	(γ)	414.0	4.6	460
	Q_I	584.5	4.8	180
	Q_{II}	637.8	4.1 (3.7)[a]	180 (250)[a]

[a] The values obtained from the E-F spectra.
[b] The value in parenthesis is at the band (B) of Z(2).
[c] The value in parenthesis is at the band (B) of Z(3).
[d] The value in parenthesis is at the band (B) of Z(4).

the E-A spectrum across the band (β) is close in shape to the second derivative of the absorption spectrum. These results suggest that the permanent electric dipole moment is induced following excitation into the band (β), as in the case of the band (B) of Z(N). It is also known that the E-A spectrum of band (β) of ZAc(4) includes the first derivative since the peak position of the absorption band (β) is different from the corresponding peak in the integral of $\int \Delta A dv$. The spectrum of $-\int \Delta A dv$ across the band (γ) does not show a peak, but gives a shape which reminds of the first derivative of the absorption band (γ), implying that $\Delta\mu$ is not zero for excitation at the band (γ).

Actually, the observed E-A spectra of ZAc(N) are reproduced quite well by assuming exciton splitting components of the Soret band, i.e., bands (α), (β) and (γ), and by assuming a linear combination of the first and the second derivatives of each absorption band. From the first and the second derivative parts, the magnitude of each of $\Delta\mu$ and $\Delta\alpha$ was determined for excitation at each splitting component. The results are shown in Table 1. The negative value of $\Delta\alpha$ shows that the magnitude of the molecular polarizability becomes smaller following excitation into the excited state.

E-A spectra in the wavenumber region less than 18000 cm^{-1}, where Q bands are located, were also simulated by assuming two absorption bands, i.e., Q_I and Q_{II}, and by assuming a linear combination between the first and the second derivatives of these absorption bands (see Figure 3). Based on the results, the magnitude of each of $\Delta\mu$ and $\Delta\alpha$ was similarly determined for the Q bands (see Table 1). Even for excitation at the Q bands, a non-zero value of $\Delta\mu$ is observed, indicating that the permanent electric dipole moment is induced even for excitation at the Q bands of phenylethynylated *meso-meso* linked porphyrin arrays.

The E-A spectra of ZAc(2) and ZAc(4) were similarly simulated with three absorption bands having a Gaussian profile in the Soret band region, i.e., bands (α), (β) and (γ), and two absorption bands in the Q band region, i.e., Q_I and Q_{II} (see Figure 4). Spectral widths of these bands were determined from the absorption bandwidth and HWHM both of the spectrum of $-\int \Delta A dv$ across each band and of the spectrum of the further integral of $\int \Delta A dv$, though some ambiguity must be admitted. Based on the comparison between the E-A spectra and the simulated spectra, the magnitude of each of $\Delta\mu$ and $\Delta\alpha$ of ZAc(2) and ZAc(4) was determined. The results are shown in Table 1.

It should be noted the value of $\Delta\alpha$ or $\Delta\mu$ shown in Table 1 were obtained by assuming that the internal field is the same as the applied field, i.e., f in Eq. (1) is unity, and that the molecular polarizability is isotropic, i.e., $\Delta\alpha_m = \Delta\overline{\alpha}$. Actual field strength on molecules is different from the applied field strength because of the dielectric properties of the environment. Then, the evaluated values of $\Delta\mu$ or $\Delta\alpha$ shown in Table 1 must be multiplied by a correction factor. Unfortunately, this factor is not known at the present, but this factor is considered to be common in all the samples of Z(N) and ZAc(N). Therefore, relative values are considered to be compared to each other. In the evaluation of $\Delta\mu$ in Table 1, the transition moment was assumed to be parallel to the linked porphyrin array for the bands (β), Q_I and Q_{II}, whereas it was assumed to be perpendicular to the linkage for bands (α) and (γ).

For the exciton splitting component of Soret band of ZAc(N) located at the longer wavelength side, i.e., band (β), a significantly large value of $|\Delta\mu|$ is observed, and the magnitude of $\Delta\mu$ increases with increasing the number of porphyrins (see Table 1). These results are essentially the same as the ones observed for the *meso-meso* linked

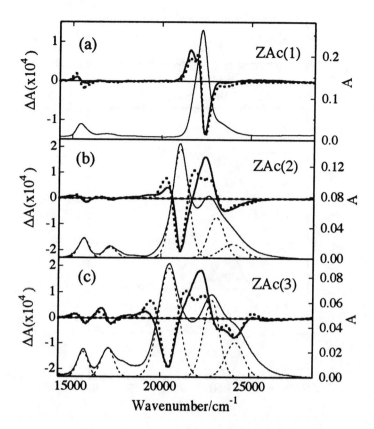

Figure 4. The thick dotted line in (a) shows the first derivative of the absorption spectrum of ZAc(1). Thick dotted lines in (b) and (c) show the E-A spectrum simulated by a linear combination between the first and the second derivatives of the absorption spectrum in ZAc(2) and ZAc(3), respectively Thin solid line and thick solid line in (a), (b) and (c) show the absorption and E-A spectra of ZAc(1), ZAc(2) and ZAc(3), respectively. Thin dotted lines in (b) and (c) show the absorption bands used for the simulation.

porphyrin arrays, i.e., Z(N).[9] Note that the magnitude of $\Delta\mu$ of Z(N) increases with increasing the number of porphyrins following excitation at the band (B). These results show that a photoinduced charge transfer efficiently occurs along the porphyrin array for excitation at the Soret band whose transition moment is along the linked array. When the results of ZAc(N) are compared with those of Z(N), it is also found that the magnitude of $\Delta\mu$ following excitation at the band (β) of ZAc(N) is smaller than the one at the band (B) of Z(N), i.e., about two-thirds in every case (see Table 1), as far as the comparison is made at the same number of the linked porphyrins. One may expect that large electron delocalization is induced by a substitution of the phenylethynyl group at both ends of the linked porphyrin array. In fact, bridges containing ethynyl groups lead to strong electronic coupling between the porphyrins.[16] Therefore, it comes as a surprise to know that the

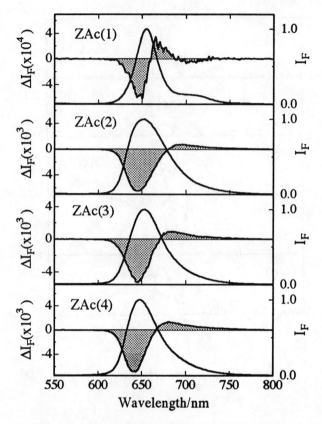

Figure 5. E-F spectra of ZAc(N) with N from 1 to 4 in a PMMA film (from top to bottom), together with the fluorescence spectra (solid line). Applied field strength was 0.75 MVcm^{-1}.

magnitude of $\Delta\mu$ of ZAc(N) is smaller than that of Z(N). It should be also noted that a non-zero value of $\Delta\mu$ is observed even for excitation into the bands whose transition moment is considered to be directed perpendicular to the linked array, e.g., at bands (α) and (γ) of ZAc(4).

E-F spectra of ZAc(1) - ZAc(4) are also observed in PMMA. The results are shown in Figure 5, together with the fluorescence spectra. Note that all the E-F spectra were obtained with excitation at the wavelength where ΔA is negligibly small, i.e., at 452.0, 432.5, 435.5 and 435.5 nm for ZAc(1) - ZAc(4), respectively. The observed E-F spectra were reproduced quite well by a linear combination of the fluorescence spectrum and its first and the second derivative spectra. The simulated results are shown in Figure 6 for ZAc(1) and in Figure 7 for ZAc(4), respectively. In addition to the Stark shift, field-induced change in Φ_F (fluorescence quantum yield) is observed especially for ZAc(N) with N $\geq$ 2, suggesting that intramolecular excitation dynamics of ZAc(N) is influenced by F. In

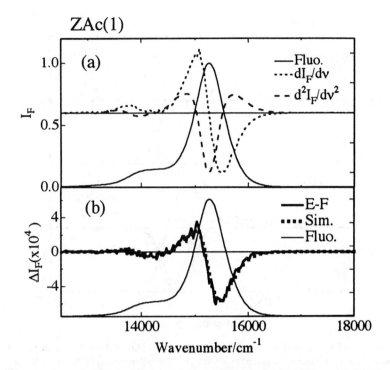

Figure 6. (a) Fluorescence spectrum of ZAc(1) (solid line), its first derivative spectrum (dotted line) and the second derivative spectrum (broken line); (b) E-F spectrum of ZAc(1) (solid line), simulated spectrum (dotted line) and fluorescence spectrum (broken line).

all the cases, ΔI_f integrated for the E-F spectrum over the full wavelength is negative, indicating that Φ_F is reduced by F. Even in ZAc(1), the integrated intensity of the E-F spectrum is not zero, but the field-induced decrease is one-order of magnitude smaller than the others. Note that $\Delta \Phi_F / \Phi_F$ is -2×10^{-4}, -3.2×10^3, -2.6×10^3 and -2.6×10^3 for ZAc(1) - ZAc(4), respectively, with a filed strength of $0.75\,\mathrm{MVcm}^{-1}$.

The field dependence of Φ_F is attributed to the field-induced change in radiative decay rate (k_r) and/or in nonradiative decay rate (k_{nr}) at the emitting state. A small decrease of Φ_F observed in ZAc(1) may be attributed to the field-induced decrease of the radiative decay rate. On the other hand, the field-induced quenching of fluorescence of ZAc(N) with $N \geq 2$, which is about one-order of magnitude larger than that of ZAc(1), may be attributed to the field-induced enhancement of k_{nr}. Thus, it is likely that the field effect on k_{nr} plays a significant role in phenylethynylated *meso-meso* linked porphyrin arrays. Based on the contribution of the derivative components for the simulated E-F spectra, the magnitude of each of $\Delta\mu$ and $\Delta\alpha$ between the fluorescing state and the ground state was also evaluated. The results are shown in Table 1. In agreement with the analysis of the E-A spectra, E-F spectra show that the Q bands also gives a non-zero value of $\Delta\mu$ for phenylethynylated *meso-meso* linked porphyrin arrays.

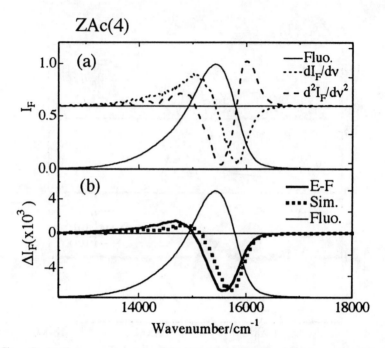

Figure 7. (a) Fluorescence spectrum of ZAc(4) (solid line), its first derivative spectrum (dotted line) and the second derivative spectrum (broken line); (b) E-F spectrum of ZAc(4) (solid line), simulated spectrum (dotted line) and fluorescence spectrum (broken line).

5. SUMMARY

External electric field effects both on absorption spectra and on fluorescence spectra of phenylethynylated *meso-meso* linked porphyrin arrays doped in a polymer film have been examined using electric field modulation spectroscopy with the number of the linked porphyrins from 1 to 4. The magnitude of the change in electric dipole moment ($\Delta\mu$) and in molecular polarizability ($\Delta\alpha$) was evaluated for excitation into each of the exciton splitting components of the Soret band and into each of the Q bands, and the results are compared with the *meso-meso* linked porphyrin arrays having no phenylethynyl substitutes.

The non-zero value of $\Delta\mu$ shows that a photoinduced charge transfer occurs along the linked porphyrin array especially following excitation into the splitting component of the Soret band which gives the transition moment along the linked array.

Field-induced quenching of fluorescence is also observed for phenylethynylated *meso-meso* linked porphyrin arrays in a PMMA film, suggesting that intramolecular nonradiative process(es) at the emitting state is significantly enhanced by an electric field.

This work was partly supported by a Grant-in-Aid for the Scientific Research (B) (2) (No. 12555239) from the Ministry of Education, Science, Sports and Culture of Japan.

References

1. See for a review, M. R. Wasielewski, Photoinduced electron transfer in supramolecular systems for artificial photosynthesis, *Chem. Rev.* **92,** 435-461 (1992).
2. A. Osuka, K. Maruyama, Synthesis of naphthalene-bridged porphyrin dimers and their orientation-dependent exciton coupling, *J. Am. Chem. Soc.* **110,** 4454-4456 (1988).
3. T. Nagata, A. Osuka, K. Maruyama, Synthesis and optical properties of conformationally constrained trimeric and pentameric porphyrin arrays, *J. Am. Chem. Soc.* **112,** 3054-3059 (1990).
4. J. L. Sessler, V. L. Capuano and A. Harriman, Electronic energy migration and trapping in quinine-substituted, phenyl-linked dimeric and trimeric porphyrins, *J. Am. Chem. Soc.* **115,** 4618-4628 (1993).
5. R. W. Wagner and J. S. Lindsey, A molecular photonic wire, *J. Am. Chem. Soc.* **116,** 9759-9760 (1994).
6. S. Anderson, H. L. Anderson, A. Bashall, M. McPartlin, J. K. M Sanders, Assembly and crystal structure of a photoactive array of five porphyrins, *Angew. Chem. Int. Ed. Engl.* **34,** 1096-1099 (1995).
7. A. Osuka, H. Shimidzu, *meso,meso*-Linked porphyrin arrays, *Angew. Chem. Int. Ed. Engl.* **36,** 135-137 (1997).
8. N. Aratani, A. Osuka and Y. H. Kim, Extremely long, discrete *meso,meso*-coupled porphyrin arrays, *Angew. Chem. Int. Ed. Engl.* **39,** 1458-11461 (2000).
9. N. Ohta, Y. Iwaki, T. Ito, I. Yamazaki, A. Osuka, Photoinduced charge transfer along a *meso,meso*-linked porphyrin array, *J. Phys. Chem.* **103,** 11242-11245 (1999).
10. J. L. Oudar, Optical nonlinearities of conjugated molecules. Stilbene derivaties and highly polar aromatic compounds, *J. Chem. Phys.* **67,** 446-457 (1977).
11. A. Nakano and A. Osuka, to be submitted.
12. S. Umeuchi, Y. Nishimura, I. Yamazaki, H. Murakami, M. Yamashita, N. Ohta, Electric field effects on absorption and fluorescence spectra of pyrene doped in a PMMA polymer film, *Thin Solid Films* **311,** 239-245 (1997).
13. W. Liptay, in: *Excited States,* edited by E. C. Lim (Academic, New York, 1974), Vol. 1, pp. 129-229.
14. S. G. Boxer, in: *The Photosynthetic Reaction Center,* edited by J. Deisenhofer and J. R. Norris (Academic Press, Sandiego 1993) Vol. 2, pp. 179-220.
15. L. Karki, F. W. Vance, J. T. Hupp, S. M. LeCours and M. J. Therien, Electronic Stark effects studies of a porphyrin-based push-pull chromophore displaying a large first hyperpolarizability: state-specific contributions to β, *J. Am. Chem. Soc.* **120,** 2606-2611 (1998).
16. J. J. Piet, P. N. Taylor, H. L. Anderson, A. Osuka and J. M. Warman, Excitonic interactions in the singlet and triplet excited states of covalently linked zinc porphyrin dimers, *J. Am. Chem. Soc.* **122,** 1749-1757 (2000).

DESIGN, STRUCTURE, AND BEHAVIOR OF INTERPOLYMER COMPLEX MEMBRANES

Sarkyt E. Kudaibergenov,[*] Larisa A. Bimendina, and Gulmira T. Zhumadilova

1. INTRODUCTION

Polyelectrolyte complex (PEC) formation reactions between oppositely charged linear polyelectrolytes have been considered in detail.[1-4] PEC formation reactions represent an interesting principle to develop PEC films,[5] membranes,[6] and microcapsules[7,8] with unique permselective characteristics towards liquids,[9] gases,[10] and ions.[11] Usually, at a low concentration of mixing polyelectrolytes PECs precipitate and the processing of thin films or membranes becomes a multistage process that includes the following steps: 1) the separation of precipitate; 2) dissolution of precipitate in a ternary mixture consisting of water, organic solvents, and neutral salts; 3) casting; 4) removal of impurities. Whereas a thin interfacial film of PEC about 20 nm thick is rapidly formed at higher polyelectrolyte concentrations (above 0.6 wt.%, or about 0.03 N).[11] This film is stoichiometric and because of its impermeability the reaction is stopped after the initial formation of the thin film.

PECs can also be formed at liquid-liquid[12] or liquid-air[13] interfaces. Layer-by-layer deposition of polyelectrolytes can build up ultrathin multilayers assemblies.[14,15] The aim of this report is to prepare hydrophilic and hydrophobic (co)polymers, to study the PEC, and membrane formation processes between cationic and anionic polyelectrolytes in aqueous solution and at the dimeric interface.

2. EXPERIMENTAL PART

2.1. Materials

2.1.1. Purification of Vinyl Ether Monomers

Commercially available vinyl butyl ether (VBE) and vinyl-2-aminoethyl ether (VAEE) monomers containing 99.5% of basic product were dried during 7-8 days with

---

* Institute of Polymer Materials and Technology, Satpaev Str.18a, 480013 Almaty, Kazakhstan.

Advanced Macromolecular and Supramolecular Materials and Processes
Edited by K. Geckeler, Kluwer Academic/Plenum Publishers, 2003

the help of annealed potash and distilled over calcium hydride three times under argon atmosphere.[16]

2.1.2. Synthesis of Water-Soluble and Organo-Soluble (Co)polymers

Copolymers of vinyl butyl ether and acrylic acid (VBE-AA), vinyl-2-aminoethyl ether and styrene (VAEE-St), acrylic acid and styrene (AA-St) were synthesized by radical copolymerization in ethanol in the presence of AIBN (c = $5 \cdot 10^{-3}$ mol·L^{-1}). The copolymers were purified by three-fold precipitation from an ethanol solution into diethyl ether. The composition of the vinyl ether-based copolymers was determined by elemental analysis and potentiometric titration.

The average-number molecular weight M_n of the (co)polymers VBE-AA, VAEE-St, and AA-St determined by ebullioscopy were between $(20-30) \cdot 10^3$. Poly(vinyl-2-aminoethyl ether) (PVAEE) with $M_n = (20-30) \cdot 10^3$ was synthesized by gamma-irradiation polymerization with the help of ^{60}Co «RXM-γ-20M» at an irradiation dose 170 rad·s^{-1} during 8 h. The viscosity-average molecular weight M_η of poly[(4-but-3-en-1-ynyl)-1-methylpiperidin-4-ol] (PVEP)[17] determined from the Mark-Kuhn-Houwink equation [η] = $1,17 \cdot 10^{-4}$ $M^{0.71}$ in methanol was $70 \cdot 10^3$.

Poly-2-vinylpyridine (P2VP) ($M_w = 3,4 \cdot 10^5$), sodium salt of polystyrenesulfonic acid (SPSS) ($M_w = 3 \cdot 10^{-5}$), poly(acrylic acid) (PAA), and poly(methacrylic acid) (PMAA) with viscosity-average molecular weights $M_\eta = 250 \cdot 10^3$ and $150 \cdot 10^3$, respectively, were purchased from Polyscience (USA) and used without purification. All organic solvents used were purified according to literature.[18] Reagent-grade sodium dodecylsulfonate (SDS) and KCl were used.

2.1.3. Preparation of Fluorescence Labeled (Co)polymers

The synthesis of fluorescence-tagged PMAA and AA-VBE was carried out by the interaction of polymers with pyrenildiazomethane (PDM) without catalyst at room or lowered temperature (Scheme 1). 0.02 g of PDM dissolved in 7.5 mL of freshly distilled dioxan was added to 10 mL of a methanol solution containing 0.5 g of PMAA. The initial dark-orange solution turned to yellow and afterwards to a lemon color during several minutes. The labeled PMAA* was purified by a two-fold precipitation from a methanol solution into diethyl ether. The mass of the dried PMAA* was 0.2 g.

0.01 g of PDM dissolved in 3.8 mL of freshly distilled dioxan was added to 7 mL of an ethanol solution containing 0.3 g of VBE-AA. The mixture was stirred during 2 hours. The labeled VBE-AA* was purified by a two-fold precipitation from dioxan into diethyl ether. The mass of labeled VBE-AA* was 0.14 g.

The amount of fluorescent label per polyacid chain was determined by UV spectroscopy at λ = 346 nm (the absorbance band of pyrene label). The concentration of pyrene labels was calculated according to the Lambert-Behr equation. It was found that PMAA* contained one pyrene label per 260 chains, VBE-AA* contained one pyrene label per 400 chains.

Nonstoichiometric ally labeled PEC (NPEC*) of definite composition φ = [PVAEE]/.[PMANa*] was prepared by mixing of aqueous solutions of PMAA* and PVAEE. The insoluble product formed in the acidic region was dissolved by the addition of an equivalent amount of sodium hydroxide.

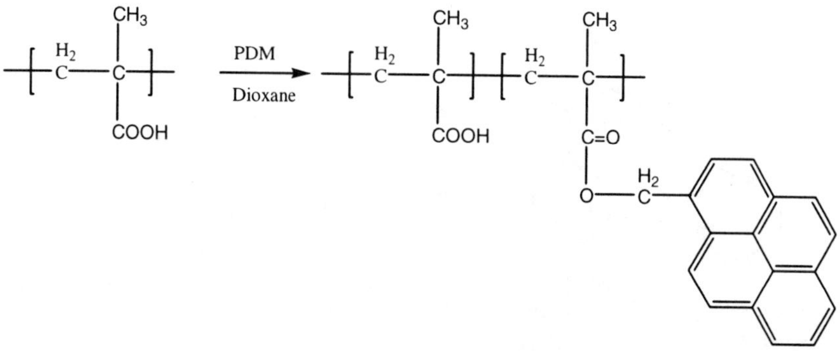

Scheme 1. Labeling of PMAA by pyrenildiazomethane (PDM).

All experiments were carried out at pH = 9.5 (in the presence of Tris-buffer) to ionize all carboxylic groups of PMAA*.

2.2. Methods

2.2.1. Study of PEC Formation on the Dimeric Interface

To study the formation of PEC on the dimeric interface, the organic solution of one polymer was layered onto the aqueous solution of another polymer, so that the surface of the aqueous solution was fully covered by the organic solution. For instance 0.18 mL benzene solution of P2VP (c = $1.25 \cdot 10^{-2}$ mol·L^{-1}) was added with a micro-syringe to 1.8 mL aqueous solution of SPSS (c = $1.25 \cdot 10^{-3}$ mol·L^{-1}) to keep the stoichiometry of reacting components at 1:1.

The quartz cuvette with a working volume of 2 mL was adjustable with respect to light beam, so that the various zones of the aqueous solution could be detected. Since SPSS has a maximal absorption peak at 225 nm, it was easy to follow by the change of concentration of SPSS with the help of UV spectrocsopy. The formation of PEC between P2VP and SPSS at the water-benzene boundary can be represented by Scheme 2.

PEC formation at the water-butanol phase boundary was carried out similarity to the P2VP-SPSS system, e.g., the n-butanol solution of VBE-AA was carefully spreaded on the water solution of PVAEE, so that the surface of the aqueous solution was fully covered by the organic solution. Changing of pH or conductivity during the complexation was fixed by pH-meter dipped into the aqueous solution without disturbing the interface layer. To determine the rate constant of the membrane formation at the interface, the change of polyelectrolyte (c_P) or proton concentration on the time τ was detected.

The dependencies of c_P - τ and pH - τ were then linearized by plotting the curves in coordinates $\ln(c_0/c_\tau)$ - τ and $\ln([OH_0]/[OH_\tau])$ - τ (where c_0 and $[OH_0]$ are the initial concentrations of polyelectrolytes, c_τ and $[OH_\tau]$ are the concentrations of polyelectrolytes at definite time τ) at the initial time period.

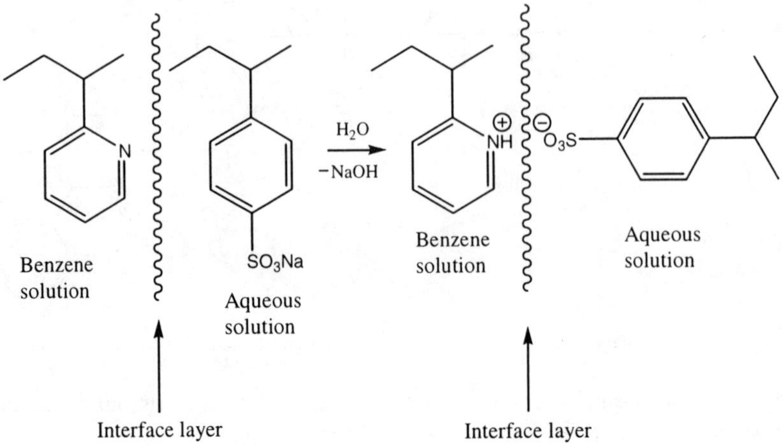

Scheme 2. Formation of PEC at the water-benzene interface.

2.2.2. Preparation and Characterization of PEC Membranes from the Aqueous Solution

PEC membranes from the aqueous solution were fabricated according to the following procedure:[5] equimolar amounts of PAA and PVAEE or AA-VBE and PVAEE (c = $5 \cdot 10^{-3}$ mol L) were dissolved in water containing 5 wt.% of formic acid. PEC solutions were poured onto the polyethylene matrix, the solvent was evaporated at room temperature, the dried films were then exposed to vacuum at 140-150°C during 1.5 h. After washing with distilled water up to neutral pH, transparent membranes were obtained. The swelling degree α of membranes was determined by a gravimetric method: $\alpha = (m-m_0)/m$, where m_0 and m are the masses of dry and in water swollen membranes, respectively.

To study the influence of pH, ionic strength, and temperature on the properties of the membranes, equilibrium-swollen samples with a diameter of 10 mm were placed in a thermostated cell containing a buffer solution. The diameter of films d_0 and d_t were measured by means of a cathetometer "V-630" (Russia), where d_0 is the initial diameter of the equilibrium-swollen sample, d_t is the change of the diameter of membrane at definite time t. The relative change of the film diameter was expressed as d_t/d_0 in dependence of the pH. The permeability of the PEC films with respect to urea was measured in a two-chamber glass cell separated by a membrane. One of the chamber contained an aqueous solution of urea (0.6 g·L^{-1}), the other distilled water.

The concentration of urea was determined a spectrophotometrically by means of a spectrophotometer "Spekol-11" (Germany) at $\lambda = 420$ nm at room temperature using a color reaction of urea with p-dimethylamino benzaldehyde.[19]

The permeability constant P was calculated by the formula: $P = \ln(c_o/\Delta C)\cdot V/2At$, where c_o is the initial concentration of urea $(g\cdot L^{-1})$, $\Delta C = c_o-c_t$ is the difference of urea concentration $(g\cdot L^{-1})$ in the chambers 1 and 2 (where C_t is the concentration of urea at the defined time t (min)), A is the membrane surface (cm^2), and V is the volume of chamber (cm^3).

2.2.3. Measurement Technique

Potentiometric and conductometric titrations were carried out with a pH/conductivity meter "Mettler Toledo MPC 227" (Switzerland) at room temperature in a thermostated cell. Spectroturbidimetric titrations were provided with the help of spectrophotometer «SPEKOL-11» (Germany) at the wavelengths $\lambda = 340$ and 420 nm and at room temperature. Spectrophotometric measurements were performed with a «Specord-M40» (Germany) at $\lambda = 225$ nm.

Fluorimetric titration was conducted with a spectrofluorimeter "Jobin Yvon-3CS" (France) at a wavelength of excitation 343 nm and registration 395 nm, respectively. IR spectra of the samples were recorded on a JASCO FT/IR-5300 spectrophotometer with KBr pellets. The band resolution was 4 cm^{-1}.

Thermogravimetric (TG) and differential thermal analysis (DTA) were carried out a SEIKO TG/DTA 220 thermal analyzer at a heating rate 20 °C/min in a nitrogen atmosphere. DSC traces were obtained using a SEIKO DSC 120 thermal analyzer at a heating and cooling rate 10 °C/min under argon. The viscosity of PEC was measured in an Ubbelohde viscometer at 298 ± 0.1 K.

The degree of conversion for the PEC formation in aqueous solution was calculated from the potentiometric data using the formula[20] $c_s = \theta\cdot c_0 = c_b/V_0 + [H^+] - [H^+]_{PA}$, where c_0 is the initial concentration of any polyelectrolyte $(mol\cdot L^{-1})$, c_s is the concentration of ionic contacts $(mol\cdot L^{-1})$, c_b is the concentration of alkaline spent for the titration of the reaction mixture $(mol\cdot L^{-1})$, V_0 is the volume of reaction mixture (L), and $[H^+]$ is the concentration of protons in the solution $(mol\cdot L^{-1})$. $[H^+]_{PA}$ was calculated from: $[H^+]_{PA} = \sqrt{K_0} \cdot c_0$, where K_0 is equal to 4.8 for VBE-AA.

3. RESULTS AND DISCUSSION

3.1. Formation of Interpolyelectrolyte Complexes and Membranes in Aqueous Solution

Fig. 1 illustrates the interaction of labeled PMAA* and VBE-AA* with PVAEE in an aqueous and aqueous-salt solution. In both cases, the formation of PEC is accompanied by a decrease of the luminescence intensity, the effectiveness of which is determined by the amount of PVAEE, e.g, by the composition of nonstoichiometric PEC (NPEC). The dependence of the relative intensity of luminescence of an aqueous solution of NPEC* ($\varphi = $ [PVAEE]/[PMANa*] = 0.5) on the concentration of NaCl shows that an increase of the salt concentration leads to a decrease of the luminescence intensity of the NPEC* solution up to the equilibrium state I/I_0.

It is connected with a conformational state of the NPEC particles under the action of salt. The double decrease of the value of luminescence intensity corresponds to the composition of NPEC* $\varphi = 0.5$ and indicates the uniform distribution of polybases on the whole polyacid chain. The selective binding of polybase with pyrene labels was not observed. Therefore, PVAEE is a fluorescence quencher. NPEC* is stable up to an ionic strength of $\mu = 1.0$, i.e. an increase of luminescent intensity does not occur. The absence of a "salting out" effect allows to study the kinetics of the interaction of PVAEE and PMAA* in a wide range of ionic strength. Kinetic curves of the PEC formation between PVAEE and PMAA* show that the fluorescent intensity decreased sharply, that corresponds to the rapid formation of some amount of polyionic contacts.

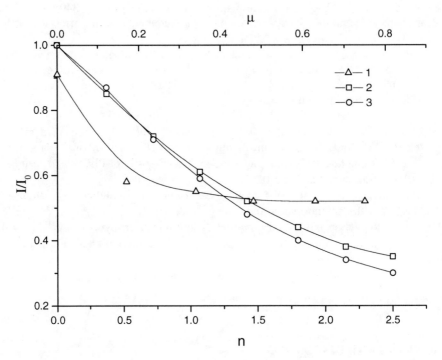

Figure 1. Dependencies of the relative intensity I/I_0 on the molar ratio of initial polyelectrolytes n = [PVAEE]/[PMAA*] (curve 3), n = [PVAEE]/[VBE-AA*] (curve 2), and ionic strength of the solution (curve 1) (n = [PVAEE]/[PMAA*] = 0.5). [PMAA*] = [VBE-AA*] = $8.6 \cdot 10^{-3}$ mol·L^{-1}. $\lambda_{excit.}$ = 346 nm, $\lambda_{registr.}$ = 395 nm.

Then the system reaches the equilibrium state. The decrease of the relative fluorescent intensity I/I_0 in pure water is higher than that in salt solutions. This is probably connected with the preparation methodology of PECs. When PVAEE is added to an acidic solution of PMAA*, only those labeled PMAA* chains, which are arranged on the surface of compact coils, are involved in complexation reaction. Most of pyrene groups are probably displaced inside of the coils.

The addition of a base to the reaction medium lead to the dissolution of NPEC*. However, the system is still in "frozen" state and there is no exchange between NPEC* particles. This resulted in a high value of fluorescent intensity because the system is in unequilibrated state. By the addition of a small amount of salt is enhancd inter- and intramolecular rearrangement of interacting polyelectrolytes and leads to a considerable decrease of luminescent intensity (Figure 1).

Kinetic data obtained for PMAA* and PVAEE were similar to literature values.[21,22] Therefore one can suppose that the lower limit of the rate constant of bimolecular reactions between polyelectrolytes K_D is close to the order of 10^9 L·mol^{-1}·s^{-1}. In other words, this value is close to the rate constant of diffusion collisions between macromolecular coils.

Figue 2 combines the viscometric and spectroturbidimetric titration curves for a system consisting of VBE-AA and PVAEE in aqueous solution. Extremums of these curves correspond to the formation of stoichiometric complexes (1:1) independent of the mixing order. For the systems consisting of VBE-AA and PVEP the formation of stoichiometric PEC in water was also observed.

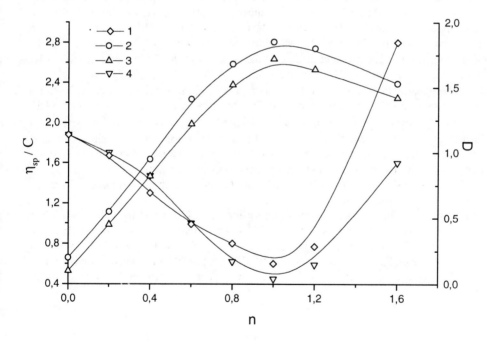

Figure 2. Viscometric (curves 1,4) and spectroturbidimetric (curve 2,3) titration curves for [PVAEE]/[VBE-AA] system in water. [VBE-AA] = 3:97 mol.% (curves 1,3) and 13:87 mol.% (curves 2,4). [PVAEE] = [VBE-AA] = 5·10^{-3} mol·L^{-1}.

3.2. Formation of Interpolyelectrolyte Complexes on the Dimeric Water-Benzene and Water-Butanol Interface

Previously we have shown the formation of PECs between P2VP and sodium SPSS (SDS, metal salts) at a water-benzene phase boundary.[12] It seems that the kinetics and mechanism of the PEC formation at the dimeric interface is different from the process taking place in aqueous solution. According to the literature,[23] the reaction zone or the thickness of the interface adsorption layer is 15-35 nm. The formation of an interface adsorption layer depends on: a) diffusion of macromolecules from the volume into the interface; b) energetic barrier of transfer of macromolecules to the surface part; c) surface diffusion of macromolecules; d) conformational transition (or orientation of hydrophilic and hydrophobic groups) of macromolecules on the interface adsorption layer. Taking into account that the interaction of macromolecules occurs only at the interface adsorption layer one can suppose that the kinetics of PEC formation will be determined by the transfer of macromolecules to the organic or water phase. The relative rate of these processes, as well as energetic and entropy factors depend on the nature of the liquid-liquid phases, temperature, and conformational state of interacting macromolecules.

Figure 3 represents the change of concentration of SPSS with time (time interval is 30 s), when the benzene solution of P2VP was layered on the aqueous solution of SPSS. The sharp decrease of SPSS concentration during 5 min is probably connected with IPC film formation process on the dimeric interface. The further slow decrease of SPSS concentration is probably due to the additional adsorption of SPSS chains on the membrane surface. The gradient change of the concentration of SPSS after 20 min $3 \cdot 10^{-5}$ mol/L that corresponds to 5% of conversion.

The linearization of the initial kinetic curves in coordinates $\ln(c_0 - c_t / c_0 - c_\infty)$ against time t (where c_0 is the concentration of SPSS after 30 s, c_t is the concentration of SPSS at time t, c_∞ is the concentration of SPSS at infinite time) gives apparently the first-order kinetics of PEC film formation that is equal to $1.7 \cdot 10^{-3}$ $L \cdot mol^{-1} \cdot s^{-1}$. The slope of curve, which corresponds to the membrane formation process (between 0.5-5 min), is 20 times higher than that of macromolecular adsorption (between 5-25 min).

The composition of PEC determined from the bend of the potentiometric titration curves is summarized in Table 1.

It is interesting to note that the composition of PEC formed by PVEP and VBE-AA depends on temperature. At 298 and 323K the formation of [VBE-AA]/[PVEP] = 2:1 and 4:1 is observed. This is probably accounted for the existing of LCST (303K) for PVEP[24]. At T<LCST the conformation of PVEP is expanded and the additional binding of PVEP occurs. At T>LCST, PVEP has a compact structure and cannot be more adsorbed.

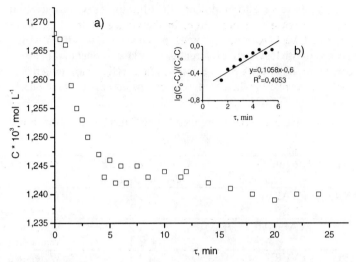

Figure 3. Change of the concentration of an aqueous solution of SPSS upon the addition of a benzene solution of P2VP with time (a) and kinetic curve of the IPC membrane formation (b) at an initial stage (0.5-5 min). [SPSS] = $1.25 \cdot 10^{-3}$ mol·L^{-1}; [P2VP] = $1.25 \cdot 10^{-2}$ mol·L^{-1}. λ = 225 nm.

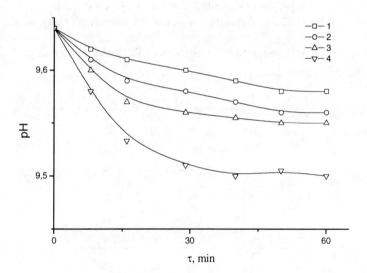

Figure 4. Change of pH of an aqueous solution of VBE-AA upon layering of a n-butanol solution of PVAEE at T = 288 (curve 1), 293 (curve 2), 298 (curve 3), and 303 K (curve 4). [PVAEE] = $1 \cdot 10^{-2}$ mol·L^{-1}. [VBE-AA] = $1 \cdot 10^{-2}$ mol·L^{-1}.

As seen from Figure 4 the formation of PEC films on the water-butanol interface is enhanced with an increase of temperature. In the course of the PEC formation, the carboxylic groups of VBE-AA interact with the amine groups of PVAEE, while the bulky butyl groups of VBE-AA are preferentially replaced in butanol phase. In the case of PAA/VAEE-St or PVEP/AA-St, the amine (or acrylic acid) moieties tend to be dipped into the aqueous phase, whereas styrene groups into the benzene phase.

Table 1. Survey of polyelectrolytes employed for PEC formation on the dimeric interface

Polyelectrolyte mol.%		Composition of PEC, [cationic]:[anionic]	Dimeric interface
anionic	cationic		
VBE-AA (55:45 mol.%)	PVAEE	2:1	Water-butanol
	PVEP	2:1	Water-butanol
PAA	VAEE-St (27:73)	1:2	Water-benzene
PMAA		1:1	Water-benzene
AA-St (15:85)	PVEP	1:1	Water-benzene
SPSS	P2VP	1:1	Water-benzene

As a result, the transfer of protons from carboxylic to amine groups takes place. However, the interface layer limits a fully mixing of the components and the reaction rate is probably controlled by the membrane formation process. Table 2 summarizes the rate constants of the PEC film formation as well as the activation energy of this process.

Table 2. Rate constants and activation energies of PEC film formation processes on a dimeric interface

Interacting system (dimeric interface)	T (K)	$K \cdot 10^4$ ($L \cdot mol^{-1} \cdot s^{-1}$)	E ($kJ \cdot mol^{-1}$)
VBE-AA/PVAEE (water-butanol)	288	0,9	40,7
	293	1,2	
	298	1,8	
	303	2,0	
PAA/VAEE-St (water-benzene)	288	9,2	22,5
	293	11,3	
	298	13,2	
	303	14,4	
	308	16,8	
VAEE-St/SDS (water-benzene)	288	20,0	13,1
	293	21,5	
	298	23,6	
	303	26,1	
	308	28,0	
VAEE-St/Cu^{2+} (water-benzene)	288	34,2	4,4
	293	35,2	
	298	36,1	
	303	37,6	
	308	38,4	

IR spectra of thin films of VBE-AA/PVAEE and VBE-AA/PVEP derived from the butanol-water interface show asymmetric vibrations of COO⁻ ions at 1563 cm⁻¹ due to the transfer of protons from carboxylic to primary amino groups. At the same time, the intensive band of C=O groups at 1708 cm⁻¹ is strongly weakened and appeared as a shoulder. The absorption band of protonated amino groups is observed at 2957 cm⁻¹. It follows from the IR spectra of VBE-AA/PVAEE that the formation of PEC films is due to electrostatic interaction between NH_3^+ and COO⁻. One of the characteristic bands of PVEP, that is very sensitive to the complexation process, is the so-called "Bohlman" band, which appears at 2800 cm⁻¹. Its appearance is connected with the availability of trans fragments between the free axial lone electron pairs of the nitrogen atom and axial CH-bonds in the alpha-position in relation to the nitrogen atom of the heterocycle.

The protonization of nitrogen atoms during the complexation leads to the disappearance of the Bohlman bands owing to the disturbance of electron interactions in this fragment. Thus, the binding of protons from the carboxylic groups of VBE-AA to nitrogen atoms of PVEP should change the position and intensity of the Bohlman bands. The appearance of a new band at 2720 cm⁻¹ reflects the protonization of tertiary amine groups of PVEP. The degree of protonization of nitrogen atoms calculated from the Bohlman band is equal to 12%. It probably reflects the conversion degree of the PEC film formation. While the conversion degree of the PEC formation between PVEP and PAA, determined from the potentiometric titration data in aqueous solution, corresponds to 34%, a calculated value from the "Bohlman" band of solid PEC up to 50% was obtained.

3.3. Complexation of VAEE-St with Sodium Dodecylsulfonate and Copper (II) Ions on the Water-Benzene Phase Boundary

Polymer-surfactant composite films were prepared by casting the solution of polycomplexes in common halogenated organic solvents.[25] By means of DSC and EPR it was shown that the films are composed of lipid multilammelae, in which lipids aggregate in a manner similar to a lipid bilayer. Polymer-surfactant membranes were also designed at the air-water interface.[13] The composition of the VAEE-St/SDS complexes depends on the concentration of SDS and is determined by the critical micellar concentration (CMC). At $c_{CMC} > c_{SDS}$ the composition of complex is equimolar, at $c_{SDS} > c_{CMC}$ the complex particles are enriched by SDS molecules and precipitate. The complexation of VAEE-St with Cu^{2+} is accompanied by the formation of an interfacial blue thin film with the maximal absorbance at $\lambda = 680\text{-}700$ nm. The shift of C-O-C and NH bands to a lower frequency region confirms simultaneously the participation of oxygen and nitrogen atoms in the complexation reaction. The parameters of the ESR spectra ($g_\perp = 2.033$, $g_\parallel = 2.260$, $A_\parallel = 170$ Oe) suggest the participation of copper(II) ions in five-membered chelate cycles, as shown in Scheme 3.

3.4. Thermal Properties and Permeability of PEC Membranes

The membranes prepared on the dimeric interface from the system PAA/VAEE-St and VBE-AA-2/PVAEE were exposed to 140-150 °C during 1.5 h.

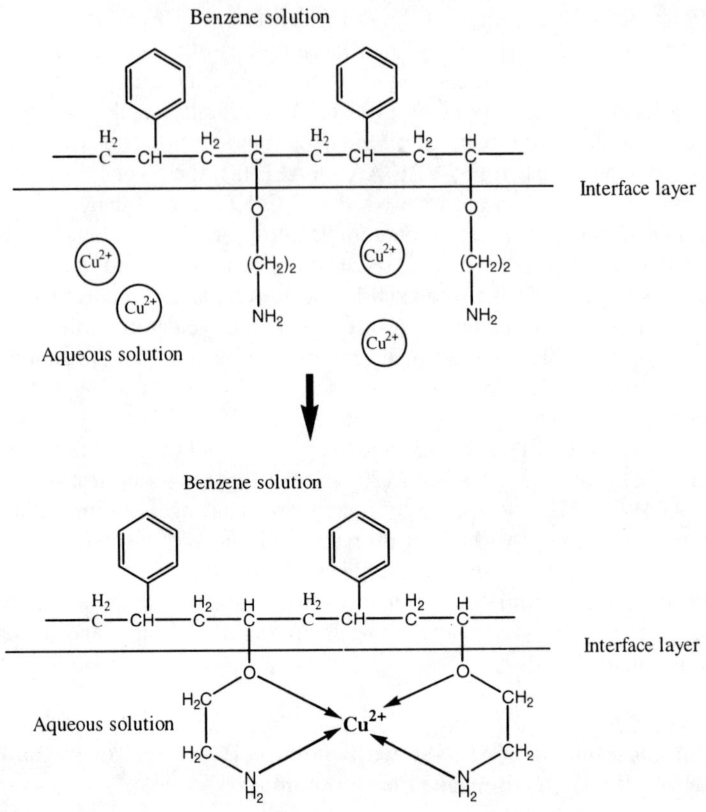

Scheme 3. Formation of copper complexes at the water-benzene interface.

The IR-spectra of films after heating contained the characteristic C=O peak at 1716 cm^{-1} and a new band at $\nu = 1653$ cm^{-1} (amid I). The latter can be connected with the formation of amide groups ~CO-NH~ as a result of the dehydration of intermolecular salt bonds: ~COO$^-$ $\cdots$ NH$^+_3$~ $\rightarrow$ ~CO-NH~ + H$_2$O. The possibility of formation of covalent amide bonds in PEC was shown for PAA-PEI and chitosan-PAA complexes.[9,26] PEC membranes prepared from VBE-AA/PVAEE and PAA/VAEE-St swell in water up to $\alpha = 3$ and 2, respectively. The degree of swelling decreased with increasing temperature exposition. Preliminary results showed that the PEC membranes VBE-AA/PVAEE formed on the butanol-water interface are stable up to 400 °C and had a T$_g$ at 20 and 70 °C.

The semipermeable properties of PAA/PVEP and VBE-AA/PVEP membranes have been checked with respect to urea transport.[27] Figure 5 shows the concentration change of urea with time during the dialysis experiments. The permeability constants (P·10^3 cm·min^{-1}) of membranes are equal to 3.4 and 2.3, respectively.

The decrease of constant values in the case of copolymer is probably connected with the destruction of the complementarity of reacting macromolecules. The membranes are stable at the interval of pH = 2.5-10.5. The minimum swelling degree of VBE-AA/PVEP membranes corresponding to pH ≈ 6.5 is due to Coulomb attractions between long sequences of acidic and basic groups of polyelectrolytes.

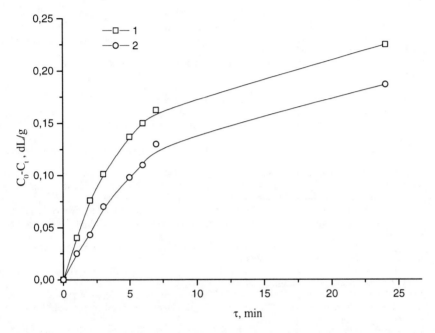

Figure 5. Change of the urea concentration with time for PEC membranes derived from [PVEP]/[VBE-AA] (curve 1) and [PVEP]/[PAA] (curve 2).

Considerable membrane swelling at pH < 6.5 and pH > 6.5 is accounted for electrostatic repulsion between positively and negatively charged groups of polyelectrolytes in acidic and basic regions respectively. A temperature factor enhances the swelling degree of PEC membranes. In water-methanol mixtures the swelling degree passes through a maximum that is connected with the change of the thermodynamic quality of the solvent with respect to hydrophilic and hydrophobic parts of membranes.

4. CONCLUSION

The aggregation structure in the two-dimensional ultrathin films assembled from the dimeric interface is different from three-dimensional solid PEC films derived from the common solvents. The thickness of the two-dimensional ultrathin films will probably be comparable to the dimension of polymer chains, because individual macromolecular components from both phases have the tendency to interfacial complexation.

5. ACKNOWLEDGEMENT

The authors acknowledge an INTAS-1746 grant for financial support.

6. REFERENCES

1. E. A. Bekturov and L. A. Bimendina, Interpolymer complexes, *Adv. Polym. Sci.* **41**, 99-147 (1981).
2. E. Tsuchida and K. Abe, Interactions between macromolecules in solution and intermacromolecular complexes, *Adv. Polym. Sci.* **45**, 1-123 (1982).
3. A. B. Zezin and V. A. Kabanov, A new class of water-soluble polyelectrolyte complexes, *Usp. Khimii* **56**, 1447-1483 (1982).
4. B. Philipp, H. Dautzenberg, K.-J. Linow, J. Koetz, W. Dawydoff, Polyelectrolyte complexes. Recent development, open problems, *Prog. Polym. Sci.* **24**, 91-172 (1989).
4. R. I. Kaljuzhnaya, A. R. Rudman, N. A. Vengerova, E. F. Razvadovskii, B. S. El'tzefon, and A. B. Zezin, Formation conditions and properties of membranes prepared from the polyelectrolyte complexes based on weak polyelectrolytes, *Vysokomol. Soedin. Ser. A*, **17**, 2786-2792 (1975).
5. P. M. Bungay, H. K. Lonsdale, and M. N. Pinho *Synthetic membranes: science, engineering and application* (Reidel, Boston, 1983).
6. H. Dautzenberg, U. Schuldt, D. Lerche, H. Woehlecke, and R. Ehwald, Size exclusion properties of polyelectrolyte complex microcapsules prepared from sodium cellulose sulphate and poly(diallyldimethylammonium chloride), *J. Membr. Sci.* **162**, 165-171 (1999).
7. N. S. Yoon, K. Kono, and T. Takagishi, Permeability control of poly(methacrylic acid)-poly(ethyleneimine) complex capsule membrane responding to external pH, *J. Appl. Polym. Sci.* **55**, 351-357 (1999).
8. S. Y. Nam and Y. M. Lee, Pervaporation and properties of chitosan-poly(acrylic acid) complex membranes, *J. Membr. Sci.* **135**, 161-171 (1997).
9. J. M. Levasalmi and T. J. McCarthy, Poly(4-methyl-1-pentene)-supported polyelectrolyte multilayer folms: Preparation and gas permeability, *Macromolecules* **30**, 1752-1757 (1997).
10. H. J. Bixler and A. S. Michaels, Polyelectrolyte Complexes, in: *Encyclopedia of Polymer Science and Technology*, edited by H. F. Mark, N. M. Bikales, Ch. G. Overberger, G. Menges, (John Wiley & Sons, New York, 1969), pp. 765-780.
11. S. E. Kudaibergenov, R. E. Khamzamulina, E. A. Bekturov, L. A. Bimendina, V. A. Frolova, and M. Zh. Askarova, Polyelectrolyte complex formation on a dimeric interface, *Macromol. Chem. Rapid Commun.* **15**, 943-947 (1994).
12. T. Seki, A. Tohnai, T. Tamaki, and K. Ueno, *J. Chem. Soc. Chem. Commun.* 1876-1880 (1993).
13. P. Bertrand, A. Jonas, A. Laschewsky, and R. Legras, Ultrathin polymer coatings by complexation of polyelectrolytes at interfaces: suitable materials, structure and properties, *Macromol. Rapid. Commun.* **21**, 319-348 (2000).
14. X. Arys, A. M. Jonas, A. Laschewsky, and R. Legras, Supramolecular polyelectrolyte assemblies, in: *Supramolecular Polymers*, edited by A. Ciferri (Marcel Dekker Inc. New York, 2000).
15. V. B. Mikhantiev, O. N. Mikhantieva, *Vinyl Ethers of Glycols*, (Voronezh State Univresity, Voronezh, 1984).
16. S. E. Kudaibergenov, V. B. Sigitov, V. Zh. Ushanov, Interpolyelectrolyte complexes of poly[4-(but-3-en-1-ynyl)-1-methylpiperidin-4-ol] with poly(carboxylic acids), *Makromol. Chem. Phys.* **198**, 183-191 (1997).
17. A. Vaisberger, E. Proskauer, J. Riddik, E. Toops, *Organicheskie Rastvoriteli*, (Mir, Moscow, 1958).
18. G. W. Watt and I. D. Chrisp, Spectrophotometric method for determination of urea, *Anal. Chem.* **26**, 452-453 (1954).
19. V. B. Rogacheva and A. B. Zezin, Polyelectrolyte complexes, in: *Fizika i Khimiya Polimerov*, (Khimiya, Moscow, 1973), pp. 3-30.
20. K.N. Bakeev, V. I. Izumrudov, A. B. Zezin, and V. A. Kabanov, Kinetics and mechanism of polyelectrolyte complex formation, *Dokl. Acad. Nauk SSSR* **299**, 1405-1408 (1988).
21. K. N. Bakeev, V. I. Izumrudov, A. B. Zezin, V. A. Kabanov, S. I. Kuchanov, and A. B. Zezin, Kinetics and mechanism of exchange at interpolymer complex, *Dokl. Acad. Nauk SSSR* **300**, 132-135 (1988).
22. B. N. Tarasevich, V. N. Izmailova, Dynamics of formation of interphase adsorption layers of gelatin on liquid interface, *Vestnik Mosk. Universiteta, Ser. Khim.* **39**, 405-407 (1998).

23. S. E. Kudaibergenov, S. S. Saltybaeva, and E. A. Bekturov, Conformational behaviour of linear and crosslinked poly(4-but-3-en-1-ynyl-1-methylpiperidin-4-ol) in solution, *Makromo. Chem.* **194**, 2713-2717 (1993).
24. K. Taguchi, S. Yano, K. Hiratano, N. Minoura, Y. Okahata, Viscoelastic properties of lipid surfactant/polymer composite films, *Macromolecules* **21**, 3336-3338 (1988).
25. V. B. Rogacheva, N. V. Grishina, A. B. Zezin, and V.A. Kabanov, Intermacromolecular amidization in dilute aqueous solution of polyelectrolyte complex of polyacrylic acid and linear polyethylene imine, *Vysokomol. Soedin. Ser. A:* **25**, 1530-1535 (1983).
26. G. T. Zhumadilova, A. D. Gazizov, L. A. Bimendina, and S. E. Kudaibergenov, Properties of polyelectrolyte complex membranes based on some weak polyelectrolytes, *Polymer* **42**, 2985-2989 (2001).

OPTIMISATION OF MEMBRANE MATERIALS
FOR PERVAPORATION

François Schué,[1*] Houssain Qariouh,[1] Rossitza Schué,[1] Nabil Raklaoui,[1] and Christian Bailly[2]

1. INTRODUCTION

Pervaporation is used world-wide as a membrane process for the separation of liquid mixtures. During a pervaporation experiment, the feed mixture[1] remains in contact with one side of a dense polymeric membrane, while the permeate is removed in the vapour state at the opposite side by vacuum or gas sweeping. Thus, the permeate undergoes a phase change, from liquid to vapour, during its transport though the membrane.

The technique has been developed on an industrial level for applications, where the separation by other more conventional methods is difficult or impossible, e.g, for azeotropic mixtures, close-boiling organic mixtures organic mixtures, or thermosensitive products. The developments in pervaporation processes have been reviewed in some recent publications.[2-4] One of the main industrial interests in this field is the removal of water from a possible substitution of fossil fuels by the use of ethanol as a biomass energy source.[5-6]

The case in which the pervaporation process has the most marked economic advantage over conventional separation methods, is the dehydration of the ethanol water azeotropic mixture (4.4% water). Many studies were performed on hydrophilic membranes with the purpose of understanding the parameters influencing the flux (J) and selectivity (α et β). Because a single polymer does not usually possess the optimum properties for a given separation, different polymers were frequently combined in the from of polymer blends,[7-8] and block- or graft-copolymers.[8-12]

[1] Université de Montpellier II, Science et Techniques de Languedoc, Laboratoire de Chimie Macromoléculaire, Place E. Bataillon, 34095 Montpellier Cedex 5, France.
[2] General Electric Plastics BV, Box 117, 4600 AC Bergen op Zoom, The Netherlands.

Advanced Macromolecular and Supramolecular Materials and Processes
Edited by K. Geckeler, Kluwer Academic/Plenum Publishers, 2003

More recently, copolymer networks and interpenetrating polymer networks (1PNs) have also been used for the same purpose.[13-16] The mass transport through each of these polymer blends can be controlled and adjusted by changing the composition, morphology, and the swelling behavior. The network structures have the additional advantage that the chemical crosslinks may be used to control the degree of swelling and to improve the dimensional stability of the membrane.

2. FUNDAMENTALS OF PERVAPORATION

2.1. Solution-Diffusion Model

In pervaporation operation, only the dense layer of the membrane contributes to separation of the mixture. Mass transport in pervaporation is generally described by a solution-diffusion mechanism[2] which can be devided into three steps: (1) solution of the liquid feed penetrant molecules at the upstream surface; (2) diffusion through the membrane; (3) desorption of the permeate in vapor form at the downstream surface of the membrane. Thus the selectivity and permeation rate are governed by solubility and diffusivity of each component of the feed mixture to be separated. Solubility is a thermodynamic property and diffusivity is a kinetic property. Vaporisation on the permeate side of the membrane is generally considered to be a fast, non-selective step, if the partial pressure is kept low, and does not have a major influence on the overall process.

The equations obtained from this model can describe the transport of a single component, but for binary liquid transport the equations become more complicated because coupling effects occur because of strong interaction between the penetrant molecules and the membrane. A more quantitative description of the actual situation can be given by the calculation of the deviation coefficient ε introduced by Drioli et al.[17]

$$\varepsilon_i = \frac{J_i}{J_0} = \frac{J_i}{J_i^0 . X_i}$$

where J_i and J_i^0 are, the permeation rates of component i (mol m^{-2} h^{-1}) during the permeation of the binary mixture and for the single component i respectively, J_i^I the ideal permeation rate of component i (i.e. in the absence of coupling effects), and X_i the mol fraction of component i in the feed mixture.

J_{H2O} and J_{EtOH} were calculated as follows:

$$J_{H2O} = J_{tot} \frac{Y_{H2O}}{Y_{H2O}M_{H2O} + (1 - Y_{H2O})M_{EtOH}}$$

and

$$J_{EtOH} = J_{tot} \frac{1 - Y_{H2O}}{Y_{H2O}M_{H2O} + (1 - Y_{H2O})M_{EtOH}}$$

in which J_{tot} represents the experimental measured total flux ($Kg\ m^{-2}h^{-1}$), Y_{H2O} the mol fraction of water in the permeate, and M the molar mass ($kg\ mol^{-1}$) of the solvents. Only if ε_i exceeds unity, do the interactions between the membrane and the solvents accelerate the permeation of a component i, compared to the ideal permeation situation, in which the solubility or diffusivity of this component is independent from that of the other.

2.2. Diffusion Coefficients

From the Fick's law of diffusion, the diffusion flux can be expressed as follows (neglegting cross terms):[11]

$$J = -D\ (dC/dx)$$

where J is the permeation flux per unit area ($mol\ m^{-2}\ s^{-1}$), D is the diffusion coefficient ($m^2\ s^{-1}$), c is the concentration of the permeant ($mol\ m^{-3}$), and x is the diffusion length (m). For simplicity, we assume that the concentration profile along the diffusion length is linear. The diffusion coefficient can be calculated by

$$D = J.e / (c_{in} - c_{out})$$

The upstream concentration (c_{in}) of the permeate in the membrane can be calculated from the sorption equilibrium relationship. The downstream concentration (c_{out}) can be assumed to be zero because the downstream pressure was very small (in this experiment, inferior to 1 mm Hg). Although the membrane thickness may change in the pervaporation experiments, we can assume a constant membrane thickness. Indeed, from the low swelling extent measurements (< 10% in our case), the volume diffrences between the swollen membrane and the dry one was negligeable (at most 3-4%)

3. DESIGN AND CHOICE OF MEMBRANE MATERIALS FOR PERVAPORATION

For the pratical application of pervaporation, the membrane must have a high permeation rate and a large separation factor. To obtain a higher permeation rate, an improved permeability coefficient is necessary, generally, diffusion of small molecules through a dense membrane is farred, and the solubility of a compound in a polymer is governed by the chemical affinity betwen the penetrant and the membrane.

Therefore, when the difformer of molecular size of two components to be separated in a mixture is large, permeation through the membrane many be preferentail for the small component, although the solubility of the big component in the membrane is large.

For this reason, many polymers are preferentially permeable to water rather than organic components because the water molecule is much smaller than the organic molecules.

Some of the most investigate polymers for pervaporation are given in Table 1.

4. APPLICATIONS OF THE PERVAPORATION PROCESS

Theoretically, the pervaporation technique can be applied of the separation of any type of liquid mixture by changing the nature of the membrane in all concentration ranges. In practice, most research efforts have been concentrated on the separation of azeotropic mixtures, close boiling point mixtures, recovery or removal of trace substances, and to replace the equilibrium of chemical reactions. In Table 2 some application of pervaporation and their state of the art are listed.

Such industrial companies as Deutsche Carbone (GFT), BP International (Kalsep), Laurgi, Semps, MTR, Separex, Mitsui, Daicell, and Tokuyama Soda, are taking part in the industrial applications of the pervaporation process. Of these, Deutsche Carbone GmbH (the former GFT mbH) equipment.

5. EXPERIMENTAL

5.1. Materials

The polyetherimide used in this study is a product of General Electric Plastics, The Netherlands, and is marked under the trade name Ultem 1000. It has a glass transition temperature of 217 °C and its chemical structure is constituted with the repeat unit shown. NMP was used as solvent for the preparation of the membrane casting solution. Water was distilled before use.

5.2. Membrane Preparation

Polyetherimide membranes were prepared by two different methods:

1) The polyetherimide material was dissolved in NMP to form a homogeneous solution of 12-14 wt%. The polymer solution was cast at ambient atmosphere on Pyrex glass plates. The cast films were then quickly placed into an oven for 1.5h at 120°C. The dried membrane was subsequently dislodged with water from the glass plate, and dried in a vacuum oven for at least 50 h at 60°C to remove extraneous residual solvents in the membrane. The thicknesses of the dense homogeneous membranes obtained by this method were measured with a micrometer.

Table 1. Polymers for pervaporation

Polymer	Application
Cellulose and derivatives	Extraction of water from an aqueous solution of ethanol, separation of benzene/cyclohexane mixtures
Chitosan	Extraction of water from an aqueous solution of ethanol
Collagen	Extraction of water from an aqueous solution of alcohols and acetone
Cuprophane	Extraction of water from an aqueous solution of ethanol
Ion exchange resins (Nafion,etc,)	Extraction of water from an aqueous solution of ethanol, pyridine
LDPE	Separation of C8 isomers
NBR	Separation of benzene/n-heptane
Nylon-4	Extraction of water from an aqueous solution of ethanol
PA	Extraction of water from an aqueous solution of ethanol, acetic acid
PA-co-PE	Separation of dichloroethane/trichloroethylene mixtures
PAA	Extraction of water from an aqueous solution of ethanol, acetic acid
PAN	Extraction of water from an aqueous solution of ethanol
PAN-co-AA	Extraction of water from an aqueous solution of ethanol
PB	Extraction of 1-propanol, ethanol from an aqueous solution
PC	Extraction of water from an aqueous solution of ethanol, acetic acid
PDMS filled with silicatite. molecular sieves, etc.	Extraction of alcohols from an aqueous solution, separation of butanol from butanol/oleyl alcohol mixture
PEBA	Extraction of alcohols and phenol from an aqueous solution, recovery of natural aromas
PI	Extraction of water from an aqueous solution of ethanol, acetic acid; separation of benzene/cyclohexane mixtures
Plasma polymerized fluorine-containing polymers	Extraction of ehtanol from aqueous solution
Plasma polymerized PMA	Separation of organic/organic mixtures
Polyion complexes	Extraction of water from an aqueous solution of ethanol
Polysulfones	Extraction of water from an aqueous solution of ethanol,acetic acid
POUA	Separation of benzene/n-hexane mixtures
PP	Separation of xylene isomers
PPO	Extraction of water from an aqueous solution of alcohols, separation of benzene/cyclohexane mixtures
PTMSP/PDMS composite	Extraction of ethanol from an aqueous solution
PTMSP and derivatives	Extraction of ethanol from an aqueous solution

Table 2. Application of pervaporation processes

Applications	State of the art
General applications:	
Dehydration:	
Alcohols: MeOH, EtOH, PrOH, BuOH	Industrially realized
Ketones: acetone, methyl ethyl keton	IndustriallyrealizedAcids:acetic
acid, chlorinated acetic acid,	Pilot test
sulfiric acid, sulfonic acid, hydrochloric acid	
Amines: pyridine, cyclohexamine	Pilot test
Extraction of organics from aqueous solutions:	
Removal of toxic components: hydrocarbons,	Partially Industrially
halogenated hydrcarbons, amines, phenols	realized
Recovery of natural aromas in agro-food industry	Pilot test
Recovery of trace components: perfumery	Pilot test
Azeotropes, close boiling-point mixtures:	
EtOH/water, i-PrOH/water	Industrially realized
Pyridine /water	Pilot test
Organic mixtures:	
Aliphatic/Aromatic: benzene/hexane,	Laboratory tested
benzene/ethanol, benzene/cyclohexane	
Isomers: butanols, C8 isomers (xylenes, ethylbenzene) Laboratory tested	
Halogenated Hcs:	Laboratory tested
dichloroethane/trichloroethylene	
Others: alcohol/ester, chlorohydrocarbons	Laboratory tested
Hybrid applications:	
PV aimed chemical/biochemical reactors:	
Esterification: ethyl acetate, butyl acetate	Laboratory tested,
	pilot tested
Continuous membrane fermenter	Laboratory tested
separator (CSFR):	
EtOH, MeOH, perfumery, analcoholic beverage	
Processing of vegetable/animal fats	Laboratory tested
	Pilot tested
Fine chemistry: synthesis of dimethylurea	Laboratory tested,

2) Another type of polyetherimide membranes investigated for the dehydration of water-ethanol mixtures are dense composite polyetherimide membranes, which are synthesised by electrodeposition of the polymer on a stainless steel mesh as a cathode. Indeed, tertiary amine groups attached to the polymer chain are obtained by opening the imide cycle of the polymer, dissolved in NMP, with a secondary-tertiary amine (N-methylpiperazine) at high temperature. These groups were protonated by an aqueous acid (lactic acid), forming the positively charged substituent required for the emulsification in water and the cataphoretic deposition of the polymer. Finally, the quasi-total re-imidisation reaction by elimination of the amine was achieved by multistep heating of the coating at elevated temperature.

5.3. Illustrative Example of Electrodeposition Deposition

Polyetherimide Ultem 1000 (80 g), N-methylpyrrolidone (165.3 g) and aceto-phenone (20.6 g) were charged in a 500-ml reaction flask. The polymer was dissolved with stirring at approximately 85-90°C under a blanket of nitrogen. Once the polymer was completely dissolved (approximately 4 h), a mixture of 18.9 g N-methyl piperazine and 61.8 g acetophenone was added over a period of 2 hours. Vigourous stirring was maintained throughout. The temperature was kept at 85-90°C during the addition. After the addition was complete, the mixture was stirred and warmed to 110°C and held at that temperature for 2.5 h. The resulting solution of modified polymer was then used to prepare the electrophoretic deposition emulsion as detailed below.

5.98 g of acetophenone and 1.48 g of 50% aqueous lactic acid were added to 30 g of the modified polymer solution. The mixture was stirred vigorously while 78 g of deionized water was added slowly. The stirred mixture became quite viscous as the gradual water addition continued, then thinned out as the addition of water was completed.

All electrodeposition experiments were carried out at constant applied voltages using a CONSORT E425. A multimeter with a data storage option (Metrix) was used to monitor current decay.

The milky emulsion was placed in a constant temperature bath at room temperature equipped with a teflon stirbar and submersible magnetic stirrer. Aluminium test pieces (40 mm x 40 mm) were pretreated (electrodes were cleaned of grease) and placed into the emulsion along with an aluminium anode measuring 10 x 10 mm. The distance betwen the cathode and the anode was ca. 40 mm. The test piece was removed, rinsed rapidly in deionised water and placed in a warm (60°C) dry chamber to evaporate solvents from the film. After 60 h, the test piece was removed from the chamber. Inspection showed a uniform, smooth coating. The coating was baked in several steps in a oven (EUROTHERM818P, THERMOLYTE) to effect the reimidisation reaction.

Nuclear magnetic resonace (NMR) spectra were recorded on a Bruker AC 250 instrument (250 MHz). Differential scanning calorimetry (DSC) curves were recorded on a Mettler instrument DSC 30 with a thermal analysis controller TA 4000. The yield of modification determined by ^{1}H-NMR is equal to the ratio of the integration of protons $-CH_2$ (d) of the amine to that $-CH_3$ (a) of bisphenol A.

5.4. Pervaporation Experiments

Pervaporation of aqueous ethanol solutions through the membrane was carried out by an ordinary pervaporation apparatus. The permeation cell consisted of two detachable stainless steel parts. The upper part was a cylindrical chamber which served as a feed reservoir. The membrane was mounted in the lower part of the cell, and the effective membrane area for permeation was 19.6 cm^2. Vaccum was applied to the downstream side of the membrane. The permeate vapor was initially condensed with liquid nitrogen and collected in one of two cold traps, and then the cold trap was switched to the other after a permeation steady-state was reached. The weight of the condensed vapor obtained over a given period of time was used for the calculation of the permeation flux.

The permeate was analysed by means of a Karl Fischer Coulometer (Metrohm, KF 684). This technique allows a fast absolute measurement with a reproducibility of approximately 0.5%.

In all the experiments, about 100 ml feed liquid was charged to the permeation cell, and the feed solution was well stirred by using a magnetic stirrer. The operation temperature was controlled by running thermostatted water through a jacket that covered the permeation unit. The downstream pressure was maintained below 1 mm Hg with a vacuum pump.

The permeation rate (J_p) could be calculated by the following equation :

$$J_p = \frac{m_p}{S.\Delta t} \ (kg/m^2.h)$$

where m_p denotes the weight of permeate collected in the cooling trap, S the effective membrane surface, and Δt the permeate time. The permeation rates were normalised to a membrane thickness of 35 μm ($J_{35\mu m}$), by assuming the flux to be inversily proportional to the membrane thickness.

The separation factor, α, is defined as :

$$\alpha_{H2O/EtOH} = \frac{Y_{H2O}/Y_{EtOH}}{X_{H2O}/X_{EtOH}}$$

where Y_{H2O}, and Y_{EtOH} denote weight fractions of water and ethanol in the permeate and X_{H2O}, and X_{EtOH} denote weight fractions of water and ethanol in the feed.

Another dimensionless parameter was used, the enrichment factor β, which is defined by the ratio of the weight fractions of the faster permeant in the permeate and in the feed:

$$\beta = \frac{Y_{H2O}}{X_{H2O}}$$

5.5. Swelling Measurements

A piece of dried membrane was immersed in the water-ethanol mixtures of known composition at 40°C. After 24 h immersion with shaking, the membrane was removed, the surface was quickly wipped off with a filter paper, and the wet weight of the membrane was measured on a semimicro balance. The swelling experiments were repeated for each solution until sorption equilibrium was reached.

The degree of swelling G is calculated by:

$$G = \frac{m_s - m_0}{m_0}.\frac{100}{\rho}$$

where m_0 and m_S respectively denote the weight of the dry and swollen membrane, the density of the liquid mixture.

The initial composition of water/ethanol mixtures and their composition after the sorption measurements were determined using the Karl Fischer instrument.
The contribution of each component of the mixture to the total swelling can be determined as follows:

$$C^m_{H2O} = \frac{(C^0_{H2O} - C^e_{H2O}) \times (V_{mixture} \times \rho)}{m_0} \quad \text{(g water / g dry membrane)}$$

and $C^m_{EtOH} = G - C^m_{H2O}$ (g ethanol / g dry membrane)

where C^0_{H2O} and C^e_{H2O} respectively denote the weight fraction of water in the initial mixture and at the equilibrium (g/g of mixture).

6. RESULTS AND DISCUSSION

Pervaporation characteristics of composite polyetherimide membranes were obtained by cataphoresis. Figure 1 shows the variation of the total permeation rate and separation factor versus the feed composition of water/ethanol mixtures through polyetherimide membranes.

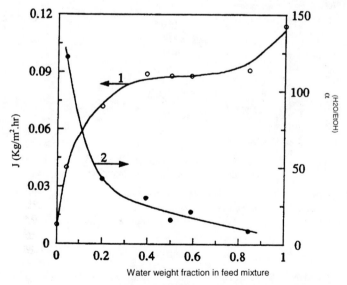

Figure 1. Effect of feed composition on the flux J and selectivity through polyetherimide membrane at 40°C. The membrane wase prepared by cataphoretic electrodeposition of the polymer on a stainnless steel mesh as support (50 μm x 50 μm).

It can be seen from the data presented that a trade-off exists between the total flux and selectivity: when the selectivity increases the permeability decreases and vice versa; but in any case, the membrane is still preferentially selective towards water over the whole concentration range. The separation factor, $\alpha_{H2O/EtOH}$, increases with decreasing water weight fraction in the feed and reached a maximum value of about 118, when the water mole fraction is equal to 0.044 (water/ethanol azeotrope). The high selectivity towards water through these membranes could be due to specific physico-chemical interactions, for instance the selective hydrogen-bonding interactions between water and imide groups of the polymer.

On the other hand, the low values of the permeation rate obtained through these membranes on the overall water concentration in the feed (less than 100 g m^{-2} h) correspond to the ones given in the literature, and represent until now the principle inconvenient of the application of the polyimides in pervaporation processes.

Figure 2 depicts the partial permeation rates of water and ethanol through the Ultem 1000 membrane. The non-monotonicity of the curves, especially when the feed composition is near to that of the azeotrope, indicates that strong interactions between the penetrants and the membrane result in coupled transport for this membrane.

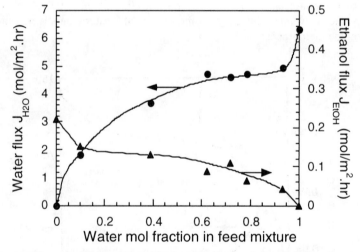

Figure 2. The permeation rates of pure components water and ethanol ($J_{i35\mu m}$) through polyetherimide membranes versus the composition of the feed mixture at 40°C. The membranes were prepared by cataphoretic electrodeposition of the polymer on a stainless steel mesh as a support (50 μm x 50 μm).

According to the solution-diffusion model, the selectivity during pervaporation is governed by both the sorption and the diffusion rates of the components at the same time. For this reason, complementary studies such as sorption measurements and diffusion coefficients (see Tables 2 and 4 in the section below) have been performed in order to obtain more informations about the Ultem1000/water/ethanol systems. It was found that the sorption contribution to the separation mechanism is nearly negligible, and the preferential permeation of water through the Ultem 1000 membrane is therefore attributed mainly to the diffusion process.

Indeed, the results obtained for selective sorption measurements of the membranes in water, ethanol, and water/ethanol azeotrope are approximately similar and with low values of degrees of swelling (1.38, 1.44, and 1.95, respectively). This means that the preferential water permeation may be due to a high water diffusion coefficient (D_{H2O} = 6.12 10^{-11} m^2 s^{-1}) in comparison with those of ethanol and azeotrope (0.58 10^{-11} and 1.57 10^{-11} m^2 s^{-1}, respectively). Finally, we can conclude that the selectivity to water can therefore be explained by, first, favorable interactions between ethanol and the membrane and the larger molecular size of ethanol, causing a low rate of diffusion for ethanol, and secondly, favorable interactions between water/ethanol leading to a coupled flux (formation of associated molecules between components), and causing a low rate of diffusion for water/ethanol azeotrope.

6.1. Investigation of a Coupling Effect in Polyimide Membranes

A more quantitative description of the actual permeation of this binary liquid mixture system (better understanding of the permeation deviation of the mixture from pure component permeation such as ideal permeation), and of the thermodynamic interactions between the penetrants and the membrane can be given by the calculation of the deviation coefficients ε_{H2O} and ε_{EtOH}. For the Ultem 1000 membrane, these deviation coefficients are drawn in Figure 3 as a function of feed composition.

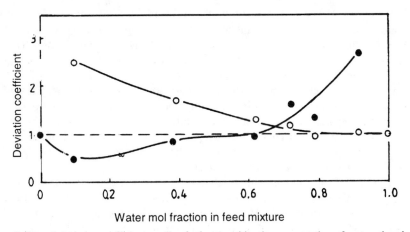

Figure 3. Plot of deviation coefficients *versus* feed composition in pervaporation of water-ethanol mixtures through a polyetherimide membrane et 40°C.

High values of ε_{H2O} are obtained over the entire composition range, especially when the concentration of water is low, indicating a large positive deviation of the ideal permeation rate. In other words, the interactions accelerate water permeation and slow down ethanol permeation. A maximum value of deviation coefficient of ε_{H2O} (= 2.52), which coresponds to a minimum value of ε_{EtOH} (= 0.50), is reached for the polyetherimide membrane at a feed composition near that of the water/ethanol azeotrope. This means that a strong interaction occurs between polymer/ethanol leading to a higher possibility for water permeation. This can explains the higher selectivity obtained for the polyetherimide membrane towards water ($\alpha = 118$).

When the mole fraction of water increases, its deviation coefficient decreases and tends to unity when its mole fraction is 0.8. Also, when the water mole fraction is greater than 0.6, the deviation coefficients for both penetrants are not less than unity, meaning that enhanced transport occurs for both components. This can be explained by the fact that the interactions such as the plasticising effect on the membrane or the formation of associated molecules between both permeants, which are partially immobilised in the membrane, cause a mutual drag. This can also explain the lower selectivity in this region.

6.2. Effects of Temperature on the Membrane Perfomances

The temperature dependence of the permeation rate can be expressed by the Arrhenius-type relationship[18]:

$$ J = A \exp(-E_p / RT) $$

where the value of the apparent activation energy for permeation (E_p) depends on both the activation energy for diffusion and the heat of solution. These parameters will be changed as a result of change in solubility and in chain segment mobility, respectively.

The membranes Ultem 1000 obtained by casting solution were tested for the dehydration of the water-ethanol azeotrope as well as for the permeation of the pure components water and ethanol at elevated temperatures. The experimental results are illustrated in Figure 4. In general, an increasing temperature leads to a higher permeation flux and a lower separation factor. The temperature dependance of permeation flux for water, ethanol, and the water/ethanol azeotrope seems to follow an Arrhenius type of relation in the temperature range of interest, whereas the temperature effect on the separation factor is less significant (Figure not shown).

The increase in the permeability with temperature is generally attributed to the increasing polymer mobility and increasing free volume in the polymer structure and to the decreasing interactions between the permeants. Indeed, if the logarithms of the fluxes are plotted against the reciprocal of absolute temperature, straight lines (R = 0.999) are obtained, the slopes of which give a value for E_p (given in Table 3).

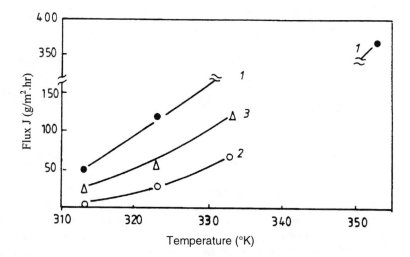

Figure 4. Effet of temperature on flux through a polyetherimide membrane. Curve 1 water; curve 2 ethanol; curve 3 azeotrope water ethanol.

Table 3. Apparent activation energy through polyetherimide membranes

Liquid	E_p , kJ mol^{-1}
H_2O	45.0
EtOH	92.0
H_2O/EtOH azeotrope	72.4

A major point is that the apparent activation energy for water permeation is twice lower than that for ethanol, suggesting a high separation efficiency of this membrane, which corroborates readily the selectivity factor $\alpha_{H2O/EtOH}$ obtained.

On the other hand, the low value of the apparent activation energy obtained for water in comparison with the one for the water-ethanol azeotrope indicates that interactions between the penetrants (water and ethanol) take place and result in coupled transport for the membrane (water permeation was inhibited in the presence of ethanol).

6.3. Investigation of Modified Polyetherimide Membranes by N-Methylpiperazine in Pervaporation Process: Correlation Between Structure and Properties

In the pervaporation of water-ethanol mixture, polyetherimide exhibited a high permselectivity towards water but a low permeability. Thus, with the hope to increase the permeability while at least maintaining the permselectivity unchanged, polyetherimide was dissolved in NMP and reacted with a secondary-tertiary amine (N-methylpiperazine) at high temperature, to obtain tertiary amine groups attached to the polymer chain.

The chemical structure of the polymer, which is a poly(etherimide-amide), deposited at the cathode on a stailess steel mesh as a support is, as follows:

It is clear that the modification of the poly(etherimide) by N-methylpiperazine creates a complicated chemical structure of the polymer. Indeed, the partial disparition of imide groups preferentially interact with water as described before, with the incorporation at an equal extent of hydrophilic groups into the polymer chain such as -NHCO-, $>$C$=$O, and -NR$_3$ groups, which also preferentially interact with water by hydrogen bond interaction leading to a high permselectivity towards water. These groups can also increase the intrasegmental mobility and the macromolecular flexibility leading to an increase in diffusivity selectivity. On the other hand, the incorporation of bulky and hydrophobic substituents such as N-methyl-piperazine cycles can tend to open the matrix with a restriction in macromolecular density, resulting in raising the permeability and reducing the selectivity of the resultant polymer. As a consequence, all these groups can sighficantely affect the pervaporation properties of these membranes.

In order to elucidate the effect of these groups on pervaporation characteristics, pervaporation experiments, sorption equilibrium measurements, and calculation of the diffusion coefficients were carried out for these membranes. The results are illustrated in Tables 4-6.

The degree of swelling for single components and the water/ethanol azeotrope of the modified Ultem 1000 membranes is much larger than the one of unmodified Ultem 1000 membrane. Further, it increased with increasing degree of modification. The solubility of ethanol and water-ethanol azeotrope is higher than that of water in modified Ultem 1000 membranes. This means that the membranes, a priory, have a higher affinity towards ethanol than water. However, the comparison of the ratios of amounts of water and ethanol absorbed in the membranes swollen in the water/ethanol azeotrope (C^m_{H2O}/ C^m_{EtOH}) are approximately identical for unmodified and modified Ultem 1000 at 70% (0.42 and 0.32, respectively) (see Table 4).

This means that these membranes are less affected by the partial desappearance of imide groups and the appearance of pronouced polar groups such as amide, carbonyl, and amine groups. The high values of swelling extent of modified Ultem 1000 can, therefore, be attributed to a reduction in intermolecular density of the polymer. This results from chain flexiblity restrictions due to the bulky piperazine cycles attached in the anhydride part of polymer (intramolecular rigidity), and due to an increase of the intermolecular spaces also provided by the size of the amine.

Table 4. Sorption equilibrium measurements of the water/ethanol azeotrope, water and ethanol in unmodified and modified polyetherimide by N-methylpiperazine

Polymer		Swelling (%)				
		Azeotrope			Water	Ethanol
	G_{az}	C^m_{H2O}	C^m_{EtOH}		G_{H2O}	G_{EtOH}
Ultem 1000 unmodified	1.95	0.58	1.37		1.38	1.44
Ultem 1000 20% of imide cycles modified	4.20	-	-		0.90	3.23
Ultem 1000 70% of imide cycles modified	10.01	2.41	7.59		2.26	9.66

It can be seen from the results shown in Table 5 that the membranes demonstrate the typical pervaporation behavior. That is, the flux increased and the selectivity decreases with the modification extent. The selectivity values $\alpha_{H2O/EtOH}$ fell significantely from 118 to 10 for unmodified and modified polyetherimide membrane at 70%, while the flux increased from 40 to 61 g m^{-2} h.

On the other hand, in binary systems as well as in tertiary systems, the diffusion coefficients of water and water-ethanol azeotrope decreased (D_{H2O} and D_{az}), while the one of ethanol increased (D_{EtOH}) (see Table 6). This large diffusion capacity of the polymer towards ethanol could be explained by an increases in intrasegmental spacing of the polymer ("free space", V_F), and also due to a change in the hydrophilic/hydrophobic balance of the polymer, resulting from the incorporation of hydrophobic bulky substituent (N-methylpiperazine) on the polymer backbone. The hydrophobic character of this group is, therefore, believed to be principally responsible of the large sorptive capacity of the membrane towards the water-ethanol mixture, which can also explain the low values of the diffusion coefficient of water. This result is in a good agreement with the selectivity obtained.

The correlation between thermodynamic and kinetic parameters as illustrated in Tables 4 and 6 shows that the preferential sorption of water and ethanol vary in the same way (increase with modification), while the diffusion coefficients vary in the opposite trend (D_{H2O} decreased and D_{EtOH} increased with modification). This suggests that the difference observed in the permselectivity reflects the difference in diffusivity selectivity. Therefore, it could be concluded that the effect of diffusivity selectivity play a dominant role in determining the pervaporation performance of the modified polyetherimide membranes.

Table 5. Pervaporation characteristics (J_{35mm}, β, α) of unmodified and modified poly-etherimide membranes by N-methylpiperazine

Polymer	J_{total} kg/m^2.h	J_{H2O} kg/m^2.h	J_{EtOH} kg/m^2.h	$\beta_{H2O/EtOH}$	$\alpha_{H2O/EtOH}$
Ultem 1000 unmodified	0.040	0.114	0.010	20	118
Ultem 1000 20% of imide cycles modified	0.049	0.026	0.029	10	20
Ultem 1000 70% of imide cycles modified	0.061	0.048	0.087	7	10

Table 6. Diffusion coefficients of water/ethanol azeotrope, water and ethanol in unmodified and modified polyetherimide by N-methylpiperazine

Polymer	Azeotrope water/ethanol*			Water	Ethanol
	D_{az} 10^{-11} m^2/s	$D_{p,H2O}$ 10^{-11}, m^2/s	$D_{p,EtOH}$ 10^{-11}, m^2/s	D_{H2O} 10^{-11}, m^2/s	D_{EtOH} 10^{-11}, m^2/s
Ultem 1000 unmodified	1.57	4.35	0.39	6.12	0.58
Ultem 1000 20% of imide cycles modified	0.89	-	-	2.21	0.67
Ultem 1000 70% of imide cycles modified	0.47	0.57	0.43	1.62	0.69

*D_{az} : diffusion coefficient of the azeotrope water/ethanol mixture.

$D_{p,i}$: diffusion coefficient of component i in the azeotrope water/ethanol mixture.

D_i : diffusion coefficient of the pure component i.

6.4. Investigation of a Surface Thermodynamic Approach in Polymer Membranes

A surface thermodynamic approach will be considered as a possible method for a better understanding of structure-property relationship of the modified polyetherimide membranes. This approach will be tested and compared with the pervaporation performance of these membranes by using the data measured in this study. The surface energy of polymer was calculated by using the general formula developed by Owens and Wend and Neumann[20]:

$$(1 + \cos\theta)\gamma_L = 2\sqrt{(\gamma_{sd} \cdot \gamma_{Ld})} - 2\sqrt{(\gamma_{sp} \cdot \gamma_{Lp})}$$

The interfacial tension between liquid and polymer surface is given in terms of the contact angle Θ by:

$$\gamma_{sL} = \gamma_s - \gamma_L \cos\theta$$

Now, consider a hypothetical drop of ethanol, 2, on a polymer membrane surface, s, surrounded by water, 1; the work of adhesion to separate unit areas of membrane, s, and water, 1, in the presence of ethanol, 2, is given by:

$$W_{(s1)2} = \gamma_{12} + \gamma_{s2} - \gamma_{s1}$$

Also, the cohesive energy for water and ethanol is given by:

$$W_c = 2\gamma_{12}$$

Defining the interfacial free energie between water, 1, and polymer, s, in the presence of ethanol, 2, as:

$$\Delta G_{(s1)2} = W_{(s1)2} - W_c = \gamma_{s2} - \gamma_{s1} - \gamma_{12}$$

where a positive value $\Delta G_{(s1)2} > 0$ implies preferential permeation of water over ethanol[26]; γ_{ij} values for solutes and polymer were calculated according to Neumann[25], where

$$\gamma_{ij} = (\gamma_i^{1/2} - \gamma_j^{1/2})^2 / (1 - 0.015\gamma_i^{1/2}\gamma_j^{1/2}).$$

As shown in Table 5, large values of contact angle were obtained with water, while these with ethanol and the water/ethanol azeotrope are low and less affected by the modification extent. Also, the wettabilty of polymer surface with water increased with increasing modification extent.

The high wettabilty of polymer surface with ethanol and water-ethanol azeotrope suggest a good interaction between these liquids and the polymer which results in their preferential sorption in modified polymers. This result can therefore explain the large values obtained in the swelling equilibrium extent as mentioned above .

On the other hand, the surface energy of polymer increased with increasing degree of modification (46.85 and 51.75 mN m^{-1} for unmodified and 100% for modified polyetherimide, respectively), which can be ascribed to the increase in $\gamma_{s,p}$ due to a change in the hydrophilic/hydrophobic balance of the polymer by incorporation of some polar groups such as -NHCO-, $>$C$=$O, and -NR$_3$ into the polymer chain by modification with N-methylpiperazine. This result could be predicted if we compare the values of surface tension of water, ethanol, and polymer. Indeed and generally speaking, a liquid of high surface tension (in our case water: 72.8 mN m^{-1}) would be expected to spread on the polymer surface of higher surface tension (in our case its a polyetherimide modified at 100%: 51.75 mN/m).

Table 7. Contact angle and surface energy measurements of unmodified and modified polyetherimide by N-methylpiperazine

Polymer	Contact angle O (°)			Surface energy of polymer γ_S (mN/m)		
	Water	Ethanol	Azeotrope	$\gamma_{s,d}$	$\gamma_{s,p}$	γ_s
Ultem 1000 unmodified	93	8.5	19	46.70	0.15	46.85
Ultem 1000 20% of imide cycles modified	87.5	10	17	45.18	0.88	46.06
Ultem 1000 70% of imide cycles modified	68	12	16	43.70	7.21	50.38
Ultem 1000 100% of imide cycles modified	63	12.5	15	41.54	10.21	51.75

In addition, the small values of $\Delta G_{s,2}$ indicate a good wettability of polymers surface by ethanol, and the larges negatives values $\Delta G_{s,1}$ indicate a preferential water transport (permeability) through the polymers surface, despite the loss of selectivity towards water caused by modification. Also, it could be seen from Table 6 that large values of $\Delta G_{(s,1)2}$ were obtained wich suggset a worse phase separation between water, 1, and ethanol, 2, in the presence of polymer, s. Although this approach cannot a priori predict which solute preferentially permeates the polymer material, the smaller value of $\Delta G_{(s,1)2}$ of unmodified polyetherimide (-8.88 mN m^{-1}) indicates a preferential water transport.

On the other hand, the larger negative values obtained for $\Delta G_{(s,2)1}$ indicate coexistence of water and ethanol into polymer surface, and suggest a good interaction between ethanol, 2, and polymer, s, in the presence of water, 1.

This effect increased with increasing modification extent. That can explain the loss in selectivity towards water due to the modification of polymer by N-methylpiperazine.

Table 8. Interfacial free energy in binary system Ultem 1000/liquid and in ternary system Ultem 1000/water/ethanol. (1) water; (2) ethanol; (s) polymer

Polymer	Interfacial free energy in binary system $\Delta G_{s,L}$ (mN/m)		Interfacial free energy in ternary system $\Delta G_{(s,i)j}$ (mN/m)	
	$\Delta G_{s,1}$	$\Delta G_{s,2}$	$\Delta G_{(s,1)2}$	$\Delta G_{(s,2)1}$
Ultem 1000 unmodified	-76.58	-0.26	-8.88	-62.74
Ultem 1000 20% of imide cycles modified	-69.65	-0.35	-15.93	-55.72
Ultem 1000 70% of imide cycles modified	-45.72	-0.51	-40.02	-39.63
Ultem 1000 100% of imide cycles modified	-40.00	-0.55	-45.80	-25.85

7. CONCLUSION

In this work, a three-pronged approach was used to evaluate polyetherimide membranes for the removal of water from water-ethanol solutions. Besides pervaporation performance (flux and selectivity), sorption and diffusion measurements were obtained, and the following conclusions can be drawn from the observations made in this study:

1. According to the solution-diffusion model, selective diffusion of water appears to dominate the pervaporation performance of polyetherimide membranes rather than the solubility.
2. The separation factor was less affected by an increase in temperature, whereas the corresponding permeation flux increased significantly.
3. Permeability rates of single components through the membrane and the calculation of deviation coefficients clearly demonstrate the presence of coupling effects in polyetherimide membranes.
4. The treatment of polyimide with an amine affects the main-chain of polymer structure by reducing the concentration of imide rings and introducing of a number of hydrophilic groups such as -NHCO-, $>$C$=$O, and -NR$_3$. This results in an increase in flux and a decrease in selectivity towards water.
5. Finally, a relatively consistent correlation was found between the surface energetic parameters approach and the pervaporation performance of the polymers studied.

8. ACKNOWLEDGEMENTS

We gratefully acknowledge the permission granted by John Wiley and Sons Ltd. on behalf of the SCI to incorporate in this manuscript the Figures 1 to 4 and the Tables 1 to 6 from the paper published in *Polym. Int.* **48**, 171-180 (1999), entitled, "Sorption, diffusion, and pervaporation of water/ethanol mixtures in polyetherimide membranes" by Houssain Qariouh, Rossitza Schué, Francois Schué, and Christian Bailly.

9. REFERENCES

1. P. Aptel and J. Néel, in *Synthetic Membranes*: Science, Engineering and Applications, edited by P. M. Bungay, H. K. Lonsdale and M. N. de Pinho, Reidel, Dordrecht, 1988, p. 403.
2. R. Y. M. Huang (ed.), *Pervaporation Membrane Separation Processes, Membrane Science and Technology, Series 1*, Elsevier, Amsterdam, 1991.
3. S. Zhang and E. Drioli, *Sep. Sci. Technol.* **30** 1 (1995).
4. T. M. Aminabhavi, R. S. Khinnavar, S. B. Harogoppad, U. S. Aithal, Q. T. Nguyen, and K. C. Hansen, *J. Macromol. Sci., Rev. Macromol. Chem. Phys.* **34** 139 (1994).
5. U. Sander and P. J. Soukup, *Membr. Sci.* **36**, 463 (1988).
6. J. Frenneson and G. Tragardh, *Chem. Eng. Commun.* **45**, 277 (1986).
7. Q. T. Nguyen, *Synth. Polym. Membr., Proc. Microsymp. Macromol.* **29**, 479 (1986).
8. Q. T. Nguyen, I. Noezar, R. Clément, C. Streicher, and H. Brueschke, *Polym. Adv. Technol.* **8**, 477 (1997).
9. R. E. Kesting, Synthetic Polymeric Membranes, 2nd edn. John Wiley & Sons, Chichester, 1985, Chapt. 2.
10. R. Y. M. Huang and Y. F. Xu, *Eur. Polym. J.* **24**, 927 (1988).
11. P. Aptel, N. Challard, J. Cuny, and J. Néel, *J. Membr. Sci.* **1**, 271 (1976).
12. H. Tanisugi and T. Kotaka, *Polym. J.* **17**, 499 (1985).
13. J. A. Kerres and H. Strathmann, *J. Appl. Polym. Sci.* **50**, 1405 (1993).
14. H. Okuno, T. Okado, A. Matsumoto, M. Oiwa, and T. Uragami, *Sep. Sci. Technol.* **27**, 1599 (1992).
15. Y. K. Lee,T. M. Tak, D. S. Lee, and S. C. Kim, *J. Membr. Sci.* **52**, 157 (1990).
16. Q. T. Nguyen, C. Léger, P. Billard, and P. Lochon, *Polym. Adv. Technol.* **8**, 487 (1997).
17. E. Drioli, S. Zhang, and A. Basile, *J Memb Sci.* **81**, 43 (1993).
18. R. Y. M. Huang, and N. R. Javis, *J. Appl. Polym. Sci.* **14**, 2341 (1970).
19. J. Belana, J. C. Canadas, J. A. Diego, M. Mudarra, R. Diaz-calleja, S. Friederichs, C. Jaimes, and M. J. Sanchis, *Polym. Int.* **46**, 11 (1998).
20. A. W. Neumann, R. J. Good, C. J. Hope, and M. Sejpal, *J. Colloid Interface Sci.* **49**, 291 (1974).

BIOFUEL CELLS BASED ON MONOLAYER-FUNCTIONALIZED BIOCATALYTIC ELECTRODES

Eugenii Katz and Itamar Willner[*]

1. INTRODUCTION

Biofuel cells use biocatalysts for the conversion of chemical energy to electrical energy.[1-3] As most organic substrates undergo combustion with the evolution of energy, the biocatalyzed oxidation of organic substances by oxygen at two-electrode interfaces provides a means for the conversion of chemical to electrical energy. Abundant organic raw materials such as methanol or glucose can be used as substrates for the oxidation processes at the anode, whereas molecular oxygen or H_2O_2 can act as the substrate being reduced at the cathode. The extractable power of a fuel cell (P_{cell}) is the product of the cell voltage (V_{cell}) and the cell current (I_{cell}) (Eq. 1).

Although the ideal cell voltage is affected by the difference in the formal potentials of the oxidizer and fuel compounds ($E^{\circ'}_{ox} - E^{\circ'}_{fuel}$), irreversible losses in the voltage (η) as a result of kinetic limitations of electron transfer at the electrode interfaces, ohmic resistances and concentration gradients, lead to decreased values of the cell voltage (Eq. 2).

$$P_{cell} = V_{cell} \times I_{cell} \tag{1}$$

$$V_{cell} = (E^{\circ'}_{ox} - E^{\circ'}_{fuel}) - \eta \tag{2}$$

Similarly, the cell current is controlled by the electrode sizes, the ion permeability and transport rate across the membrane separating the catholyte and anolyte

---

[*] Institute of Chemistry, The Hebrew University of Jerusalem, Jerusalem 91904, Israel.

Advanced Macromolecular and Supramolecular Materials and Processes
Edited by K. Geckeler, Kluwer Academic/Plenum Publishers, 2003

compartments of the biofuel cell, and specifically, the rate of electron transfer at the respective electrode surfaces.

These different parameters collectively influence the biofuel cell power, and for improved efficiencies the V_{cell} and I_{cell} values should be optimized.

Biofuel cells can use biocatalysts, enzymes or whole cell organisms in two different routes.[1-3] (i) The biocatalysts can generate the fuel substrates for the cell by biocatalytic transformations or metabolic processes. (ii) The biocatalysts may participate in the electron transfer chain between the fuel substrates and the electrode surfaces. That is, microorganisms or redox enzymes facilitate the electron transfer between the fuel substrate or the oxidizer and the electrode interfaces, thereby enhancing the cell current. Most of the redox enzymes lack, however, direct electron transfer features with conductive supports, and a variety of electron mediators (electron relays) were used for electrical contacting of the biocatalysts and the electrode.

Our laboratory is extensively involved in the functionalization of electrode surfaces with monolayers and multilayers consisting of redox enzymes, electrocatalysts and bioelectrocatalysts that stimulate electrochemical transformations at the electrode interfaces.[4,5] The assembly of electrically contacted bioactive monolayer electrodes could be advantageous for biofuel cell applications as the biocatalyst and electrode support are integrated. Here we report on novel biofuel cell configurations based on the biocatalytic monolayer structures integrated with cathodes and anodes of biofuel cells.

2. MONOLAYER-FUNCTIONALIZED ELECTRODES AS BIOELECTRO-CATALYTIC ANODES OR CATHODES IN BIOFUEL CELLS

2.1. An Anode Based on the Bioelectrocatalyzed Oxidation of NAD(P)H

The nicotinamide redox cofactors (NAD^+ and $NADP^+$) play important roles in biological electron transport, acting as carriers of electrons and activating the biocatalytic functions of dehydrogenases, the vast majority of redox enzymes.[6] Application of $NAD(P)^+$-dependent enzymes (e.g. lactate dehydrogenase, EC 1.1.1.27; alcohol dehydrogenase, EC 1.1.1.71; glucose dehydrogenase, EC 1.1.1.118) in biofuel cells allows the use of many organic materials such as lactate, glucose or alcohols as fuels. The biocatalytic oxidation of these substrates requires the efficient electrochemical regeneration of $NAD(P)^+$-cofactors in the anodic compartment of the cells. The biocatalytically produced NAD(P)H cofactors participating in the anodic process should transport electrons from the enzymes to the anode and the subsequent electrochemical oxidation of the reduced cofactors regenerates the biocatalytic functions of the system.

In an aqueous solution at pH 7.0, the thermodynamic redox potential ($E^{\circ\prime}$) for $NAD(P)^+/NAD(P)H$ is ca. -0.56 V (vs. SCE),[6] that is sufficiently negative for the anode operation. Electrochemistry of NAD(P)H has been extensively studied at different electrodes and it has been demonstrated that the electrochemical oxidation process is highly irreversible and proceeds with large overpotentials (η) (ca. 0.4 V, 0.7 V and 1 V vs. SCE at carbon, Pt and Au electrodes, respectively).[7,8] Strong adsorption of NAD(P)H and $NAD(P)^+$ (e.g. on Pt, Au, glassy carbon, and pyrolytic graphite) generally poisons the electrode surface and inhibits the oxidation process.

Furthermore, NAD(P)$^+$ is an inhibitor of the direct oxidation of NAD(P)H, and adsorbed NAD(P)H can be oxidized to undesired products that lead to the degradation of the cofactor and to biologically-inactive products (e.g. NAD$^+$-dimers). Thus, non-catalyzed electrochemical oxidation of NAD(P)H cannot be realized in biofuel cells.

For the efficient electrooxidation of NAD(P)H, mediated electrocatalysis is necessary.[7,8] Mediators immobilized on electrodes have been applied for regeneration of NAD(P)$^+$. Different immobilization techniques were applied for the preparation of these modified electrodes: the mediator molecules were directly adsorbed onto electrode surfaces, incorporated into polymer layers, or covalently linked to functional groups on electrode surfaces.[8] The covalent coupling of redox mediators to self-assembled mono-layers on Au-electrode surfaces has an important advantage for preparation of multi-component organized systems.[9]

Pyrroloquinoline quinone, PQQ, (1) has been covalently attached to amino groups of a cystamine monolayer assembled onto a Au-surface (Figure 1(A)). The resulting electrode demonstrated good electrocatalytic activity for NAD(P)H oxidation, particularly in the presence of Ca^{2+}-cations as promoters (Figure 1(B)).[10] A quasi-reversible redox-wave at the formal potential, $E^{o'} = -0.155$ V (vs. SCE at pH = 8.0) is observed, corresponding to the two-electron redox process of the quinone units (Figure 1(B), curve a).

Coulometric analysis of the quinone redox-wave indicates that the PQQ surface coverage on the electrode corresponds to $\Gamma_{PQQ} = 1.2 \times 10^{-10}$ mol $\cdot$ cm^{-2}, a value that is typical for monolayer coverage. The electron transfer rate constant was found to be $k_{et} = 8$ s^{-1}.

Figure 1(B), curve b, shows the cyclic voltammogram of the PQQ-functionalized electrode upon addition of NADH, 10 mM, in the presence of Ca^{2+}-ions. An electrocatalytic anodic current is observed in the presence of NADH, implying the effective electrocatalyzed oxidation of the cofactor, Eqs. (3)-(4).

$$NADH + PQQ + H^+ \rightarrow NAD^+ + PQQH_2 \qquad (3)$$

$$PQQH_2 \rightarrow PQQ + 2H^+ + 2e^- \text{ (to anode)} \qquad (4)$$

In view of high cost of NAD(P)$^+$/NAD(P)H cofactors, practical applications require their immobilization together with the enzymes. Nonetheless the covalent coupling of natural NAD(P)$^+$ cofactors to an organic support results in a substantial decrease of their functional activity. Mobility of the cofactor is vital for its efficient interaction with enzymes, and thus attention was paid to the synthesis of artificial analogs of the NAD(P)$^+$ cofactors carrying functional groups separated from the bioactive site of the cofactor by spacers.[11,12]

The spacer is usually linked to N-6 position of the NAD(P)$^+$ molecule, and should provide some flexibility for the bioactive part of the cofactors, allowing them to be associated with the enzyme molecules. Structure/activity relationships of the artificial functionalized NAD(P)$^+$-derivatives have been studied with different enzymes and the possibility to substitute the natural NAD(P)$^+$ cofactor with these artificial analogs has been demonstrated.[11,13]

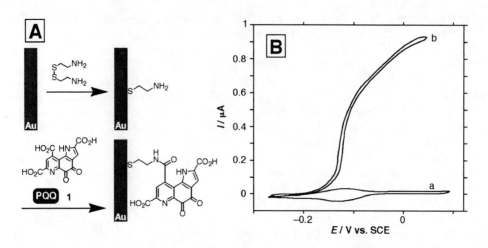

Figure 1. (A) Assembly of the PQQ-modified Au-electrode. (B) Cyclic voltammograms of a Au-PQQ electrode (geometrical area 0.2 cm², roughness factor ca. 1.5) in the presence of: (a) 0.1 M Tris-buffer, pH 8.0, (b) 10 mM NADH and 20 mM Ca^{2+}. Recorded at a scan rate of 1 mV·s⁻¹.

An efficient electrode that acts as an anode in the presence of an NAD(P)⁺-dependent enzyme should include three integrated, electrically-contacted components: The NAD(P)⁺-cofactor that is associated with the respective enzyme and a catalyst that allows the efficient electrocatalytic regeneration of the cofactor. Electrodes functionalized with cofactor monolayers (e.g., NAD⁺-monolayers) demonstrate the ability to form stable affinity complexes with their respective enzymes.[14-16] These interfacial complexes can be further crosslinked to produce integrated bioelectrocatalytic matrices consisting of the relay-units, the cofactor and the enzyme molecules. Electrically contacted biocatalytic electrodes of NAD⁺-dependent enzymes have been organized by the generation of affinity complexes between a catalyst/NAD⁺-monolayer and the respective enzymes.[15,16]

A PQQ monolayer covalently linked to an amino-functionalized nicotinamide adenine dinucleotide, N^6-(2-aminoethyl)- NAD⁺ (**2**) was assembled onto a Au-electrode. The resulting monolayer-functionalized electrode binds NAD⁺-dependent enzymes (e.g. L-lactate dehydrogenase, LDH, EC 1.1.1.27) by affinity interactions between the cofactor and the biocatalyst (Figure 2(A)).

These enzyme-electrodes electrocatalyze the oxidation of their respective substrates (e.g. lactate). Two-dimensional crosslinking of the enzyme layer associated with the PQQ/NAD⁺-cofactor monolayer, using glutaric dialdehyde, generates a stable, integrated, electrically-contacted, cofactor-enzyme electrode.

Figure 2(B) shows the electrical responses of a crosslinked layered PQQ/NAD⁺/-LDH electrode in the absence (curve a) and the presence (curve b) of lactate, whereas Figure 2(B), inset, shows the respective calibration curve corresponding to the amperometric output of the integrated LDH layered electrode at different lactate concentrations. This system exemplifies a fully integrated rigid biocatalytic matrix composed of the enzyme, NAD⁺-cofactor, and catalyst.

The complex between the NAD^+-cofactor and LDH aligns the enzyme on the electrode support thereby enabling the effective electrical communication between the enzyme and the electrode, while the PQQ-catalytic sites provide the electrocatalytic units for the regeneration of NAD^+.

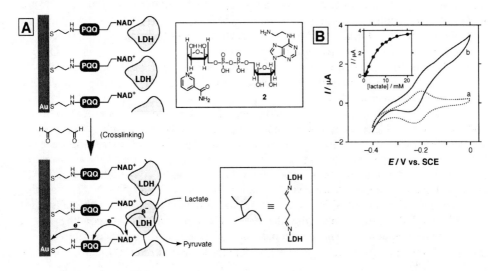

Figure 2. (A) The assembly of an integrated lactate dehydrogenase monolayer-electrode by the crosslinking of an affinity complex formed between the lactate dehydrogenase, LDH, and a PQQ-NAD$^+$ monolayer-functionalized Au-electrode. (B) Cyclic voltammograms of the integrated crosslinked PQQ-NAD$^+$/LDH electrode (geometrical area ca. 0.2 cm^2, roughness factor ca. 15): (a) In the absence of lactate. (b) With lactate, 20 mM. Data recorded in 0.1 M Tris-buffer, pH 8.0, in the presence of 10 mM CaCl$_2$, under Ar, scan rate, 2 mV·s^{-1}. Inset: Amperometric responses of the integrated electrode at different concentrations of lactate upon the application of a constant potential corresponding to 0.1 V vs. SCE.

2.2. An Anode Based on the Bioelectrocatalyzed Oxidation of Glucose by Glucose Oxidase Reconstituted on a FAD/PQQ-Monolayer-Functionalized Electrode

The electrical contacting of redox-enzymes that defy direct electrical communication with electrodes can be established by using synthetic or biologically-active charge- carriers as intermediates between the redox-center and the electrode.[4] The overall electrical efficiency of an enzyme-modified electrode depends not only on electron transport properties of the mediator, but on the different transfer steps occurring in the biocatalytic assembly on the electrode. In order to accomplish the superior electron contacting, the mediator should be selectively placed in an optimum position between the redox-center and the enzyme periphery. In the case of surface-confined enzymes, the orientation of the enzyme-mediator assembly with respect to the electrode should also be optimized. A novel means for the establishment of electrical contact between the redox-center of enzymes and their environment based on a reconstitution approach has recently been demonstrated.[17]

The organization of a reconstituted enzyme aligned on an electron-relay-FAD monolayer was recently realized by the reconstitution of an apo-enzyme on a surface functionalized with a relay-FAD monolayer (Figure 3(A)).[18,19]

Pyrroloquinoline quinone, PQQ, (1) was covalently linked to a base cystamine monolayer at a Au-electrode, and N^6-(2-aminoethyl)-FAD (3) was then attached to the PQQ redox-relay units. Apo-glucose oxidase (apo-GOx) was then reconstituted onto the FAD units of the PQQ-FAD- monolayer architecture to yield an immobilized biocatalyst on the electrode with a surface coverage of 1.7×10^{-12} mol $\cdot$ cm^{-2}. The apo-GOx was obtained by the extraction of the native FAD-cofactor from glucose oxidase (EC 1.1.3.4). The resulting enzyme- reconstituted PQQ-FAD-functionalized electrode reveals bio-electrocatalytic properties.

Figure 3(B) shows cyclic voltammograms of the enzyme electrode in the absence and the presence of glucose (curves a and b, respectively). When the substrate is present, an electrocatalytic anodic current is observed, implying electrical contact between the reconstituted enzyme and the electrode surface. The electrode constantly oxidizes the PQQ site located at the protein periphery, and the PQQ-mediated oxidation of the FAD-center activates the bioelectrocatalytic oxidation of glucose, Eqs. (5)-(7). The resulting electrical current is controlled by the recycling-rate of the reduced FAD by the substrate.

Figure 3(B, inset) shows the derived calibration curve corresponding to the amperometric output of the enzyme-reconstituted electrode at different concentrations of glucose. The resulting current densities are unprecedentedly high (300 mA $\cdot$ cm^{-2} at 80 mM of glucose).

$$\text{FAD} + \text{glucose} + 2\text{H}^+ \rightarrow \text{FADH}_2 + \text{gluconic acid} \tag{5}$$

$$\text{FADH}_2 + \text{PQQ} \rightarrow \text{FAD} + \text{PQQH}_2 \tag{6}$$

$$\text{PQQ H}_2 \rightarrow \text{PQQ} + 2\text{H}^+ + 2\text{e}^- \text{ (to anode)} \tag{7}$$

Control experiments revealed that reconstituted GOx on an electrode-FAD assembly (lacking the PQQ component) does not exhibit direct electron-transfer communication with the electrode surface, demonstrating that the PQQ relay unit is, indeed, a key component in the electro-oxidation of glucose.[18,19] The electron-transfer turnover rate of GOx with molecular oxygen as the electron acceptor corresponds to ca. 600 s^{-1} at 25°C. Using an activation energy of 7.2 kcal$\cdot$mol^{-1}, the electron-transfer turnover rate of GOx at 35°C is estimated to be ca. 900 s^{-1}.[18,19]

A densely-packed monolayer of GOx (ca. 1.7×10^{-12} mol $\cdot$ cm^{-2}) that exhibits the theoretical electron-transfer turnover rate is expected to yield an amperometric response of ca. 300 mA $\cdot$ cm^{-2}. This indicates that reconstituted GOx on the PQQ-FAD monolayer exhibits an electron-transfer turnover with the electrode of similar effectiveness to that observed for the enzyme with oxygen as a natural electron acceptor. Thus, the high current output of the resulting enzyme- electrode is preserved in the presence of O$_2$ in the solution.

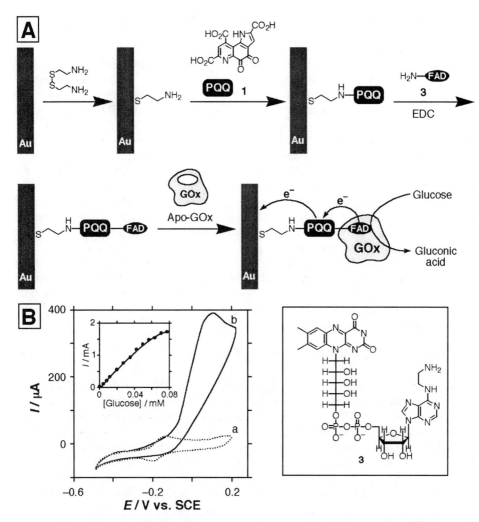

Figure 3. (A) The surface-reconstitution of apo-glucose oxidase on a PQQ-FAD monolayer assembled on a Au-electrode (geometrical area ca. 0.4 cm², roughness factor ca. 20). (B) Cyclic voltammograms of the glucose oxidase-reconstituted PQQ-FAD-functionalized Au-electrode: (a) In the absence of glucose. (b) With glucose, 80 mM. Recorded in 0.1 M phosphate buffer, pH 7.0, under Ar, at 35°C, scan rate, 5 mV s⁻¹. Inset: Calibration curve corresponding to the current output (measured by chronoamperometry, E = 0.2 V vs. SCE) of the PQQ-FAD-reconstituted glucose oxidase enzyme-electrode at different concentrations of glucose.

2.3. A Microperoxidase-11 Bioelectrocatalytic Cathode for the Reduction of Peroxides

Hydrogen peroxide is a strong oxidizer ($E°' = 1.535$ V vs. SCE), yet its electro-chemical reduction proceeds with a very high overpotential.[20]

Different materials were applied to catalyze the electrochemical reduction of H_2O_2.[21,22] Bioelectrocatalyzed reduction of H_2O_2 was accomplished in the presence of various peroxidases (e.g. horseradish peroxidase, EC 1.11.1.7).[23]

Microperoxidase-11, MP-11, (**4**) is an oligo- peptide consisting of 11 amino acids and a covalently linked Fe(III)-protoporphyrin IX heme site.[24] The oligopeptide is obtained by the controlled hydrolytic digestion of cytochrome c and represents the structure of the active-site microenvironment of cytochrome c. MP-11 reveals several advantages over usual peroxidases: It has much smaller size, high stability and exhibits direct electrical communication with electrodes since its heme is exposed to the solution.

The MP-11 was covalently linked to a cystamine monolayer self-assembled on a Au-electrode.[25] The MP-11 (**4**) structure suggests two different modes of coupling of the oligopeptide to the primary cystamine monolayer: (i) Linkage of the carboxylic functions associated with the protoporphyrin IX ligand to the monolayer interface. (ii) Coupling of carboxylic acid residues of the oligopeptide to the cystamine residues. These two modes of binding MP-11 to the monolayer reveal similar formal potentials, $E^{o'} = -0.40$ V vs. SCE, (Figure 4(A)). The electron transfer rates of the two binding modes of MP-11 were kinetically resolved, using chronoamperometry.[26] The binding modes appear at an approximately 1:1 ratio and the interfacial electron transfer rates to the heme-sites linked to the electrode by the two binding modes are 8.5 s^{-1} and 16 s^{-1}. Coulometric analysis of the MP-11 redox wave, corresponding to the reversible reduction/oxidation of the heme, Eq. (8), indicates a surface coverage of 2×10^{-10} mol $\cdot$ cm^{-2}.

$$[\text{heme-Fe(III)}] \underset{-e}{\overset{+e}{\rightleftarrows}} \quad [\text{heme-Fe(II)}] \tag{8}$$

Figure 4(B) shows the cyclic voltammograms of the MP-11-functionalized electrode recorded at positive potentials in the absence of H_2O_2 (curve a) and in the presence of added H_2O_2 (curve b). The observed electrocatalytic cathodic current indicates the effective electrobiocatalyzed reduction of H_2O_2 by the functionalized electrode. It should be noted that the electrocatalytic current for the reduction of H_2O_2 in aqueous solutions is observed at much more positive potentials than the MP-11 redox potential registered in the absence of H_2O_2 (Cf. Figures 4(A) and 4(B)). The reason for this potential-shift is the formation of Fe(IV) intermediate species in the presence of H_2O_2, Eqs. (9)-(11). Control experiments reveal that no electroreduction of H_2O_2 occurs at the bare Au electrode within this potential window.

$$[\text{heme-Fe(III)}] + H_2O_2 \rightarrow [\text{heme-Fe(IV)=O}]^{\cdot+} + H_2O \tag{9}$$

$$[\text{heme-Fe(IV)=O}]^{\cdot+} + e^- \text{ (from cathode)} + H^+ \rightarrow [\text{heme-Fe(IV)-OH}] \tag{10}$$

$$[\text{heme-Fe(IV)-OH}] + H^+ + e^- \text{ (from cathode)} \rightarrow [\text{heme-Fe(III)}] \tag{11}$$

Enzymes,[27] particularly peroxidases,[23] can function in non-aqueous solutions. A horseradish peroxidase-modified electrode was applied for the biocatalytic reduction of organic peroxides in non-aqueous solvents.[28] The biocatalytic activity of enzymes, particularly of horseradish peroxidase,[29] is, however, usually lower (sometimes by an order of magnitude) in organic solvents, compared to aqueous solutions.

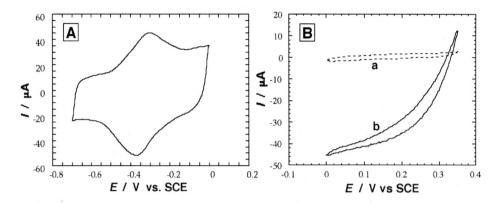

Figure 4. (A) Cyclic voltammogram of the MP-11-modified Au-electrode (geometrical area ca. 0.2 cm^2, roughness factor ca. 15) in 0.1 M phosphate buffer, pH 7.0, under Ar atmosphere, scan rate 50 mV · s⁻¹. (B) Cyclic voltammograms of the MP-11-modified electrodes recorded at positive potentials in 0.1 M phosphate buffer, pH 7.0, scan rate 10 mV · s⁻¹, (a) without H_2O_2, (b) in the presence of 5 mM H_2O_2.

Micro- peroxidase-11 monolayer-modified electrodes demonstrated high activity and stability for the electrocatalytic reduction of organic hydroperoxides in non-aqueous (acetonitrile and ethanol) solutions as compared to the reduction of H_2O_2 in an aqueous solution.[30]

In order to perform a biocatalytic cathodic reaction in a medium immiscible with an aqueous solution, the MP-11-modified electrode was studied[31] in a dichloromethane electrolyte (Figure 5, inset). A quasi-reversible redox wave of the heme center of MP-11 was observed at $E°' = -0.30$ V (vs. aqueous SCE) in dichloromethane (0.05 M tetrabutyl-ammonium tetrafluoroborate, TBATFB). Coulometric assaying of the redox wave indicates a surface coverage of ca. 3×10^{-10} mol · cm⁻². Figure 5 shows the cyclic voltammograms of the MP-11-functionalized electrode in the absence of an organic peroxide (curve a), and in the presence of added cumene peroxide (**5**) (curve b). The observed electrocatalytic cathodic current indicates the effective bioelectrocatalyzed reduction of cumene peroxide by the functionalized electrode. The sequence of electron transfers leading to the reduction of the peroxide is summarized in Eqs. (12)-(13).

$$[\text{heme-Fe(II)}] + \text{Cumene peroxide } (\textbf{5}) \rightarrow [\text{heme-Fe(III)}] + \text{Cumene alcohol } (\textbf{6}) \quad (12)$$

$$[\text{heme-Fe(III)}] + e^- \text{ (from cathode)} \rightarrow [\text{heme-Fe(II)}] \quad (13)$$

It should be noted that the MP-11 bioelectrocatalyzed reduction of organic peroxides in non-aqueous solutions does not include the intermediate formation of the Fe(IV) species and it proceeds at the MP-11 potential corresponding to Fe(III)/Fe(II) redox transformation (Eq. 8).

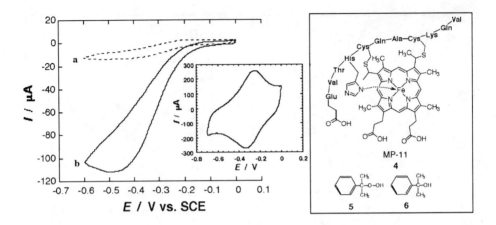

Figure 5. Cyclic voltammograms of the MP-11-functionalized Au-electrode (geometrical area ca. 0.4 cm^2, roughness factor ca. 20): (a) In the absence of a peroxide. (b) In the presence of cumene peroxide, 5×10^{-3} M. Potential scan rate, 5 mV · s^{-1}. Inset: Cyclic voltammogram of the MP-11-monolayer-modified Au electrode in the absence of cumene peroxide. Potential scan rate, 50 mV · s^{-1}; Ar atmosphere; electrolyte composed of a dichloromethane solution with 0.05 M TBATFB. Structures of MP-11 (**4**), cumene peroxide (**5**) and cumene alcohol (**6**) are shown.

2.4. A Cathode Consisting of Cytochrome C/Cytochrome Oxidase Assembly for the Bioelectrocatalyzed Reduction Of Oxygen

Electrochemical reduction of oxygen proceeds with a very large overpotential (e.g. at ca. -0.3 V vs. SCE at a Au electrode, pH = 7).[20] Thus, the reduction of oxygen at the cathode requires a catalyst in order to be used in fuel cells. Many electrocatalytic and bioelectrocatalytic systems have been reported to decrease the overpotential for the reduction of O_2.[32,33] Organized layered enzyme systems were never used, however, for this purpose. The four-electron transfer reduction of O_2 to water, without formation of peroxide or superoxide species, is a major challenge for the future development of biofuel cell elements, since such reactive intermediates would degrade the biocatalysts in the system. Accordingly, we examined the possibility to assemble a layered cytochrome c / cytochrome oxidase electrode as a biocatalytic interface for the concerted four-electron reduction of O_2 to H_2O.[34,35]

Cytochrome c that includes a single thiol group in the 102-cysteine-residue (yeast iso- 2-cytochrome c from *Saccharomyces cerevisiae*) was assembled as a monolayer on a Au- electrode by the covalent linkage of the thiol functionality to the maleimide monolayer- modified electrode as outlined in Figure 6.

The Cyt c monolayer reveals electrical contact with the electrode. The quasi-reversible cyclic voltammogram, E°' =0.03 V vs. SCE, (Figure 7(A)) indicates that the resulting heme-protein exhibits direct electrical contact with the electrode as the result of the structural alignment of the heme-protein on the electrode.

Coulometric assay of the redox-wave indicates protein coverage of 8×10^{-12} mol $\cdot$ cm^{-2}. Taking into account the Cyt c diameter as ca. 4.5 nm,[36] this surface coverage corresponds to a densely packed monolayer.

Cytochrome c (Cyt c) acts in various biological transformations as an electron relay that mediates electron transfer to redox enzymes.[37,38] The electron transfer proceeds via the formation of an inter-protein complex. The association constant between the Cyt c monolayer and cytochrome oxidase, COx, was determined[14] to be $K_a = 1.2 \times 10^7$ M^{-1}, and an integrated Cyt c/COx layered electrode was prepared as outlined in Figure 6.[34,35]

The Cyt c monolayer electrode was interacted with COx to generate the affinity complex on the surface. The resulting layered complex was crosslinked with glutaric dialdehyde to yield an integrated, electrically-contacted, electrode. The similar Cyt c/COx assembly was organized on a Au-quartz crystal microbalance surface. Microgravimetric analyses indicates that the surface coverage of COx on the base Cyt c monolayer is ca. 2×10^{-12} mol $\cdot$ cm^{-2}. This surface density corresponds to an almost densely packed monolayer of COx.

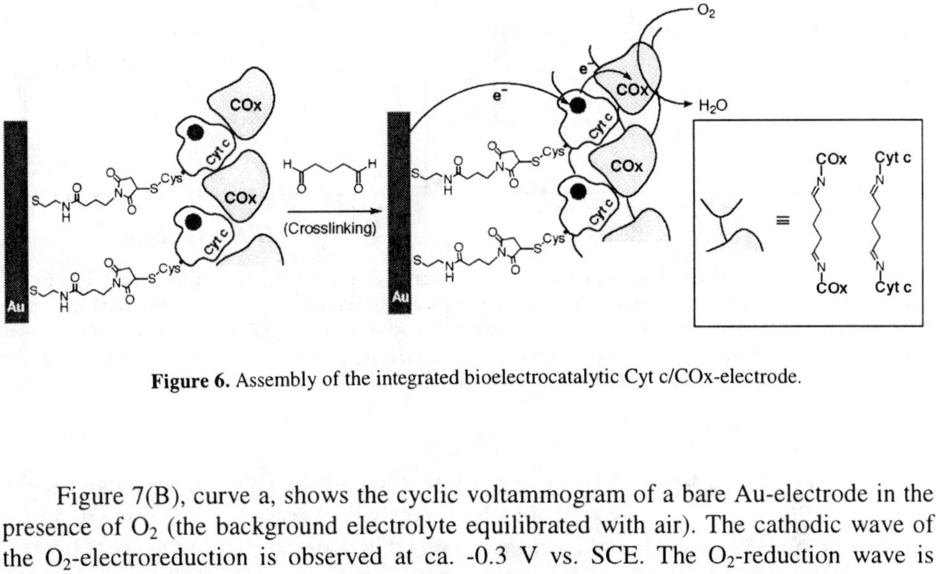

Figure 6. Assembly of the integrated bioelectrocatalytic Cyt c/COx-electrode.

Figure 7(B), curve a, shows the cyclic voltammogram of a bare Au-electrode in the presence of O$_2$ (the background electrolyte equilibrated with air). The cathodic wave of the O$_2$-electroreduction is observed at ca. -0.3 V vs. SCE. The O$_2$-reduction wave is negatively shifted in the presence of the Cyt c monolayer electrode (Figure 7(B), curve b), implying that the heme-protein layer is inactive as a biocatalyst for the reduction of O$_2$. In fact, the Cyt c monolayer enhances the overpotential for the reduction of dioxygen due to the hydrophobic blocking the electrode surface.

Figure 7(B), curve c, shows the cyclic voltammogram of the layered Cyt c/COx crosslinked electrode in the presence of O$_2$. An electrocatalytic wave is observed at ca. -0.07 V (vs. SCE), indicating that the Cyt c/COx layer acts as a biocatalytic interface for the reduction of dioxygen. In a control experiment, a COx monolayer was assembled on the Au-electrode without the base Cyt c layer. The COx monolayer lacks electrical contact with the electrode and no bio- electrocatalytic activity towards the reduction of O$_2$ is observed. Thus, the effective bio- electrocatalyzed reduction of O$_2$ by the Cyt c/COx interface originates from the direct electrical communication between the Cyt c and the electrode and the electrical contact in the crosslinked Cyt c/COx assembly.

The electron transfer to Cyt c is followed by electron transfer to COx, which acts as an electron storage biocatalyst for the concerted four-electron reduction of O_2, Eqs. (14)-(16). (The two-electron reduction of O_2 yields H_2O_2, while a concerted four-electron reduction of O_2 generates H_2O).

$$Cyt\ c_{ox} + e^- \text{ (from cathode)} \rightarrow Cyt\ c_{red} \tag{14}$$

$$4\ Cyt\ c_{red} + COx_{ox} \rightarrow 4Cyt\ c_{ox} + COx_{red} \tag{15}$$

$$COx_{red} + O_2 + 4H^+ \rightarrow COx_{ox} + 2H_2O \tag{16}$$

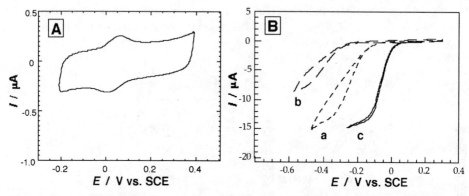

Figure 7. (A) Cyclic voltammogram of the Cyt c monolayer-modified Au-electrode (geometrical area ca. 0.2 cm², roughness factor ca. 1.5) measured under argon, potential scan rate 100 mV·s⁻¹. (B) Cyclic voltammograms obtained in an O_2-saturated electrolyte solution at: (a) A bare Au-electrode. (b) The Cyt c monolayer-modified Au-electrode. (c) The Cyt c/COx assembly-modified Au-electrode; potential scan rate 10 mV · s⁻¹. The experiments were performed in 0.1 M phosphate buffer, pH 7.0.

Rotating disk electrode (RDE) experiments were performed to estimate the electron transfer rate constant for the overall bioelectrocatalytic process that corresponds to the reduction of O_2.[35] The number of electrons involved in the reduction of O_2, n = 3.9 ± 0.2, and the electrochemical rate constant, $k_{el} = 5.3 \times 10^{-4}$ cm · s⁻¹, were found from the Koutecky-Levich plot.

The overall electron transfer rate constant, $k_{overall} = k_{el} / \Gamma_{COx} = 6.6 \times 10^5$ M⁻¹ · s⁻¹, was calculated taking into account the surface density of the bio- electrocatalyst, $\Gamma_{COx} = 2 \times 10^{-12}$ mol · cm⁻². To determine the limiting step in the bio-electrocatalytic current formation, the experimental diffusion-limited current density, 11.6 μA·cm⁻², was compared to the calculated current density that assumes the primary electron transfer to Cyt c as the rate-limiting step. Taking into account the electron transfer rate constant to Cyt c, 20 s⁻¹, and the surface density of Cyt c, $\Gamma_{Cyt\ c} = 8 \times 10^{-12}$ mol · cm⁻², the calculated current density of the system is 15.5 μA · cm⁻².

Since the calculated current density is only slightly higher than the experimental value, we assume that the primary electron transfer process from the electrode to the Cyt c monolayer is the limiting step in the overall bioelectrocatalytic reduction of O_2.

Thus, optimization of the primary electrochemical step could enhance the whole biocatalytic process at the cathode.

3. BIOFUEL CELLS BASED ON LAYERED ENZYME-ELECTRODES

The previous sections have addressed the supramolecular engineering of electrodes with layered biomaterials and the assembly of separate biocatalytic anodes and cathodes. For the design of biofuel cell elements, it is essential to couple the cathode and anode units into an integrated united device. The integration of the units is not free of limitations. The oxidizer may react with the biocatalyst relay, or cofactor units at the anode interface, thus decreasing or prohibiting the biocatalyzed oxidation of the fuel substrate.

Furthermore, for synchronous operation of the biofuel cell the charge compensation between the two electrodes must be attained, and the flow of electrons in the external circuit must be compensated by cation-transport in the electrolyte solution. To overcome these limitations the catholyte and anolyte solutions may be compartmentalized. Alternatively, the respective bioelectrocatalytic transformations at the electrodes may be sufficiently efficient that interfering components do not perturb the cell operation.

Nonetheless, in any biofuel cell the bioelectrocatalytic transformation or the transport process is a rate-limiting step controlling the cell efficiency. The mechanistic characterization and understanding of the biofuel cell performance provides then a means for the further optimization of the cell efficiency.

3.1. A Biofuel Cell Based on Pyrroloquinoline Quinone and Microperoxidase-11 Monolayer-Functionalized Electrodes

The bioelectrocatalyzed reduction of H_2O_2 and oxidation of NADH by MP-11 and PQQ monolayer-modified electrodes, respectively, enable to design a biofuel cell using H_2O_2 and NADH as the cathodic and anodic substrates (Figure 8).[39]

For the optimization of the biofuel cell element, the potentials of the biocatalytic functionalized electrodes as a function of the cathodic and anodic substrate concentrations were determined vs. the reference electrode (SCE).

Figure 9(A) shows the potential of the PQQ-electrode at different concentrations of NADH (curve a) and the potential of the MP-11-electrode at different H_2O_2 concentrations (curve b). The potentials of the PQQ monolayer-electrode and the MP-11-functionalized electrode are negatively shifted and positively shifted as the concentrations of NADH and H_2O_2 are elevated, respectively. The potentials of the electrodes reveal Nernstian-type behavior reaching saturation at high substrate concentrations (ca. 1×10^{-3} M). From the saturation potential values of the PQQ- and MP-11-functionalized electrodes an open-circuit voltage of the cell that corresponds to ca. 0.3 V is estimated.

Taking into account the surface density of the catalysts (1.2×10^{-10} mol $\cdot$ cm^{-2} and 2×10^{-10} mol $\cdot$ cm^{-2} for PQQ and MP-11, respectively), their interfacial electron transfer rate constants (ca. 8 s^{-1} and ca. 14 s^{-1} for PQQ and MP-11, respectively) and the number of electrons participating in a single electron transfer event (2 and 1 for PQQ and MP-11, respectively), one may derive the theoretical limit of the current densities that can be extracted by the catalytically active electrodes.

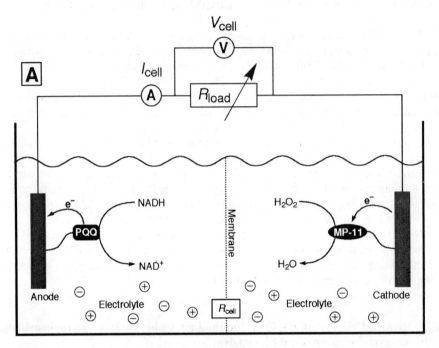

Figure 8. Schematic configuration of a biofuel cell employing NADH and H_2O_2 as fuel and oxidizer substrates and PQQ- and MP-11-functionalized-electrodes as catalytic anode and cathode, respectively.

These values are ca. 185 $\mu A \cdot cm^{-2}$ and 270 $\mu A \cdot cm^{-2}$ for the PQQ and MP-11 modified electrodes, respectively. The biofuel cell performance was examined at the concentration corresponding to 1×10^{-3} M of each of the two substrates: fuel and oxidizer. The cell voltage rises upon increasing the external load resistance and levels off to a constant value of ca. 310 mV at ca. 50 kΩ. Upon the increase of the load resistance the cell current dropped and reached almost zero at the resistance of ca. 50 kΩ.

Figure 9(B) shows the current-voltage behavior of the biofuel cell at different external loads. The cell yields the short-circuit current, I_{sc}, and open-circuit voltage, V_{oc}, of ca. 100 μA and 310 mV, respectively. The short-circuit current density was ca. 30 $\mu A \cdot cm^{-2}$ that is almost one order of magnitude less than the theoretical limits for the catalyst-modified electrodes. Thus, the interfacial kinetics of the biocatalyzed transformations at the electrodes is, probably, not the limiting step that controls the resulting current.

The power extracted from the biofuel cell ($P_{cell} = V_{cell} \cdot I_{cell}$) is shown in Figure 9(B), inset, for different external loads. The maximum power corresponds to 8 μW at an external load of 3 kΩ.

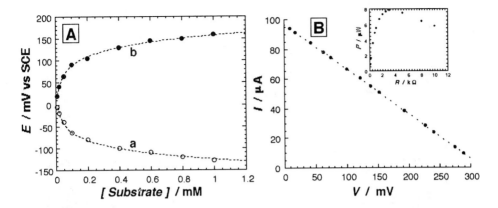

Figure 9. (A) Potentials of: (a) The PQQ-functionalized Au-electrode as a function of NADH concentration. (b) The MP-11-functionalized Au-electrode as a function of H_2O_2 concentration. The potentials of the modified electrodes were measured vs. SCE. (B) Current-voltage behavior of the PQQ-anode/MP-11-cathode biofuel cell measured at different loading resistances. Inset: Electrical power extracted from the biofuel cell at different external loads.

The ideal voltage-current relationship for an electrochemical generator of electricity is rectangular.[22] The linear dependence observed for the biofuel cell has a significant deviation from the ideal behavior and yields the fill factor of the biofuel cell of $f \approx 0.25$ (Eq. 17).

$$f = P_{cell} \times I_{sc}^{-1} \times V_{oc}^{-1} \tag{17}$$

The theory of the various types of overvoltage which produce non-rectangular V_{cell}-I_{cell} relationships has been addressed in detail by Vetter[20] and Delahay.[40] In the present case, the observed deviation results from mass transport losses reducing the cell voltage below its reversible thermodynamic value.[22]

It should be noted that in this study NADH is used as the fuel. In real biofuel cell the NADH fuel should be generated *in situ* from a respective substrate and an NAD^+-dependent dehydrogenase (e.g. alcohol or lactate acid in the presence of alcohol dehydrogenase or lactate dehydrogenase, respectively).

3.2. Bifuel Cells Based on Glucose Oxidase and Microperoxidase-11 Monolayer-Functionalized Electrodes

The bioelectrocatalyzed reduction of H_2O_2 by the MP-11 monolayer electrode and the effective bioelectrocatalyzed oxidation of glucose by the reconstituted GOx-monolayer- electrode allow us to design biofuel cells using H_2O_2 and glucose as the cathodic and anodic substrates, respectively, (Figure 10).[41]

For the optimization of the extractable power from the biofuel cell element, the potentials of the monolayer modified-electrodes as a function of the concentration of the cathodic and anodic substrates, were determined vs. the SCE reference electrode. The MP-11 monolayer electrode acts as the cathode, whereas the GOx monolayer electrode is the anode of the biofuel cell element.

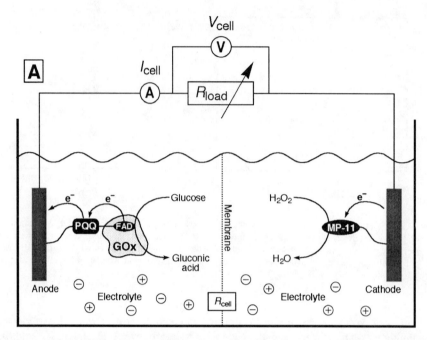

Figure 10. Schematic configuration of a biofuel cell employing glucose and H_2O_2 as a fuel and an oxidizer, respectively. PQQ-FAD/GOx and MP-11-functionalized Au-electrodes act as the biocatalytic anode and cathode, respectively.

Figure 11(A) shows the potentials of the GOx monolayer electrode at different concentrations of glucose (curve a) and the potentials of the MP-11 monolayer electrode at different concentrations of H_2O_2 (curve b). The potentials of the GOx monolayer electrode and of the MP-11 monolayer electrode are negatively shifted and positively shifted as the concentrations of the glucose and H_2O_2 are elevated, respectively. The potentials of the electrodes reveal Nernstian-type behavior, reaching saturation at a high concentration of the substrates. For the specific system, the saturation potentials of the anode and cathode are reached at ca. 1×10^{-3} M of glucose and ca. 1×10^{-3} M of H_2O_2, respectively.

From the saturated potential values of the GOx and MP-11 monolayer electrodes, the theoretical limit of the open-circuit voltage of the cell is estimated to be ca. 320 mV. The short- circuit current, I_{sc}, generated by the cell is 340 μA. Taking into account the geometrical electrode area (0.2 cm^2) and the electrode roughness factor (ca. 15), the current generated by the cell can be translated into the current density, ca. 114 μA · cm^{-2}. The theoretical limit of the current density extractable from the MP-11 monolayer electrode is ca. 270 μA · cm^{-2} (MP-11 surface coverage × interfacial electron transfer rate × Faraday constant).

For the GOx monolayer electrode the maximum extractable current density was estimated to be ca. 200 μA · cm^{-2}.

This value is based on the surface coverage of the reconstituted GOx on the electrode, 1.7×10^{-12} mol · cm^{-2} and the turnover-rate of the enzyme, ca. 600 s^{-1}. Thus, the observed short-circuit current density of the cell is probably controlled and limited by the bioelectrocatalyzed oxidation of glucose. This suggests that increasing the GOx content associated with the electrode could enhance the current density of the biofuel cell element and the extractable power.

The biofuel cell performance was examined at the concentration corresponding to 1 $\times 10^{-3}$ M of each of the two substrates: fuel and oxidizer. The cell voltage increases as the external load resistance is elevated and at an external load of ca. 50 kΩ it levels off to a constant value of ca. 310 mV. Upon increasing the external load, the current drops and it is almost zero at an external load of 100 kΩ.

Figure 11(B) shows the current–voltage behavior of the biofuel cell at different external loads. The linear dependence observed for the biofuel cell has significant deviation from the ideal rectangular behavior and the fill factor of the biofuel cell corresponds to ca. 0.25. The observed deviation results from mass transport losses reducing the cell voltage below its reversible thermodynamic value.

The power extracted from the biofuel element shown in Figure 11 (inset), for different external loads. The maximum power corresponds to 32 μW at an external load of 3 kΩ. It should be noted that the biofuel cell voltage and current outputs are identical under Ar and air. This originates from the effective electrical contact of the surface-reconstituted GOx with the electrode support, as a result of its alignment. This makes the GOx- monolayer-electrode insensitive to oxygen.[18,19]

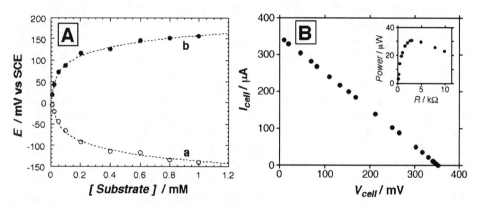

Figure 11. (A) Potential of the PQQ–FAD/GOx-modified Au-electrode as a function of glucose concentrations (a). Potential of the MP-11-functionalized Au-electrode as a function of H$_2$O$_2$ concentrations (b). Potentials were measured vs. SCE. (B) Current–voltage behavior of the GOx-anode/MP-11-cathode biofuel cell at different external loads. Inset: electrical power extracted from the biofuel cell at different external loads.

The stability of the biofuel cell was examined at the optimal loading resistance of 3 kΩ as a function of time.[41] The power decreases by ca. 50% after ca. 3 h of the cell operation. This decrease in the cell power output could originate from the depletion of the fuel substrate, inter-penetration of the fuel and oxidizer into the respective counter compartments and, eventually, the degradation of the biocatalysts that are included in the cell. The current decreases by ca. 50% within 3 h of operation of the cell.

At the same time interval, the cell voltage appears to be stable. Integration of the current output within these time interval yields the charge that passes through the cell during the time of operation. We calculate that ca. 25% of the original fuel-substrate concentration had been depleted upon operation of the cell for ca. 3 h. Thus, 25% of the total decrease in the current output can be attributed to the consumption of the fuel-substrate loaded in the cell. Recharging the cell with the fuel substrate and oxidizer could compensate for this decrease in the current output.

Charge transfer process in fuel cells across the interface of two immiscible electrolyte solutions can provide an additional potential difference between cathodic and anodic reactions due to the potential difference on the liquid/liquid interface that operates as a membrane separating the catholyte and anolyte. Many different systems with liquid/-liquid interfaces have been studied using numerous experimental approaches.[42]

The application of two immiscible solvents that exhibit perspectives for enhancing the biofuel cell output has not been used previously. The reduction of cumene peroxide in dichloromethane electrocatalyzed by the MP-11-monolayer electrode and the oxidation of glucose in aqueous solution bioelectrocatalyzed by the reconstituted GOx-monolayer-electrode enables us to design a biofuel cell using cumene peroxide and glucose as the cathodic and anodic substrates, respectively, and to operate the cell in the presence of two immiscible electrolyte solutions.[31] For the optimization of the biofuel cell element, the potentials of the monolayer- modified electrodes as a function of the concentration of the cathodic and anodic substrates were determined vs. the aqueous SCE reference electrode.

Figure 12(A) shows the potential of the GOx-monolayer electrode at different concentrations of glucose in the aqueous electrolyte solution (curve a) and the potential of the MP-11-monolayer electrode at different concentrations of cumene peroxide in the dichloromethane electrolyte solution (curve b). The potentials of the GOx-monolayer electrode and of the MP-11-monolayer electrode are negatively shifted and positively shifted, respectively, as the concentrations of the glucose and cumene peroxide are elevated.

The potentials of the electrodes reveal Nernstian-type behavior, showing a logarithmic increase and reaching saturation at high concentrations of the substrates. For our specific system, the saturation potential values of the anode and cathode are reached at ca. 1×10^{-3} M of glucose and 1×10^{-3} M of cumene peroxide, respectively. From the saturated potential values of the GOx- and MP-11-monolayer electrodes, the theoretical limit of the open-circuit voltage of the cell is estimated to be ca. 1.0 V.

It should be noted that the potentials extrapolated to zero concentrations of the substrates show a large difference, ca. 700 mV, which results from the potential jump at the liquid-liquid interface. The phase separation of the fuel and oxidizer is the origin for the enhanced efficiency of the cell. The cell reveals an open-circuit voltage of ca. 1.0 V and a short-circuit current density of ca. 830 $\mu A \cdot cm^{-2}$. The maximum power output of the cell is 520 μW at an optimal loading resistance of 0.4 kΩ (Figure 12(B)).

3.3. A Non-Compartmentalized Biofuel Cell Based on Glucose Oxidase and Cytochrome C / Cytochrome Oxidase Monolayer-Electrodes

The next generation of biofuel cells could utilize complex, ordered enzyme or multi-enzyme systems immobilized on both electrodes, and eventually could eliminate the need for compartmentalization of the anode and the cathode.

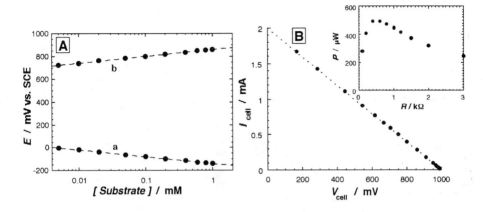

Figure 12. (A) Potential of: (a) The PQQ-AD/GOx-modified Au-electrode as a function of glucose concentration in 0.01 M phosphate buffer, pH 7.0, and 0.05 M TBATFB. (b) The MP-11-functionalized Au-electrode as a function of cumene peroxide concentration in a dichloromethane solution, 0.05 M TBATFB. Potentials were measured vs. aqueous SCE. (B) Current–voltage behavior of the biofuel cell at different external loads. Inset: Electrical power extracted from the biofuel cell at different external loads. The biocatalytic cathode and anode (ca. 0.8 cm^2 geometrical area, roughness factor ca. 1.3) were assembled in a thin-layer electrochemical cell with the distance between the electrodes 5 mm.

Tailoring of efficient electron transfer at the enzyme-modified electrode could enable specific biocatalytic transformations that compete kinetically with any chemical reaction of the electrode or biocatalysts with interfering substrates (e.g. substrate transport from the counter compartment, oxygen, etc.). This would enable to tailor non-compartmentalized biofuel cells where the biocatalytic anode and cathode are immersed in the same phase with no separating membrane.

In a working example, the anode described above (based on the reconstituted GOx) is connected to a cathode based on an aligned Cyt c / cytochrome oxidase couple providing the reduction of O_2 to water (Figure 13(A)).[34] Since the reconstituted GOx provides extremely efficient biocatalyzed oxidation of glucose that is unaffected by oxygen, the anode can operate in the presence of oxygen. Thus, this biofuel cell uses O_2 as an oxidizer and glucose as a fuel without the need for compartmentalization.

The cell operation was studied at different external loads (Figure 13(B)), and achieved a fill factor of ca. 40% with a maximum power output of 4 μW at an external load of 0.9 kW. The relatively low extracted power from the cell mainly originates from the small potential difference existing between the anode and cathode. The bioelectrocatalyzed oxidation of glucose occurs at the redox potential of the PQQ-electron mediator, $E^{o\prime}$ = - 0.125 V (vs. SCE at pH 7.0), whereas the redox potential of Cyt c is $E^{o\prime}$ = 0.03 V.

This yields a potential difference of only ca. 155 mV between the anode and cathode. By the application of electron mediators that exhibit more negative potentials, the extractable power from the cell could be enhanced. The major advance of the present system is its operation in a non- compartmentalized biofuel cell configuration. This suggests that the electrodes may be used as an invasive electrical energy generation device taking fuel and oxidizer (i.e. glucose and O_2) from a blood stream and providing electrical power for a pacemaker or an insulin pump.

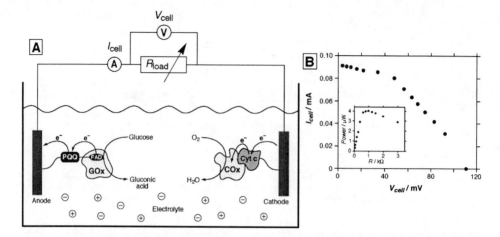

Figure 13. (A) Schematic configuration of a non-compartmentalized biofuel cell employing glucose and O_2 as fuel and oxidize, and using PQQ-FAD/GOx and Cyt c / COx-functionalized Au-electrodes as biocatalytic anode and cathode, respectively. (B) Current-voltage behavior of the biofuel cell at different external loads. Inset: Electrical power extracted from the biofuel cell at different external loads.

4. CONCLUSIONS AND PERSPECTIVES

This paper has addressed recent advances in the organization of redox enzymes as monolayer assemblies on solid conductive supports for bioenergetic applications. Specifically, the electrical contacting of the enzyme layers with the electrode supports allowed us to design novel biofuel cells. We have emphasized the basic scientific interests and the practical implications of tailoring organized protein architectures in ordered and defined nanostructures. The key elements in the integration of redox proteins with electronic transducers include the physical or chemical deposition of the proteins on the solid supports and the electrical contacting between the biomaterials and the electrodes.

Monolayer assemblies of redox-enzymes were organized on electrodes using covalent bonds and affinity interactions. Theoretical understanding of electron-transfer processes in enzymes and the availability of chemical and biological means to modify biomaterials have enabled us to improve the electrical coupling between the redox enzymes and the electrodes.

Site-specific modification of redox enzymes and surface- reconstitution of enzymes, represent novel and attractive means to align and orient biocatalysts on electrode surfaces. The effective electrical contact of aligned proteins with electrodes suggests that future efforts should be directed to develop structural mutants of redox-proteins to enhance their electrical communication with electrodes. The stepwise nanoengineering of the electrode surfaces with relay-cofactor-biocatalyst units by organic synthesis principles allows us to control the electron transfer cascades in the assemblies. By tuning the redox-potentials of the synthetic relays or of the biocatalytic mutants, enhanced power outputs from the biofuel cells may be envisaged.

The configurations of the biofuel cells discussed in this paper can be extended to other redox enzymes and fuel substrates allowing numerous technological applications including invasive biofuel cells that could operate inside of a human body providing electrical power for implanted devices (e.g. pacemakers, insulin pumps, etc.).

5. ACKNOWLEDGEMENTS

This research is supported by the German-Israeli project cooperation program (DIP). The support of the Max-Planck Award for International Cooperation (I.W.) is acknowledged.

6. REFERENCES

1. W. J. Aston and A. P. F. Turner, Biosensors and biofuel cells, *biotech. Gen. Eng. Rev.* **1**, 89-120 (1984).
2. C. van Dijk, C. Laane, and C. Veeger, Biochemical fuel cells and amperometric biosensors, *Recl. Trav. Chim. Pays-Bas* **104**, 245-252 (1985).
3. G. Tayhas, R. Palmore, and G. Whitesides, Microbial and Enzymatic Biofuel Cells, in: *Enzymatic conversion of biomass for fuels production*, edited by M. E. Himmel, J. O. Baker, and R. P. Overend (Am. Chem. Soc., Washington, DC, 1994), Chapter 14, pp. 271-290.
4. I. Willner and E. Katz, Integration of layered redox-proteins and conductive supports for bioelectronic applications, *Angew. Chem. Int. Ed.* **39**, 1180-1218 (2000).
5. I. Willner, E. Katz, and B. Willner, Electrical-contact of redox enzyme layers associated with electrodes: routes to amperometric biosensors, *Electroanalysis* **13**, 965-977 (1997).
6. A. L. Lehninger, *Biochemistry* (Worth, New York, 1975), p. 479.
7. P. N. Bartlett, P. Tebbutt, and R. C. Whitaker, Kinetic Aspects of the use of modified electrodes and mediators in bioelectrochemistry, *Prog. Reaction Kinetics* **16**, 55-155 (1991).
8. I. Katakis, and E. Dominguez, Catalytic electrooxidation of NADH for dehydrogenase amperometric biosensors, *Mikrochim. Acta* **126**, 11-32 (1997).
9. I. Willner and A. Riklin, Electrical communication between electrode and NAD(P)$^+$-dependent enzymes using pyrroloquinoline quinone - enzyme electrodes in a self-assembled monolayer configuration: design of a new class of amperometric biosensors, *Anal. Chem.* **66**, 1535-1539 (1994).
10. E. Katz, T. Lötzbeyer, D. D. Schlereth, W. Schuhmann, and H.-L. Schmidt, E lectrocatalytic oxidation of reduced nicotinamide coenzymes at gold and platinum electrode surfaces modified with a monolayer of pyrroloquinoline quinone. effect of Ca^{2+} cations, *J. Electroanal. Chem.* **373**, 189-200 (1994).
11. M. Maurice and J. Souppe, Spectral, Biochemical, and electrochemical properties of chemically modified nicotinamide adenine dinucleotides, *New J. Chem.* **14**, 301-304 (1990).
12. A. F. Bückmann and V. Wray, A simplified procedure for the synthesis and purification of N^6-(2-Aminoethyl)-NAD$^+$ and tricyclic N^6-ethanoadenine-NAD$^+$, *Biotechnol. Biochem.* **15**, 303-310 (1992).
13. J. Hendle, A. F. Bückmann, W. Aehle, D. Schomburg, and R. D. Schmid, Structure-activity relationship of adenine-modified NAD$^+$ derivatives with respect to porcine heart lactate-dehydrogenase isozyme H-4 simulated with molecular mechanics, *Eur. J. Biochem.* **213**, 947-956 (1993).
14. A. B. Kharitonov, L. Alfonta, E. Katz, and I. Willner, Probing of bioaffinnity interactions at interfaces using impedance spectroscopy and chronopotentiometry, *J. Electroanal. Chem.* **487**, 133-141 (2000).
15. E. Katz, V. Heleg-Shabtai, A. Bardea, I. Willner, H. K. Rau, and W. Haehnel, Fully integrated biocatalytic electrodes based on bioaffinity interactions, *Biosens. Bioelectron.* **13**, 741-756 (1998).
16. A. Bardea, E. Katz, A. F. Bückmann, and I. Willner, NAD$^+$-dependent enzyme electrodes: electrical contact of cofactor-dependent enzymes and electrodes, *J. Am. Chem. Soc.* **119**, 9114-9119 (1997).
17. A. Riklin, E. Katz, I. Willner, A. Stocker, and A. F. Bückmann, Reconstitution of flavoenzyme-derived apoproteins with ferrocene-modified FAD cofactor yields electroactive enzymes, *Nature* **376**, 672-675 (1995).
18. I. Willner, V. Heleg-Shabtai, R. Blonder, E. Katz, G. Tao, A. F. Bückmann, and A. Heller, Electrical wiring of glucose oxidase by reconstitution of FAD-modified monolayers assembled onto au-electrodes, *J. Am. Chem. Soc.* **118**, 10321-10322 (1996).

19. E. Katz, A. Riklin, V. Heleg-Shabtai, I. Willner, and A. F. Bückmann, Glucose Oxidase Electrodes *via* reconstitution of the apo-enzyme: tailoring of novel glucose biosensors, *Anal. Chim. Acta* **385**, 45-58 (1999).

20. K. J. Vetter, *Electrochemical kinetics* (Academic Press, New York, 1967).

21. R. R. Bessette, J. M. Cichon, D. W. Dischert, and E. G. Dow, A study of cathode catalysis for the aluminium / hydrogen peroxide semi-fuel cell, *J. Power Sources* **80**, 248-253 (1999).

22. O. 'M. Bockris and S. Srinivasan, *Fuel cells: Their electrochemistry* (McGraw-Hill, New York, 1969).

23. T. Ruzgas, E. Csöregi, J. Emneus, L. Gorton, and G. Marko-Varga, Peroxidase-modified electrodes: fundamentals and application, *Anal. Chim. Acta* **330**, 123-138 (1996).

24. P. A. Adams, Microperoxidases and iron porphyrins, in: *Peroxidases in chemistry and biology*, edited by J. Everse, K.E. Everse, and M.B. Grisham (CRC Press, Boca Raton, 1991), Vol. II, Chapter 7, pp. 171-200.

25. T. Lötzbeyer, W. Schuhmann, E. Katz, J. Falter, and H.-L. Schmidt, Direct Electron Transfer between the Covalently Immobilized Enzyme Microperoxidase MP-11 and a Cystamine-Modified Gold Electrode, *J. Electroanal. Chem.* **377**, 291-294 (1994).

26. E. Katz and I. Willner, Kinetic separation of amperometric responses of composite redox-active monolayers assembled onto au-electrodes: implication to the monolayer structure and composition, *Langmuir* **13**, 3364-3373 (1997).

27. A. M. Klibanov, Enzymatic catalysis in anhydrous organic-solvents, *Trends Biochem. Sci.* **14**, 141-144 (1989).

28. J. Li, S. N. Tan and J. T. Oh, Silica sol-gel immobilized amperometric enzyme electrode for peroxide determination in the organic phase, *J. Electroanal. Chem.* **448**, 69-77 (1998).

29. L. Yang, and R.W. Murray, Spectrophotometric and Electrochemical Kinetic Studies of Poly(ethylene glycol)-Modified Horseradish Peroxidase Reactions in Organic-Solvents and Aqueous Buffers, *Anal. Chem.* **66**, 2710-2718 (1994).

30. A. N. J. Moore, E. Katz, and I. Willner, Electrocatalytic reduction of organic peroxides in organic solvents by microperoxidase-11 immobilized as a monolayer on a gold electrode, *J. Electroanal. Chem.* **417**, 189-192 (1996).

31. E. Katz, B. Filanovsky, and I. Willner, A biofuel cell based on two immiscible solvents and glucose oxidase and microperoxidase-11 monolayer-functionalized electrodes, *New J. Chem.* **23**, 481-487 (1999).

32. J.P. Collman and K. Kim, Electrocatalytic four-electron reduction of dioxygen by iridium porphyrins adsorbed on graphite, *J. Am. Chem. Soc.* **108**, 7847-7849 (1986).

33. G. T. R. Palmore and H. Kim, Electro-enzymatic reduction of dioxygen to water in the cathode compartment of a biofuel cell, *J. Electroanal. Chem.* **464**, 110-117 (1999).

34. E. Katz, I. Willner, and A. B. Kotlyar, A non-compartmentalized glucose-O_2 biofuel cell by bioengineered electrode surfaces, *J. Electroanal. Chem.* **479**, 64-68 (1999).

35. V. Pardo-Yissar, E. Katz, I. Willner, A. B. Kotlyar, C. Sanders, and H. Lill, Biomaterial engineered electrodes for bioelectronics, *Faraday Discussions* **116**, 119-134 (2000).

36. D. S. Goodsell, and A. J. Olson, Soluble proteins - size, shape and function, *Trends Biochem. Sci.* **18**, 65-68 (1993).

37. A. E. G. Cass, G. Davis, H. A. O. Hill, and D. J. Nancarrow, The reaction of flavocytochrome b$_2$ with cytochrome c and ferricinium carboxylate. Comparative kinetics by cyclic voltammetry and chronoamperometry, *Biochim. Biophys. Acta* **828**, 51-57 (1985).

38. M. Lion-Dagan, E. Katz, and I. Willner, A bifunctional monolayer electrode consisting of 4-pyridyl sulfide and photoisomerizable spiropyran: photoswitchable electrical communication between the electrode and cytochrome c, *J. Chem. Soc., Chem. Comm.* 2741-2742 (1994).

39. I. Willner, G. Arad, and E. Katz, A biofuel cell based on pyrroloquinoline quinone and micro-peroxidase-11 monolayer-functionalized electrodes, *Bioelectrochem. Bioenerg.* **44**, 209-214 (1998).

40. P. Delahay, *Double layer and electrode kinetics* (Wiley, New York, 1965).

41. I. Willner, E. Katz, F. Patolsky, and A. F. Bückmann, A biofuel cell based on glucose oxidase and microperoxidase-11 monolayer-functionalized electrodes, *J. Chem. Soc., Perkin Trans.* 2, 1817-1822 (1998).

42. A. G. Volkov and D. W. Deamer, *Liquid-liquid interface theory and methods* (CRC, Boca Raton, 1996).

OSCILLATING EXCITATION TRANSFER IN DITHIAANTHRACENOPHANE — QUANTUM BEAT IN A COHERENT PHOTOCHEMICAL PROCESS

Iwao Yamazaki, Seiji Akimoto, Tomoko Yamazaki, Shin-ichiro Sato, and Yoshiteru Sakata[*]

1. INTRODUCTION

Since the Förster's mechanism[1] of excitation energy transfer between molecules has been formulated, many works have been reported for molecules incorporated in various types of media such as fluid solution and solid matrices. The energy transfer dynamics has been discussed in terms of the very-weak coupling limit of the Förster's theory; the interaction energy $\beta < 1$ cm^{-1} and the transfer rate $w < 10^{11}$ s^{-1}.

On the other hand, recent problem is focused on the energy transfer in highly organized molecular systems in which reacting molecules of donor and acceptor are connected chemically with close proximity and specific orientation. Molecules in such systems can be coupled to an adjacent molecule with relatively strong intermolecular interaction ($\beta > 10$ cm^{-1}) and therefore they may undergo ultrafast energy transfer reaction ($w > 10^{12}$ s^{-1}).

According to the Förster's theory,[1] the energy transfer under such condition can be expressed in terms of the *intermediate coupling* case, where the excitation energy transfer occurs accompanying with the recurrence of excitation, i.e., the excitation localized initially on donor jumps back and forth among donor and acceptor molecules. To observe the recurrence experimentally, the reaction should compete with or exceeds over the electronic dephasing, and then it can be probed more or less as a coherent process.

[*] Iwao Yamazaki, Seiji Akimoto, Tomoko Yamazaki, Shin-ichiro Sato, Department of Molecular Chemistry, Graduate School of Engineering, Hokkaido University, Sapporo 060-8628, Japan;
* Yoshiteru Sakata, The Institute of Scientific and Industrial Research, Osaka University, Mihoga-oka, Ibaraki, Osaka 567-0047, Japan.

Advanced Macromolecular and Supramolecular Materials and Processes
Edited by K. Geckeler, Kluwer Academic/Plenum Publishers, 2003

Whether or not such a predicted macroscopic quantum coherence will be observable depends on the strength of the coupling of the exciton to the heat bath.

In fact, it is found in several cases of linear and crossed-linear porphyrin arrays[2,3] that the interchromophore energy transfer rate are in $w \sim 10^{13}$ s^{-1} comparable to the rates of intramolecular energy relaxations such as internal conversion (IC), vibrational relaxation (VR), and internal vibrational redistribution (IVR) in an initially photo-excited molecule. One can expect in highly organized molecular systems to observe the electronic coherence or the recurrence of excitation on their fluorescence decays and anisotropies.

Hochstrasser's group[4,5] studied the coherent process in the excitation energy transfer between two naphthalene rings in 2,2'-binaphthyl; the anisotropy decay of polarized transient absorption exhibited a damping oscillation corresponding to the interaction energy (2β) of 41 cm^{-1}, with dephasing time of 0.2 ps. In their observation, the ainplitude of the oscillating component is rather weak because the bichromophoricc molecule having large rotational degree of freedom at the single bond which connects the two chromophores. To examine the coherent phenomenon much more precisely, it is needed to use a molecular system in which two chromophores are rigidly connected to one another.

We have studied the intramolecular excitation transfer in dithia(1,5)[3,3]-anthracenophane (DTA)[6] with a rigidly stacked pair of identical chromophores. The molecular structure is shown in Figure 1. Two anthracene rings are stacked almost in parallel (dihedral angle, 5.0°) with an interplanar distance of 3.41 Å and one of the rings rotates at 88.5° around the center of the anthracene ring.[6] The spatial arrangement of this molecular system is suitable for examining the coherent dynamics of excitation transfer. The fluorescence up-conversion method was adapted for measurement of fluorescence decays in fenitosecond time regime. The anisotropy decay exhibited a damping oscillation. We here report on the coherent energy transfer and present a discussion comparing with Hochstrasser's observation on binaphthyl.

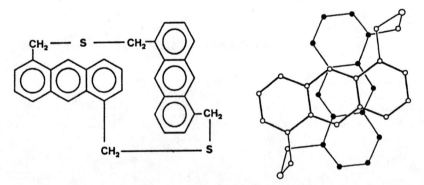

Figure 1. Molecular formula of dithiaanthracenophane (DTA) (left), and a view of DTA on leastsquares plane defined with an anthracene ring determined from X-ray analysis.

2. EXPERIMENTAL

Reaction of 1,5-di(methoxycarbonyl)anthracene with $LiAlH_4$ in THF, followed by bromination with PBr_3 in benzene gave 1,5-di(bromomethyl) anthracene in 20% yield. Coupling reaction with Na_2S was carried out in benzene-ethanol-water (44 : 9 : 1). After work-up and column chromatography (silica gel, benzene-hexane), DTA was obtained in 33% yield along with oligomer. The other possible isomer with eclipsed orientation was not found in the reaction mixture. Further detailed procedure of the synthesis of the anthracenophane are described in a previous paper.[7]

The excitation source was the second harmonics of a Ti:Sappire laser (Spectra Physics Tsunami, 816 nm, 80 MHz) pumped with a diode-pumped solid state laser (Spectra Physics, Millenia X). The output laser pulse had a pulse width of 80 fs (FWHM), and an average energy 16 nJ at 816 nm. Approximately half of the laser beam was frequency doubled in a BBO crystal for the sample excitation (0.5 nJ at 408 mn), and the remainder was used for the gating pulse.

The gating pulse passed through a variable delay line using a translation stage (Sigma STM-20X) with 1.0 mm per a step (corresponding to the delay time of 6.7 fs) under computer control. The fluorescence emission and the gating pulse were focused into a 0.5 mm BBO crystal by a 5.0 cm focal length lens. The wavelength of monitoring emission was adjusted with rotating BBO crystal in a type I phase matching geometry. The wavelength of the detection which employed a 0.5 min BBO crystal for up-conversion, was approximately 290 nm, corresponding to a fluorescence frequency of 450 nm. The sum frequency signal was both filtered and frequency selected using a monochromator (JASCO CT-10) and a single-photon counting apparatus equipped with a Hamamatsu R106UH photomultiplier.

The polarization of excitation beam was changed with rotating a 1/2 plate, i.e., parallel and perpendicular to the polarization of the excitation laser light which was polarized horizontally.

3. RESULTS

3.1 Absorption and Fluorescence Spectra

Figure 2 shows absorption and fluorescence spectra of DTA in THF. The absorption spectrum of DTA is similar to the anthracene spectrum with several vibrational bands except for spectral shift. The whole spectrum of ILa absorption with the 0-0 band at 407 nm is red-shifted -30 nm relative to that of anthracene, due to substitution with two methylene groups at α-positions of anthracene. The stationary-state fluorescence spectrm exhibits only a diff-use and broad band centered at 469 nm in THF or in cyclohexane.

Previously, we have examined the ps timeresolved fluorescence spectra (see Figure 2 of Ref. 6) in relation to the excimer formation of DTA.[6] The fluorescence spectrum in earlier time region (0 - 25 ps) exhibits a well-defined, structured spectrum (curve 1 in Figure 2) with a mirror image of the absorption, while after 50 ps it shows a broad band centered at 469 nm (curve 2 in Figure 2).

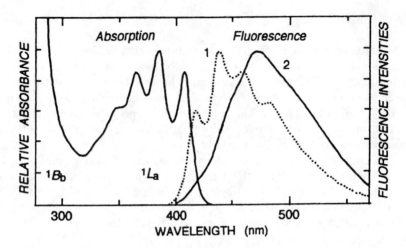

Figure 2. Absorption and fluorescence spectra of DTA in THF. Curve I is the time resolved fluorescence spectrum in 0 - 25 ps, and curve 2 is the spectrum after 1 ns.

From these observations, we assigned the broad red-shifted band to an excimer fluorescence and the structured band to a monomer one. In the present study, we are concerned with the time region shorter than 10 ps where we can confine the fluorescence decay dynamics to those of monomeric anthracene under the energy transfer interaction with another anthracene ring.

3.2 Fluorescence Decays and Anisotropy Decays

Picosecond fluorescence decay curve depends on the monitoring wavelength; the decay at 5 00 mm is almost single exponential ($\tau = 3.2$ ns), while that at 43 0 nm is biexponential ($\tau = 25$ ps and 3.2 ns). Note that the time-resolved fluorescence spectra of DTA, as mentioned above, show the spectrum of anthracene monomer in a time region of 0 - 20 ps. The lifetime of 25 ps corresponds to the transition from monomer to excimer, and that of 3.2 ns corresponds to the deactivation of the excimer state to the ground state of monomer.

Ferntosecond fluorescence decays of DTA in THF at 296 K are shown in Figure 3. DTA in THF was excited with a laser pulse (pulsewidth 70 fs, $\Delta v = 211$ cm^{-1}) at around the 0 - 0 band of $^{1}L_a$ absorption transition, 408 nm and 415 nm. When the excitation is performed at the center of 0-0 band and the fluorescence is monitored at 450 nm, the fluorescence decay exhibits oscillation in the initial time region of < 3 ps (Figure 3a). Monitoring of the fluorescence at longer wavelength gives no oscillation on the decay curve (Figs. 3b and 3c). The decay curve shown in Figure 3a was analyzed as a superposition of oscillating components and exponential decays.

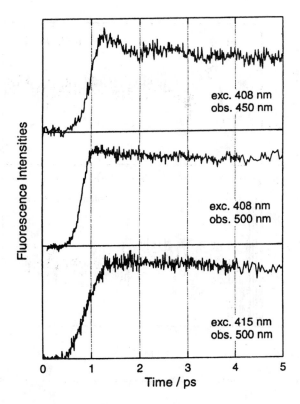

Figure 3. Fluorescence decay curves of DTA in TBF, obtained in excitation at 0-0 band of ^1L_a absorption and in different monitoring wavelengths.

The experimental curve was fitted to the following equation:

$$\rho(t) = \sum_i \alpha_i e^{-\frac{t}{\tau_i}} \cos\left(\frac{2\pi}{T_i}t + \delta_i\right) + \sum_j A_j e^{-\frac{t}{\tau_j}} \tag{1}$$

The exponential decay term, the second term of Eq. (1), can be adapted with the two exponentially decaying components with lifetimes of 25 ps and 3.2 ns obtained from the picosecond time-resolved measurement mentioned before. The oscillating components corresponding to the first term of Eq. (1) is shown in Figure 4a.

The Fourier analyses revealed three components, as listed in Table 1. Figure 4b shows the fluorescence anisotropy decay; much simpler damped oscillation appears with a period of 1.2 ps and decay constant of oscillation of 1.5 ps. This oscillating component appears commonly in the isotrpic decay and the anisotropy decay.

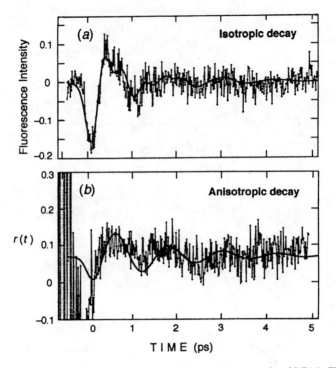

Figure 4. Isotropic fluorescence decay (a) and anisotropy decay (b) of DTA in THF.

4. DISCUSSION

Usually, molecules in condensed phase undergo solute-solvent interaction, which will change chaotically with time the electronic transition frequenciy and the energy transfer interaction. The randam and chaotic modification of the excitation energy transfer interaction causes the energy transfer to change from an ordered (coherent) to a chaotic (incoherent) process. Taking these dephasing processes into account, the experimental results can be analyzed based on a scheme of the interaction and the relaxation processes as shown in Figure 5.

The most straightforward method for analysis of the energy transfer dynamics coupled to a heat bath is to use the density matrix. The time evolution of the density matrix elements ρ_{11}, and ρ_{22}, population of states I and 2 respectively, can be expressed by using a Bloch equation.:

$$\Delta \ddot{n} + \left(\frac{1}{T_1} + \frac{1}{T_2'} \right) \Delta \dot{n} + \left(4\beta^2 + \frac{2}{T_1 T_2'} \right) \Delta n = 0 \qquad (2)$$

where Δn *is* the population difference $\Delta n = \rho_{11}, - \rho_{22}$. β is the resonance transfer inter-action energy, the phenomenological decay times T_1, for populations and T_2 for coherences are fundamental parameters for a two-level system coupled to a bath. $1 / T2 = 1 / (2T1) + 1 / T_2'$, where T_2' is the pure dephasing time (i.e., the decay of ρ_{12}),

The solution of this equation is obtained in the case of underdamped condition, $2\beta T'$ > 1 as follows:

$$\Delta n(t) \propto e^{-\left(\frac{1}{T_1}+\frac{1}{T_2'}\right)t} \cos(\omega_{osc}t + b), \qquad \omega_{osc} = \sqrt{4\beta^2 - \left(\frac{1}{T_2'}\right)^2} \qquad (3)$$

From the experimental values for w_{osc} and $(T_1)^{-1} + (T_2')^{-1}$ in Eq. (3), the values of T_2' and β for each component were obtained as summarized in Table 1.

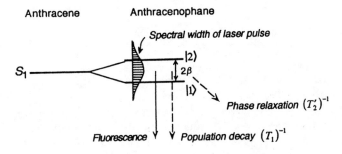

Figure 5. Diagramatic illustration of excitation-pulse width, resonance energy transfer interaction and relxation processes in the S_1 excited state of DTA.

Table 1. Components of damped oscillation in fluorescence isotropic and anisotropy decays of DTA

Components	Isotropic decay			Anisotropy decay		
	Amplitude	Period (ps)	Dephasing time T_2' (ps)	Period (ps)	Dephasing time T_2' (ps)	Interaction energy 2β (cm^{-1})
1	3.0	1.21	1.5	1.2	1.5	35.4
2	2.8	0.86	1.0			51.2
3	1.7	0.46	0.7			86.8

The fluorescence anisotropy exhibits a damped oscillation which is much simpler than those of the isotropy decay, and it can be fitted well with almost single oscillating component with a period of 1.2 ps (Figure 4b). One should note that the value of anisotropy r(t) is negative at t = 0 and then it converges on r(t) = + 0.1 after damping oscillation. Theoretical studies on time-resolved fluorescence anisotropy from a pair of chromophores coupled by an energy transfer interaction were reported.[8-11] According to Hochstrasser et al.[10], the anisotropy is expressed in the following equation:

$$r(t) = \frac{S_{\parallel}(t) - S_{\perp}(t)}{S_{\parallel}(t) + 2S_{\perp}(t)} = \frac{1}{10}\left\{1 + 3f(t) + 3e^{-\gamma t}\right\} \tag{4}$$

where

$$f(t) = \left[\frac{\gamma}{2\Omega}\sin(\Omega t) + \cos(\Omega t)\right]e^{-(2\Gamma + \gamma/2)t} \tag{5}$$

In the case of a pair of two chromophores,

$$r_D(t) = \frac{r(t) + \frac{1}{10}[3 + e^{-\gamma t} - 3f(t)]\cos^2\theta}{1 + e^{-\gamma t}\cos^2\theta} \tag{6}$$

where θ is the angle between the transition dipole moments of two chromophores.

Since θ is near 90° and cos θ ~ 0 in the present case of DTA, Eq. (4) can be used suitably in analyzing the present experiments. For the case that the pumped and probed transition dipoles are perpendicular as shown in Figure 6, the calculation of Eq. (4) leads to be r = - 0.5 at very short time before dephasing, r = - 0.2 after dephasing and r = 0.1 after population equalization. The time-dependent anisotropy (Figure 4b) corresponds to this assumed case.

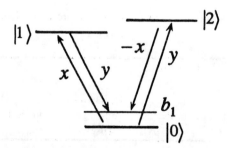

Figure 6. Energy level scheme with symmetries of states and directions and singns of transition dipole moments in the frame of a molecule with D2 symmetry.
The x- and y-axes are taken to be orthogonal to the D2 rotation axis and the z-axis coincides with it. The directions of transition dipole moment are determined VX/b by the symmetries of the vibrational state of the final electronic state.

The vibrational state which gives the transition dipoles perpendicular to the absorption transition should have a b_1 symmetry. From the MO calculation, DTA (D_2 symmetry) has several normal vibrations belonging b_1 symmetry in a range of 1100 - 1500 cm^{-1}. Therefore, the probed fluorescence is due to the transition from a level of ^1L_a (0-0) to a vibronic level of the ground state with one vitrational quantum of b_1 symmetry.

Among three oscillating components in the isotropic decay, only the component I appears on the anisotropic decay. Note that the anisotropy beating behavior is fundamental for observation of the recurrence of energy transfer between two coupled chromophores. Thus the component I is due to the excitation recurrence, and remaining two oscillatiing components arise from the vibrational coherence. It follows that the resonance energy transfer interaction 2β is ' 35 cm^{-1}, and the electronic dephasing time T_2' is 1.5 ps in DTA.

Hochstrasser et al.[5] reported the coherent excitation transfer in 2,2'-binaphthyl (CCl$_4$ solution at room temperature) probed by the polarized transient absorption. They derived the parameters of coherent dynamics from the damped oscillating anisotropy; $T_2' = 0.2$ ps and $2,6 = 41$ cm^{-1}. The dephasing time in DTA is longer by a factor of 7 than that of binaphthyl. Significantly longer dephasing time of DTA can be interpreted as arising from a rigid conformation of DTA, as is seen from the structure (Figure 1), which reduces thermal fluctuation of conformation and therefore increases the electronic dephasing time.

5. REFERENCES

1. Th. Förster, in *Modern Quantum Chemistry*, edited by O. Sinanoglu, Part III, *Action of Light and Organic Crystals*, Academic Press, New York (1965), pp. 93-137.
2. I. Yamazaki, S. Akimoto, T. Yamazaki, H. Shiratori, and A. Osuka, *Acta Phys. Polonica A* **95**, 105 (1999).
3. S. Akimoto, T. Yamazaki, 1. Yamazaki, A. Nakano, and A. Osuka, *Pure Appl. Chem.* **71**, 2107 (2000).
4. Y. P. Kim. P. Share, M. Pereira, M. Sarisky, and R. M. Hochstrasser, *.J. Chem. Phys.* **91**, 7557 (1989).
5. F. Zhu, C. Galli, and R. M. Hochstrasser, *J Chem. Phys.* **98**, 1042 (1993).
6. Y. Sakata, T. Toyoda, T. Yamazaki and 1. Yamazaki, *Tetrahedron Lett.* **33**, 5077 (1992).
7. M. Kuritani, Y. Sakata, F. Ogura and M. Nakagawa, *Bull. Chem. Soc. Jpn.* **46**, 605 (1973).
8. K. Wynne and R. M. Hochstrasser, *Chem. Phys.* **171**, 179 (1993).
9. K. Wynne, S. Gnanakaran, C. Galli, F. Zhu, and R. M. Hochstrasser, *J. Luminesc.,* **60/61**, 735 (1994).
10. K. Wynne and R. M. Hochstrasser, *J. Raman Spectrosc.* **26**, 561 (1995).
11. A. Matro and J. A. Cina, *J. Phys. Chem.* **99**, 2568 (1995).

BIODEGRADABLE BLOCK COPOLYMERS, STAR-SHAPED POLYMERS, AND NETWORKS *VIA* RING-EXPANSION POLYMERIZATION

Hans R. Kricheldorf,[*] Sven Eggerstedt, Dennis Langanke, Andrea Stricker, and Björn Fechner

1. INTRODUCTION

Over the past decades biodegradable (or more precisely "resorbable") polyesters have found rapidly increasing interest of scientists and chemical companies. Among the numerous potential applications which are under investigations certain medical and pharmaceutical applications have already proven their usefulness, for instance: devices for controlled drug release,[1-4] resorbable medical sutures,[4-6] resorbable and transparent wound dressings,[7,8] tissue separating film for surgery,[9] resorbable rods, pins, and screws for the fixation of bone fractures,[10] scaffolds for surgery,[4] and cell containers for osteosyntheses.[4] This situation has stimulated intensive research activities in the field of synthesis and characterization of biodegradable polyesters.

The most widely used strategy for the preparation of biodegradable polyesters consists of the ring-opening polymerization of lactones and cyclic diesters such as glycolide or lactide. These cyclic monomers can be polymerized via four different mechanisms: anionic mechanism, cationic mechanism, coordination-insertion mechanism and enzymatic catalysis. Numerous catalysts and initiators including various enzymes were tested with regard to their usefulness.

* Institut für Technische und Makromolekulare Chemie, Universität Hamburg, Bundesstr. 45, D-20146 Hamburg, Germany.

Advanced Macromolecular and Supramolecular Materials and Processes
Edited by K. Geckeler, Kluwer Academic/Plenum Publishers, 2003

207

In the present work we report on a new synthetic approach which is based on the "ringexpansion polymerization" of lactones and cyclic diesters by means of cyclic tin alkoxides as initiators. Both the initiation and the propagation steps follow the pattern of the coordinationinsertion mechanism known from non-cyclic tin alkoxides (and other covalent metal alkoxides) (Scheme 1).[11,12] "Ring-expansion polymerization" initiated by cyclic dibutyltin alkoxides proved to be particularly useful for preparative purposes, because ring-opening polymerization may be combined *in situ* with a broad variety of condensation reactions.[13]

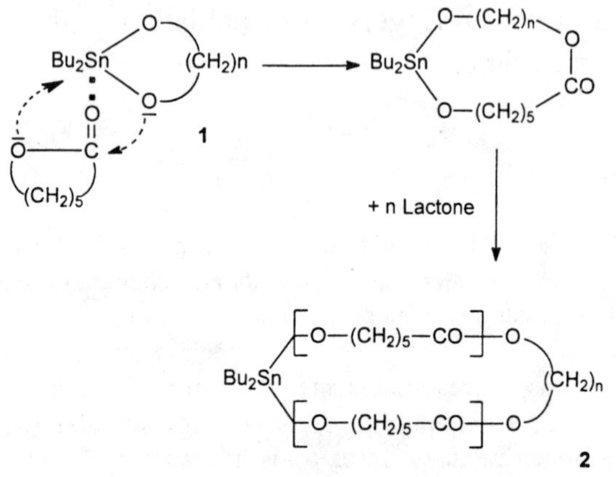

Scheme 1

2. RESULTS AND DISCUSSION

2.1. Syntheses of Block Copolymers

The ring-expansion polymerization (REP) with cyclic alkoxides, such as 1 in Scheme 1, can be performed in such a way that the molecules weight of the resulting cyclic polylactones 2 parallel to the monomer/initiator ratio (M/I). Since the Sn-bonds are still reactive after complete conversion of the monomer (M_1), another monomer M_2 may be added and a cyclic block copolymer (e.g. 3) will be formed, if transesterification can be avoided.

As illustrated in Scheme 2 for ε-caprolactone (ε-CL) and β-D,L-butyrolactone (β-D,L-Bu) the sequential copolymerization may be followed by treatment with 1,2-dimercaptoethane which yields a triblock polymer with free OH endgroup (4) without side reactions.[14] Another advantage of this approach is the synthesis of an isomeric triblock copolymer 5 by the reverse sequence of the monomer addition.

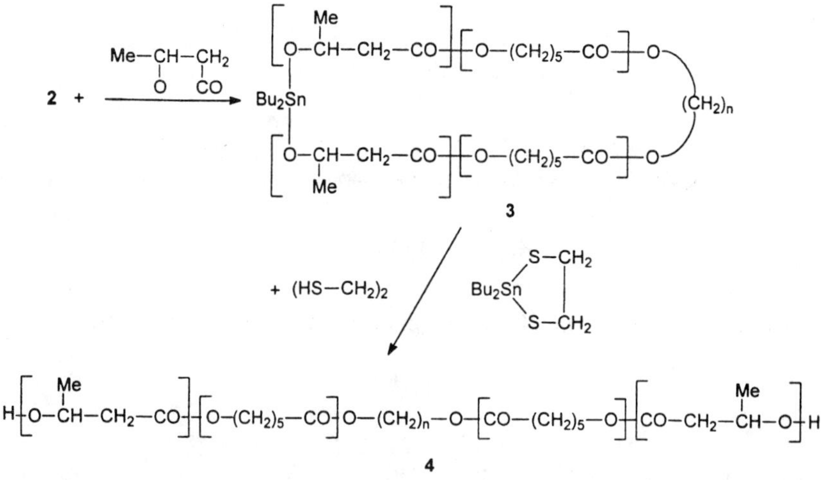

$$\text{2} + \begin{array}{c}\text{Me}-\text{CH}-\text{CH}_2\\ \ \ \ \ \ | \ \ \ \ \ |\\ \ \ \ \ \ \text{O} \ \ \ \ \text{CO}\end{array} \xrightarrow{\text{Bu}_2\text{Sn}} \text{3}$$

Me—CH—CH₂ with O and CO bridge; product **3**:

$$\begin{bmatrix}\text{O}-\overset{\text{Me}}{\text{CH}}-\text{CH}_2-\text{CO}\end{bmatrix}\begin{bmatrix}\text{O}-(\text{CH}_2)_5-\text{CO}\end{bmatrix}\text{O}$$
$$\text{Bu}_2\text{Sn} \qquad (\text{CH}_2)_n$$
$$\begin{bmatrix}\text{O}-\text{CH}-\text{CH}_2-\text{CO}\\ \quad | \\ \quad \text{Me}\end{bmatrix}\begin{bmatrix}\text{O}-(\text{CH}_2)_5-\text{CO}\end{bmatrix}\text{O}$$

3

$$+ \ (\text{HS}-\text{CH}_2)_2 \qquad \text{Bu}_2\text{Sn}\begin{array}{c}\text{S}-\text{CH}_2\\ | \\ \text{S}-\text{CH}_2\end{array}$$

$$\text{H}\begin{bmatrix}\text{O}-\overset{\text{Me}}{\text{CH}}-\text{CH}_2-\text{CO}\end{bmatrix}\begin{bmatrix}\text{O}-(\text{CH}_2)_5-\text{CO}\end{bmatrix}\text{O}-(\text{CH}_2)_n-\text{O}\begin{bmatrix}\text{CO}-(\text{CH}_2)_5-\text{O}\end{bmatrix}\begin{bmatrix}\text{CO}-\text{CH}_2-\overset{\text{Me}}{\text{CH}}-\text{O}\end{bmatrix}\text{H}$$

4

Scheme 2

An interesting and important aspect of this approach is the risk of transesterification causing a more or less complete randomization of the initially block sequence. The risk of transesterification does not only increase with higher temperatures and longer reaction times, it also depends on the sequence of the monomer addition. For instance, it is easier to avoid transesterification during the synthesis of **5** than during the synthesis of **4**. The reason is, that during the second polymerization step in the synthesis of **4** the active chain end has the choice either to attack β-D,L-Bu (which is less reactive than ε-CL) or the ε-CL units in the first block. Apparently, the difference in the reactivity of both electrophilic reaction partners is relatively small. In the second stage of the synthesis of **5** the active chain end has the choice to attack either the highly reactive c-CL or the far less reactive poly(β-D,L-)blocks, so that little transesterification occurs.

Similar observations were made for sequential copolymerizations of trimethylene carbonate (TMC) with L-lactide and various lactones.[15] The isomeric triblock copolymers **6** and **7** were obtained with little transesterification. However, when ε-CL was used as comonomer, a perfect triblock copolymer (structure **8**) was only formed with TMC as the first monomer and ε-CL as M$_2$. The reverse addition caused intensive transesterification. When L-lactide was polymerized first, no block copolymer was obtained, because the Sn-O-CH group of lactide was not reactive enough to attack the TMC. The reverse sequence of monomer addition yielded the cyclic block copolymer **9** without any transesterification. Treatment with 1,2-dimercaptoethane yielded an A-B-A triblock copolymer with two OH endgroups.[15] However, in this case, another synthetic approach was also explored namely the in situ ringopening polycondensation with a dicarboxylic acid dichloride (Scheme 3).

By means of sebacoyl chloride the multiblock copolyester **10** was obtained which contains amorphous poly(TMC) blocks with a low glass transition temperature (T$_g$), and crystalline poly(L-lactide) blocks. Consequently, stress-strain measurements revealed the typical mechanical properties of a (biodegradable) thermoplastic elastomer.[16]

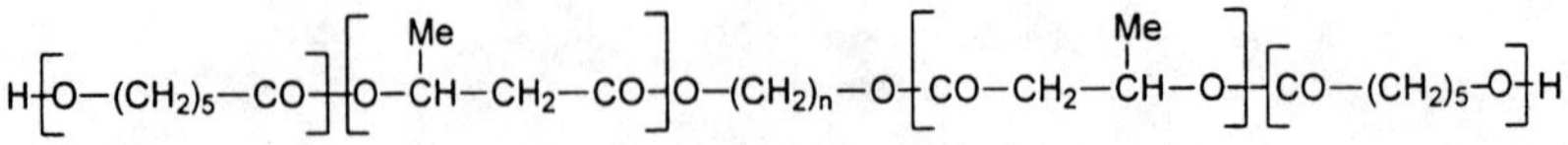

$$H-[O-(CH_2)_5-CO]-[O-\overset{Me}{\underset{|}{CH}}-CH_2-CO]-[O-(CH_2)_n-O]-[CO-CH_2-\overset{Me}{\underset{|}{CH}}-O]-[CO-(CH_2)_5-O]-H$$

5

$$H-[O-\overset{Me}{\underset{|}{CH}}-CH_2-CO]-[O-(CH_2)_3-O-CO]-[O-(CH_2)_n-O]-[CO-O-(CH_2)_3-O]-[CO-CH_2-\overset{Me}{\underset{|}{CH}}-O]-H$$

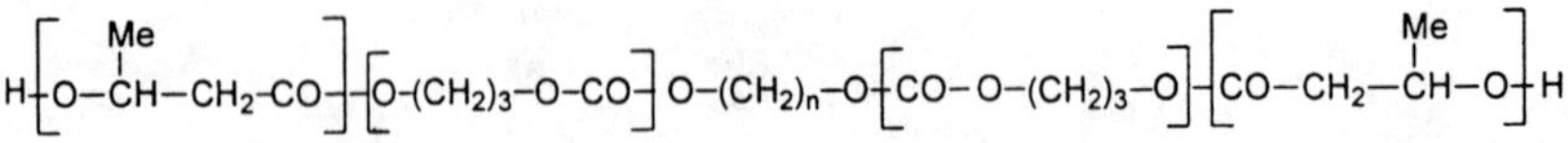

6

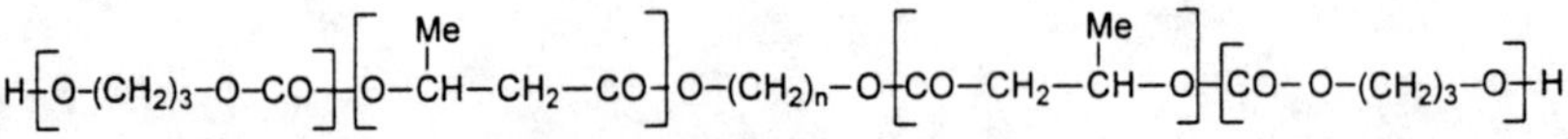

$$H-[O-(CH_2)_3-O-CO]-[O-\overset{Me}{\underset{|}{CH}}-CH_2-CO]-[O-(CH_2)_n-O]-[CO-CH_2-\overset{Me}{\underset{|}{CH}}-O]-[CO-O-(CH_2)_3-O]-H$$

7

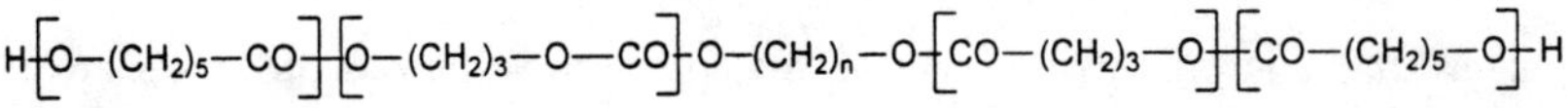

$$H-[O-(CH_2)_5-CO]-[O-(CH_2)_3-O-CO]-[O-(CH_2)_n-O]-[CO-(CH_2)_3-O]-[CO-(CH_2)_5-O]-H$$

8

$$Bu_2Sn\begin{bmatrix} [O-\overset{Me}{\underset{|}{CH}}-CO]-[O-(CH_2)_3-O-CO]-O \\ [O-\underset{|}{\overset{}{CH}}-CO]-[O-(CH_2)_3-O-CO]-O \\ \quad\quad Me \end{bmatrix}(CH_2)_n \quad + ClCO(CH_2)_8-COCl$$

9

$$\downarrow \; -Bu_2SnCl_2$$

$$\left(\begin{matrix} -CO-(CH_2)_8-CO-[O-\overset{Me}{\underset{|}{CH}}-CH_2-CO]-[O-(CH_2)_5-O-CO]-O \\ [O-\underset{|}{\overset{}{CH}}-CH_2-CO]-[O-(CH_2)_3-O-CO]-O \\ \quad\quad Me \end{matrix} \right)(CH_2)_n$$

10

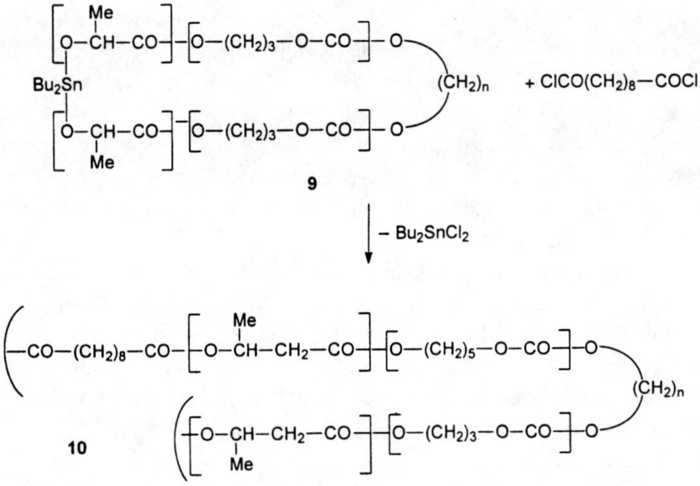

Scheme 3

Another approach yielding A-B-A triblock copolymers or multiblock copolymers is based on the finding that (poly)condensations of $Bu_2Sn(OMe)_2$ with oligo- or polyether-diols produces cyclic oligo- or polyethers (12) and never tin-containing polyethers.[17,18] These macrocycles can again serve as initiators of a ring-expansion polymerization yielding the cyclic block copolymers of structure 13. Elimination of the Bu_2Sn group with 1,2-dimercaptoethane liberates the triblock copolymers 14 (Scheme 4).

Using poly(tetrahydrofuran) diols as starting materials the triblock copolymers 16 and with β-(ω-bis(hydroxypropyl) oligosiloxanes the triblock copolymers 16 were prepared.[19] The tin-containing cyclic block copolymer, such as 13, are again reactive enough to enable ring opening polycondensations with dicarboxylic acid dichlorides. In this way, the multiblock copoly(ether ester) 17 was prepared from a commercial poly-(tetrahydrofuran)diol and ε-CL.[18] This multiblock copolymer shows again the typical mechanical properties of a thermoplastic elastomer.

Finally, the synthesis of A-B-A–triblock-copolymers containing polypeptide blocks (20, Scheme 7) should be mentioned. Starting out from the cyclic poly(E-caprolactone)s of structure 2 [20] the telechelic polylactones 18 were prepared by in situ condensation with 4–nitrobenzoylchloride. Hydrogenation yielded 4-aminobenzoyl endgroups (19) which served as initiators for the ring-opening polymerization of (x-amino acid N-carboxy-anhydrides.[21] Another series of triblock-copolymers was prepared from polylactones having two L-alanine chain ends (21).[22]

2.2. Star-Shaped Polylactones

Condensation of $Bu_2Sn(OMe)_2$ with pentaerythritol or its hydroxyethyl derivatives yield spirocycles such as 22, 23, or 24 and not gels. However, the spirocyclic tin alkoxide (22) polymerizes irreversibly upon cooling of the hot reaction mixture, and is thus not easy to handle and not attractive for preparatives purposes.[23] In contrast, the hydroxylated derivatives 23 and 24 are stable and easy to handle in inert solvents. They may serve as initiator for lactones yielding spirocyclic polylactones (25, Scheme 5). Addition of an excess of carboxylic acid chlorides allows an in situ synthesis of star-shaped polylactones having functional endgroups (26).[24] The average lengths of the star arms can be controlled via the monomer/initiator ratio. A perfect ftinctionalization of all four star arms is easier to achieve, when the spiroinitiar 24 is used for the ring-expansion polymerization.

Biodegradable gels were prepared in three ways. Firstly, the spirocyclic polylactones of structure 25 were polycondensed with various dicarboxylic dichlorides whereby gels in yields of 60-75% (after extraction) were isolated.[25] A similar strategy was followed when methyl α-D-glucoside was used as starting material for the synthesis of the initiator. The tetraacetate was reacted with $Bu_2Sn(OMe)_2$ whereby under elimination of methyl acetate the tricyclic crystalline tin derivative 27 was formed. At temperatures $\leq 80°C$ only the six-membered "tin-ring" incorporates lactones (Scheme 6), so that the tricyclic poly-lactones of structure 28 were obtained.

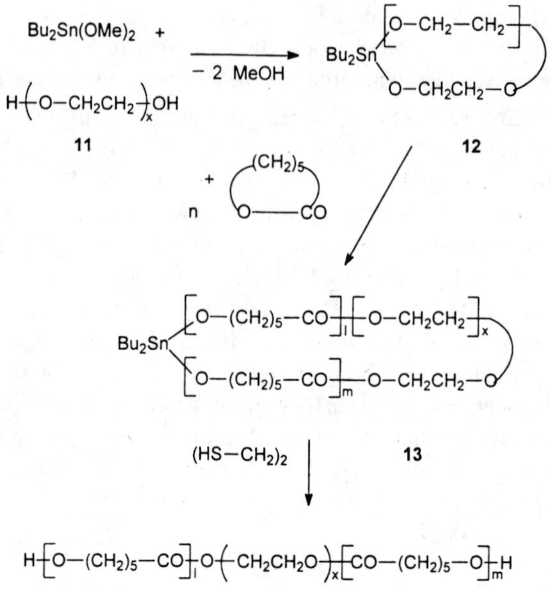

Bu$_2$Sn(OMe)$_2$ +

$\xrightarrow{\quad - 2 \text{ MeOH} \quad}$

H$\left[\text{O}-\text{CH}_2\text{CH}_2\right]_x$OH

11

$$\text{Bu}_2\text{Sn} \begin{array}{c} \left[\text{O}-\text{CH}_2-\text{CH}_2\right] \\ \text{O}-\text{CH}_2\text{CH}_2-\text{O} \end{array}$$

12

$$+ \quad n \begin{array}{c} (\text{CH}_2)_5 \\ \text{O} \longrightarrow \text{CO} \end{array}$$

$$\text{Bu}_2\text{Sn} \begin{array}{c} \left[\text{O}-(\text{CH}_2)_5-\text{CO}\right]_l \left[\text{O}-\text{CH}_2\text{CH}_2\right]_x \\ \left[\text{O}-(\text{CH}_2)_5-\text{CO}\right]_m \text{O}-\text{CH}_2\text{CH}_2-\text{O} \end{array}$$

13

(HS$-$CH$_2$)$_2$ $\quad\downarrow$

H$\left[\text{O}-(\text{CH}_2)_5-\text{CO}\right]_lO\left(\text{CH}_2\text{CH}_2\text{O}\right)_x\left[\text{CO}-(\text{CH}_2)_5-\text{O}\right]_m$H

14   (n = l + m)   Scheme 4

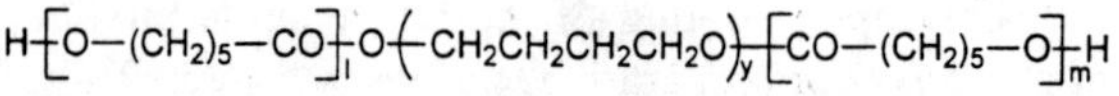

H$\left[\text{O}-(\text{CH}_2)_5-\text{CO}\right]_lO\left(\text{CH}_2\text{CH}_2\text{CH}_2\text{CH}_2\text{O}\right)_y\left[\text{CO}-(\text{CH}_2)_5-\text{O}\right]_m$H

15

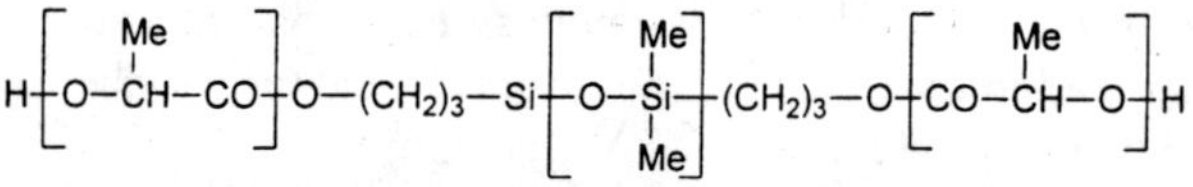

$$\text{H}\left[\text{O}-\overset{\text{Me}}{\underset{\mid}{\text{CH}}}-\text{CO}\right]\text{O}-(\text{CH}_2)_3-\text{Si}-\left[\text{O}-\overset{\text{Me}}{\underset{\text{Me}}{\text{Si}}}\right](\text{CH}_2)_3-\text{O}\left[\text{CO}-\overset{\text{Me}}{\underset{\mid}{\text{CH}}}-\text{O}\right]\text{H}$$

16

$$\left(\text{CO}-(\text{CH}_2)_8-\text{CO}\left[\text{O}-(\text{CH}_2)_5-\text{CO}\right]\text{O}\left[(\text{CH}_2)_4-\text{O}\right]_x\left[\text{CO}-(\text{CH}_2)_5-\text{O}\right]\right)$$

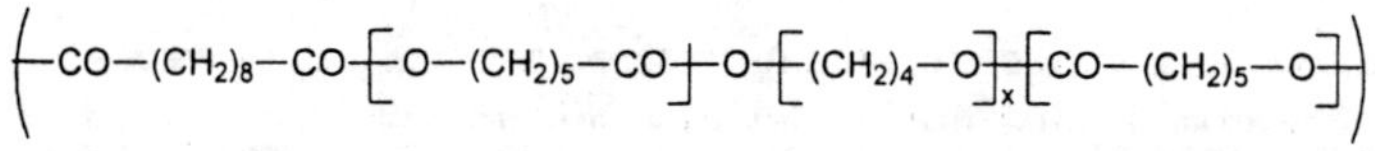

17

Scheme 4

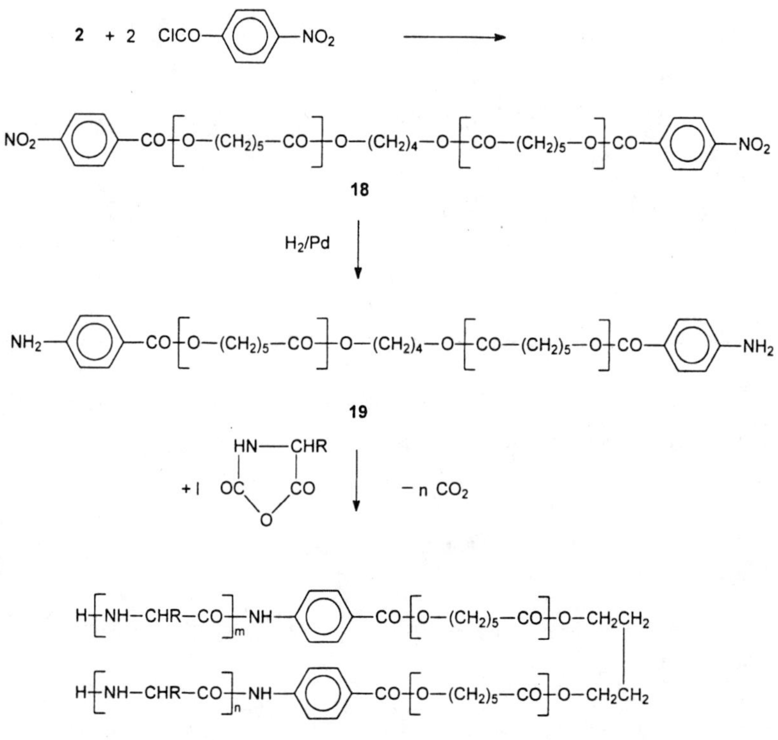

2 + 2 ClCO—⟨benzene⟩—NO$_2$ ⟶

NO$_2$—⟨benzene⟩—CO—[O—(CH$_2$)$_5$—CO]—O—(CH$_2$)$_4$—O—[CO—(CH$_2$)$_5$—O]—CO—⟨benzene⟩—NO$_2$

18

H$_2$/Pd ↓

NH$_2$—⟨benzene⟩—CO—[O—(CH$_2$)$_5$—CO]—O—(CH$_2$)$_4$—O—[CO—(CH$_2$)$_5$—O]—CO—⟨benzene⟩—NH$_2$

19

+l HN——CHR / OC   CO \ O   ↓ − n CO$_2$

H—[NH—CHR—CO]$_m$—NH—⟨benzene⟩—CO—[O—(CH$_2$)$_5$—CO]—O—CH$_2$CH$_2$

H—[NH—CHR—CO]$_n$—NH—⟨benzene⟩—CO—[O—(CH$_2$)$_5$—CO]—O—CH$_2$CH$_2$

20

Scheme 5

Me
NH$_2$—CH—CO—[O—(CH$_2$)$_5$—CO]—O—(CH$_2$)$_4$—O—[CO—(CH$_2$)$_5$—O]—CO—CH—NH$_2$
                                                                                    Me

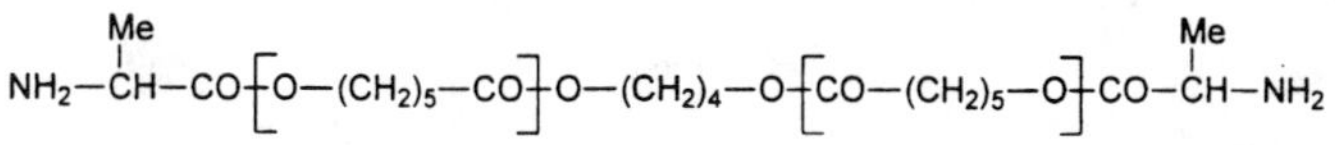

21

Scheme 5

2.3. Biodegradable Networks

In situ-polycondensation with sebacoyl chloride yielded the desired gels.[26] As evidenced by ^{1}H-NMR spectroscopy and by the extent of swelling, the segment length of these gels depend on the M/I ratio used for the ring-expansion polymerization. Furthermore, it was observed for all these gels that the T$_g$ and the melting temperature show the expected dependence on the segment length.

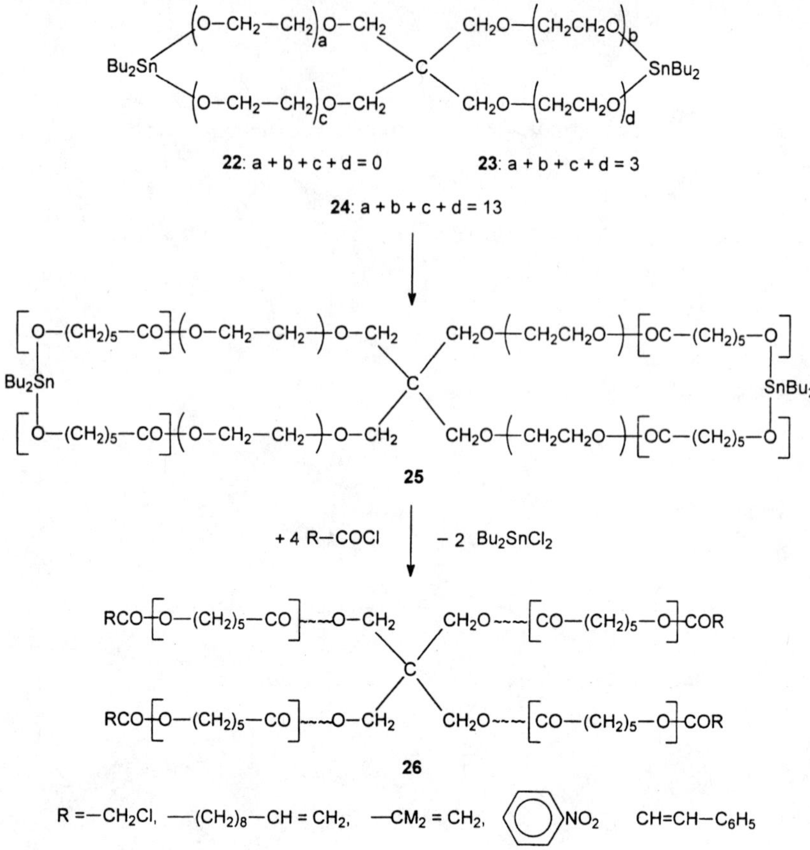

Bu_2Sn $\left(O-CH_2-CH_2\right)_a O-CH_2$ $CH_2O-\left(CH_2CH_2O\right)_b$

C

$\left(O-CH_2-CH_2\right)_c O-CH_2$ $CH_2O-\left(CH_2CH_2O\right)_d$ $SnBu_2$

22: $a + b + c + d = 0$ **23**: $a + b + c + d = 3$

24: $a + b + c + d = 13$

$Bu_2Sn \left[O-(CH_2)_5-CO\right]\left(O-CH_2-CH_2\right)O-CH_2$ $CH_2O-\left(CH_2CH_2O\right)\left[OC-(CH_2)_5-O\right]$

C

$\left[O-(CH_2)_5-CO\right]\left(O-CH_2-CH_2\right)O-CH_2$ $CH_2O-\left(CH_2CH_2O\right)\left[OC-(CH_2)_5-O\right] SnBu_2$

25

$+ 4\ R{-}COCl \quad | \quad -2\ Bu_2SnCl_2$

$RCO\left[O-(CH_2)_5-CO\right]{\sim}O-CH_2$ $CH_2O{\sim}\left[CO-(CH_2)_5-O\right]COR$

C

$RCO\left[O-(CH_2)_5-CO\right]{\sim}O-CH_2$ $CH_2O{\sim}\left[CO-(CH_2)_5-O\right]COR$

26

$R = -CH_2Cl,\ -(CH_2)_8-CH=CH_2,\ -CM_2=CH_2,$ ⬡NO_2 $CH=CH-C_6H_5$

Scheme 6

Another approach, which was elaborated in detail, is based on the ring-expansion polymerization of lactones and D,L-lactide by the cyclic initiator **1**. When the cyclic polylactones of structure **2** are condensed (*in situ*) with tri- or tetrafunctional acid chlorides, a network is formed under elimination of Bu_2SnCl_2 (Scheme 7).

Due to the aromatic protons the complete reaction of trimesoyl chloride can be detected by ^{1}H-NMR spectroscopy, so that the control of the segment length *via* the M/I ratio can be checked in the aftermath. Using D,L-lactide as monomer in combination with an aliphatic trifunctional acid chloride networks completely composed of resorbable building blocks were synthesized (Scheme 8).[27]

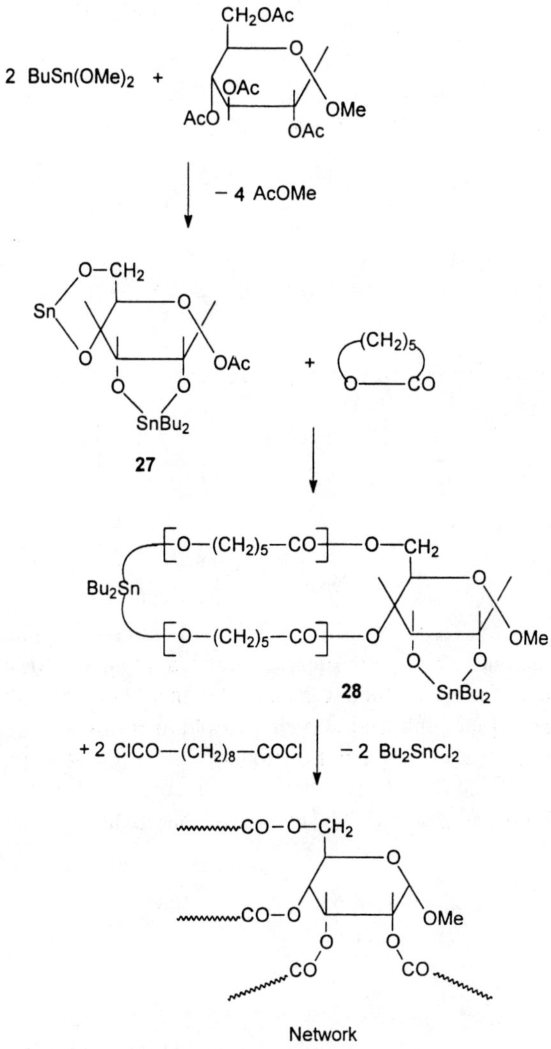
CH2OAc
2 BuSn(OMe)2 +
OAc
AcO
OMe
OAc
− 4 AcOMe
O−CH2
Sn
OAc
+
(CH2)5
O
CO
SnBu2
27
O−(CH2)5−CO
O−CH2
Bu2Sn
OMe
SnBu2
28
+ 2 ClCO−(CH2)8−COCl
− 2 Bu2SnCl2
CO−O−CH2
CO−O
OMe
CO
CO
Network

Scheme 7

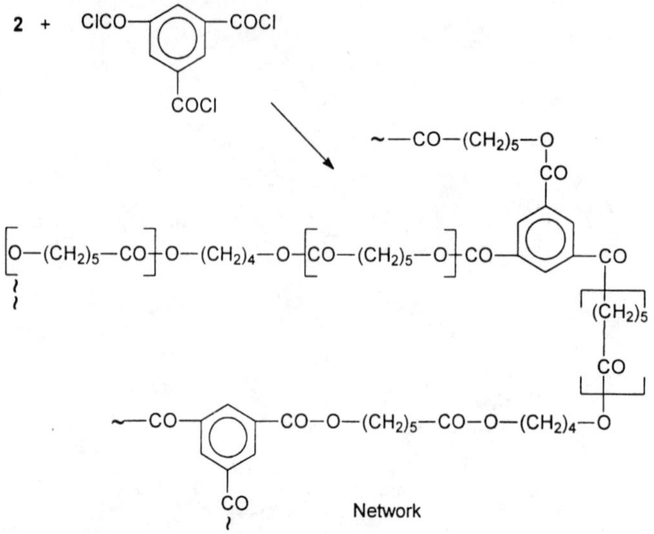

Scheme 8

3. CONCLUSION

The results summarized in this review demonstrate that ring-expansion polymerizations of lactones and cyclic diesters initiated by cyclic tin alkoxides offer an easy access to a broad variety of biodegradable polymers such as triblock copolymers, multiblock copolymers, functionalized star-shaped polymers, and networks. Particularly useful and attractive is the *in situ* combination of ring-expansion polymerization and (poly)condensation. This approach is certainly versatile enough to allow a broader variation of architecture and properties of biodegradable materials than discussed in this review.

4. REFERENCES

1. R. G. Sinclair, *Env. Sci. Techn.* **7**, 955 (1973).
2. S. Yolles, *Degradable Polymers for Sustained Drug Release*, in *Drug Delivery Systems* (R. L. Juliano, Ed.), Oxford Univ. Press, 1980.
3. N. B. Graham, D. A. Wood, *Macromolecular Biomaterials* (G. W. Hartings, and P. Ducheym, Eds.), CRC Press, Boca Raton, USA, 1984.
4. G. B. Kharas, *Polymers of Lactic Acid in Plastics from Microbes* (D. P. Mobley, Ed.), Hanser Publ., München, Wien, New York, 1994, Chapter 4.
5. A. Thiede, and B. Lünstedt in *Degradation Phenomena of Polymeric Biomaterials* (H. Planck, M. Dauner, and M. Renardy, Eds.), Springer Publ., Berlin, Heidelberg, New, York, (1999), p. 133-152.

6. T. Nakamura, Y. Shimizu, T. Matsui, N. Okumura, S.H. Hyon, and K. Nishiya, in *Degradation Phenomena of Polymeric Biomaterials* (H. Planck, M. Dauner, and M. Renardy, Eds.), Springer Publ., Berlin, Heidelberg, New York (1999), p. 153 –162.

7. Ch. Jürgens, H. R. Kricheldorf, H.-R. Kortmann, and N. Langenbecks, *Arch. Chir. Suppl.* 611 (1991).

8. H.R. Kricheldorf, I. Kreiser-Saunders, Ch. Jürgens, and D. Wolter, *Macromol. Symp.* **103**, 85 (1996).

9. Ch. Jürgens, H. R. Kricheldorf, and I. Kreiser-Saunders, Ger. Offen. 19600095 Al (1996).

10. P. Rokkanen, et al., *The Lancet*, 1422 (1985).

11. H. R. Kricheldorf, M. Berl, and N. Scharnagl, *Macromolecules* **21**, 266 (1988).

12. H. R. Kricheldorf and S. Eggerstedt, *Macromol. Chem. Phys.* **199**, 283 (1998).

13. H. R. Kricheldorf, *Macromol. Rapid Commun* **21**, 528 (2000).

14. H. R. Kricheldorf and S.-R. Lee, *Macromolecules* **28**, 6718 (1995).

15. H. R. Kricheldorf and A. Stricker, *Macromol. Chem. Phys* **200**, 1726 (1999).

16. H. R. Kricheldorf and O. Petermann, in preparation.

17. H. R. Kricheldorf and D. Langanke, *Macromol. Chem. Phys.* **200**, 1174 (1999).

18. H. R. Kricheldorf and D. Langanke, *Macromol. Chem. Phys.* **200**, 1183 (1999).

19. H. R. Kricheldorf and D. Langanke, in preparation.

20. H. R. Kricheldorf and K. Hauser, *Macromolecules* **31**, 614 (1998).

21. H. R. Kricheldorf and K. Hauser, *Biomacromolecules*, submitted (Polylactones 55.).

22. H. R. Kricheldorf and K. Hauser, in preparation.

23. H. R. Kricheldorf and S.-R. Lee, *Macromolecules* **29**, 868 (1996).

24. H. R. Kricheldorf and B. Fechner, *J. Polym. Sci., Part A, Polym. Chem.*, submitted.

25. H. R. Kricheldorf and B. Fechner, in preparation.

26. H. R. Kricheldorf and A. Stricker, *Macromolecules* **33**, 696 (2000).

27. H. R. Kricheldorf and B. Fechner, *Macromolecules*, in press.

AMPHIPHILIC POLYETHERS OF CONTROLLED CHAIN ARCHITECTURE

Andrzej Dworak,[*] Wojciech Walach, Barbara Trzebicka, Agnieszka Kowalczuk, Marzena Nowicka, and Justyna Filak

1. INTRODUCTION

The amphiphilic polymers, polymers which contain in their macromolecules both hydrophilic and hydrophobic units[1], are the base for valuable materials due to their diversified interaction with liquids. So they may act as emulsifiers, compatibilizers, "smart" materials (responding to external stimuli)[2] and many others. In order to control their properties, the hydrophilic – hydrophobic balance in the macromolecules has to be controlled. This balance depends not only upon the constitution of the chain repeating units, but also upon the art of their distribution in the chains (copolymers of controlled unit sequences), the chain topology, the size of the chains and many others. So a careful engineering of the macromolecules is necessary to obtain desired material functions.

Polyethers, especially the simplest one, the poly(ethylene oxide), are often used as building blocks of amphiphilic structures. Poly(ethylene oxide)[3] has many valuable properties, but also the disadvantage of being difficult to crosslink and to functionalize, as its chain backbone does not contain any functional groups.

Here, we want to report about the synthesis and some properties of amphiphilic polyether macromolecules of controlled chain architecture, based upon a hydrophilic analog of ethylene oxide, the 2,3-epoxypropanol-1, the glycidol.

[*] A. Dworak, B. Trzebicka, W. Walach: Polish Academy of Sciences, Institute of Coal Chemistry, 44-121 Gliwice, Poland;
A. Kowalczuk, M. Nowicka: Silesian University of Technology, Faculty of Chemistry, 44-121 Gliwice, Poland;
A. Dworak, J. Filak: University of Opole, Institute of Chemistry, 54-052 Opole, Poland.

Advanced Macromolecular and Supramolecular Materials and Processes
Edited by K. Geckeler, Kluwer Academic/Plenum Publishers, 2003

219

2. LINEAR POLYGLYCIDOL *VIA* LIVING ANIONIC POLYMERIZATION

Glycidol(2,3-epoxypropanol-1) contains a hydroxyl group next to the oxirane ring. If transferred to the macromolecule, these hydroxyl groups would make the polymer reactive, without lowering its hydrophilicity. It is however rather difficult to control the polymerization of glycidol itself. The chain branching is intrinsic to this process. In the cationic polymerization[4,5], the presence of the hydroxyl groups induces the chain growth to proceed via the activated monomer rather then via the active chain end mechanism, i.e. the activated (protonated) monomers are added to the hydroxyl groups present in the macromolecules, which causes branching. In the anionic process[6], the fast exchange of protons between the hydroxyl groups and the alcoholate anions multiplies the active sites, which leads to branched products. Moreover, the process is rather difficult to control, except when highly branched, dendrimer-like macromolecules are aimed at[7]. The molecular mass remains low.

In order to obtain linear polyglycidol chains of controlled length the hydroxyl group of the monomer has to be protected. Fitton[8] described the transformation of this hydroxyl group into an acetal group in a reaction of the monomer with ethyl vinyl ether. Spassky[9] later has shown that this monomer may be polymerized, and the protective group may be removed without any significant degradation of the polyether chain, otherwise not very stable under acid conditions (Scheme 1).

We were able to show that the polymerization of the polyglycidol so protected may be carried out in homogenous system.[10] The process is close to living. Molecular masses up to $2 \cdot 10^4$ may easily be obtained. The molecular mass distribution is relatively narrow, as indicated by the SEC chromatography (Figure 1).

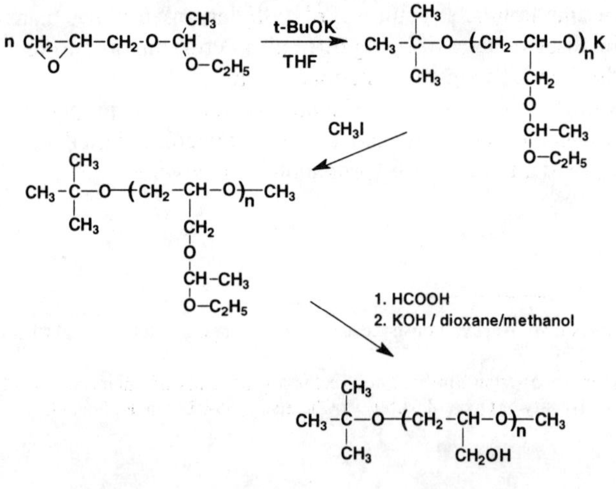

Scheme 1. Anionic polymerization of glycidol acetal and hydrolysis to linear polyglycidol.

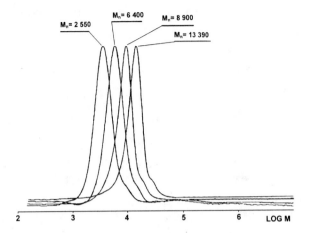

Figure 1. GPC traces (RI response, THF) of poly(glycidol acetal)[13] obtained via anionic polymerization according to scheme 1. Reprinted from Macromolecular Symposia 153, (2000) p. 233-242, Copyright 2000, with permission from Wiley-VCH.

These conditions for living polymerization of protected glycidol we applied for the subsequent synthesis of amphiphilic and hydrophilic macromolecular structures of controlled constitution and topology.

3. AMPHIPHILIC MACROMONOMERS CONTAINING STYRENE DOUBLE BOND AND HYDROPHILIC POLYGLICYDOL CHAINS – GRAFT- AND COMB-LIKE POLYMERS

The living character of the anionic polymerization of glycidol with protected hydroxyl group makes possible the synthesis of macromonomers via the termination method, which contain a polymerizable styrenic double bond and a strongly hydrophilic, polyglycidol chain. To achieve this, the polymerization of glycidol acetal was initiated with potassium t-butoxide and terminated with p-chloromethyl styrene[10] (Scheme 2). After removal of the protecting groups polyglycidol macromonomers were obtained of varying length of the hydrophilic tail.

The ^{1}H-NMR spectra yield proof that the termination of the living chains with p-chloromethyl styrene is quantitative within the limits of error.

Both series of macromonomers, the acetal macromonomers (**a** in Scheme 2) and the glycidol macromonomers (**b** in Scheme 2) were copolymerized with styrene. When co-polymerizing the macromonomer with protected hydroxyl group, styrene copolymers are obtained which contain up to 70% of polyether units. The protecting groups may be removed, which yields polystyrene–graft–polyglycidol.

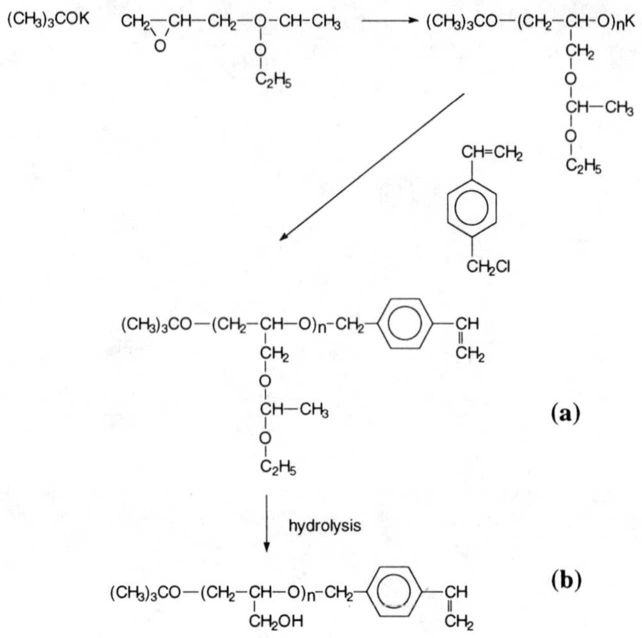

$(CH_3)_3COK$ $CH_2\!-\!CH\!-\!CH_2\!-\!O\!-\!CH\!-\!CH_3$ $\longrightarrow$ $(CH_3)_3CO\!-\!(CH_2\!-\!CH\!-\!O)_nK$

$CH_2\overset{\diagdown O \diagup}{}CH$ ... O ... C_2H_5

(scheme)

$(CH_3)_3CO\!-\!(CH_2\!-\!CH\!-\!O)_n\!-\!CH_2\!-\!\langle\bigcirc\rangle\!-\!CH$

(a)

hydrolysis

$(CH_3)_3CO\!-\!(CH_2\!-\!CH\!-\!O)_n\!-\!CH_2\!-\!\langle\bigcirc\rangle\!-\!CH$ (b)

Scheme 2. Synthesis of polyglycidol macromonomer.

The polyglycidol macromonomer (**b** in Scheme 2) acts as emulsifier. 0.1 weight-% of this macromonomer is sufficient to obtain a stable water emulsion of styrene. The emulsion is stable up to a content of the macromonomer of 10 wt.%. When a free radical initiator (sodium peroxy disulfate) is introduced, emulsion copolymerization of styrene with the macromonomer is initiated. The macromonomer is being incorporated into the polystyrene chains, thus acting as a surfmer – surfactant – monomer. The conversion of the macromonomer falls with its increasing concentration and varies from 60 to 90%.

Under proper conditions, the emulsion copolymerization of the glycidol macro-monomer with styrene leads to polystyrene microspheres[*]. Very uniform microspheres of a number average diameter D_n varying from 200 to over 600 nm and of narrow diameter distribution are obtained[11] (Table 1).

The polyglycidol is incompatible with polystyrene. The result is that the polyglyci-dol chains are expelled from the inside of the polystyrene microspheres, resulting in an outer surface, which is enriched with polyglycidol chains and their hydrophilic, reactive functional group. X-ray photoelectron spectroscopy evidences that the concentration of the polyglycidol units on the outer surface is up to an order of magnitude higher then the

---

[*] The studies of microspheres were carried out in cooperation with. T. Basinska and S. Slomkowski, Polish Academy of Sciences, Centre of Molecular an Macromolecular Studies, Lodz, Poland, who performed the characterization of the microspheres.

average content of these units in the copolymer. The hydrophilicity of the outer sphere greatly and favorably reduces the physical absorption of proteins (e.g. human serum albumin, HSA) on the surface. The concentration of physically adsorbed HSA on these microspheres in ca. 10 times lower, then the absorption of this protein on polystyrene microspheres synthesized without the polyglycidol macromonomer[11].

Table 1. Polystyrene–block–polyglycidol microspheres obtained by the emulsion copolymerization

M_n Macromonomer	[Styrene]/[Macromonomer] in feed	D_n (nm)	D_w/D_n
950	20	281	1.08
950	10	264	1.05
950	7	216	1.10
3000	100	651	1.01
3000	50	465	1.01
3000	20	353	1.02

4. HYDROPHILIC BLOCK COPOLYMERS OF ETHYLENE OXIDE AND GLYCIDOL AND HYDROPHILIC NETWORKS

Poly(ethylene oxide) is a highly hydrophilic polymer. Its networks are highly swelling gels, able to absorb large amounts of water. However, due to the lack of functional groups it is difficult to crosslink such polymer, in general, radiation or photochemical methods are used.

The living character of the anionic polymerization of glycidol initiated with alkali metal alcoholates opens the route to the synthesis of block copolymers, containing the central poly(ethylene oxide) block flanked on both sides with hydrophilic, reactive polyglycidol chains, as presented in Scheme 3.

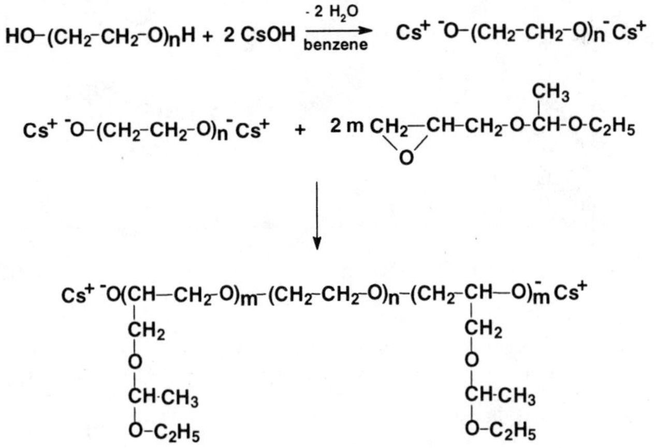

$$HO-(CH_2-CH_2-O)_nH + 2\ CsOH \xrightarrow[benzene]{-\ 2\ H_2O} Cs^+\ ^-O-(CH_2-CH_2-O)_n^-\ Cs^+$$

$$Cs^+\ ^-O-(CH_2-CH_2-O)_n^-\ Cs^+ \quad + \quad 2\ m\ CH_2-CH-CH_2-O-\overset{\overset{\displaystyle CH_3}{|}}{CH}-O-C_2H_5$$

Scheme 3. Synthesis of polyglycidol–block-poly(ethylene oxide)–block–polyglycidol.

After removal of the acetal group, triblock copolymers polyglycidol–block–poly(ethylene oxide)–block–polyglycidol containing central poly(ethylene oxide) blocks from 45 to 230 units and flanking polyglicydol chains of DP varying from 25 to 150 are obtained[12].

They are all water soluble in any ratio. The reactive hydroxyl groups were crosslinked with glutar aldehyde, yielding networks. These networks are highly swelling in water and in methanol. The equilibrium degree of swelling (Figure 2) depends upon the length of the poly(ethylene oxide) blocks – the longer the PEO blocks, the higher the swelling. Equilibrium swelling degrees over 3000% may be obtained.

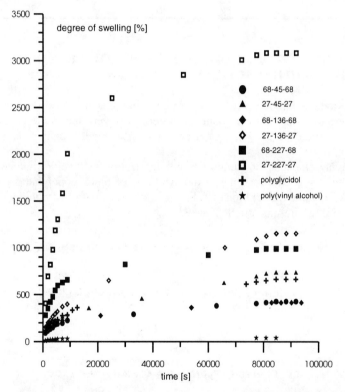

Figure 2. Swelling in water of polyglycidol–block–poly(ethylene oxide)–block polyglycidol[12]. The numbers at symbols denote the block length. Reprinted from Reactive and Functional Polymers 42, A. Dworak, G. Baran, B, Trzebicka, W. Walach, "Polyglycidol-block-poly(ethylene oxide)-block-polyglycidol: synthesis and swelling properties"p. 31-36, Copyright 1999, with permission from Elsevier Science,

5. AMPHIPHILIC BLOCK COPOLYMERS OF STYRENE AND GLYCIDOL

Carbanions are known to be efficient initiators of the polymerization of oxiranes and lead in many cases to the living growth of the polyether chains. This and the above evidenced fact of the living anionic polymerization of glycidol acetal make the synthesis of the block copolymers possible, which consist of the central, strongly hydrophobic polystyrene block and two hydrophilic, reactive polyglycidol chains.

The polymerization of styrene was initiated with potassium naphthalenide. This process is known to generate living polystyrene dianion. The polystyrene dianion initiates the polymerization of glycidol acetal, leading after hydrolysis to three-block copolymers (Scheme 4). [13]

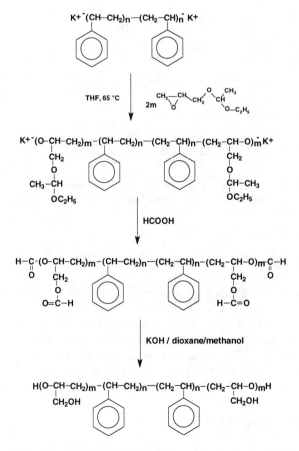

Scheme 4. Synthesis of polyglycidol–block–polystyrene–block-polyglycidol.

After removal of the protecting groups, block copolymers containing polystyrene blocks of DP = 280 to 770 flanked with two polyglycidol chains of DP = 80 to 250 result.

The macromolecules of block copolymers of very different block philicity, consisting of hydrophilic and hydrophobic blocks, tend to organize themselves in water or organic media to form micelles, where the blocks, which are compatible with the solvent, form the outside shell, keeping the micelle in solution. The micelle formation of the studied copolymers is clearly evidenced by the NMR spectra (Figure 3).

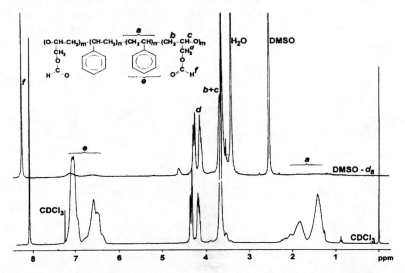

Figure 3. ^{1}H NMR spectra of polystyrene–polyglycidol copolymers[13] in DMSO and in CDCl$_3$. Reprinted from Macromolecular Symposia 153, (2000) p. 233-242, Copyright 2000, with permission from Wiley-VCH.

DMSO is a non-solvent for polystyrene, however the poly(glycidol formate) blocks are soluble in this solvent. The block copolymers are apparently soluble both in DMSO and in chloroform. In DMSO, however, the insolubility of the PS blocks makes the copolymer chains to aggregate to form micelles.

The core of these micelles consists of insoluble PS-blocks. The micelles are kept in solution by the soluble polyether chains. The interior of the micelles cannot be penetrated by dimethyl sulfoxide. The segments in the core are therefore very crowded, the 1H spin–lattice relaxation time becomes very short and the NMR line very broad, so that they cannot be seen in ^{1}H-NMR spectra.

6. ACKNOWLEDGEMENT

This work was in part supported by the Polish Committee of Scientific Research, grant no. 3T09A04518.

7. REFERENCES

1. R. S. Velichkova and D. C. Christova, *Progr. Polym. Sci.* **20**, 819 (1995).
2. C. L. McCormic (Ed.), *Stimuli Responsive Water Soluble and Amphiphilic Polymers*, ACS Symposium Series, Vol. 780, American Chemical Society, Washington DC, 2001.
3. F. E. Bailey and J. V. Koleske, *Polyethylene Oxide*, Academic Press, New York, 1976.
4. R. Tokar, P. Kubisa, S. Penczek, and A. Dworak, *Macromolecules* **27**, 320 (1994).
5. A. Dworak, W. Walach, B. Trzebicka, and Dworak, *Macromol. Chem. Phys.* **196**, 1963 (1995).
6. E. J. Vandenberg, *J. Polym. Sci. Polym. Chem.* **23**, 915 (1985).
7. A. Sunder, R. Hanselmann, H. Frey, and R. Muhlhaupt, *Macromolecules* **30**, 5602 (1999).
8. A. Fitton, J. Hill, D. Jane, and R. Miller, *Synthesis*, 1140 (1987).
9. D. Taton, A. Leborgne, M. Sepulchre, and N. Spassky, *Macromol. Chem. Phys.* **195**, 139 (1994).
10. A. Dworak, I. Panchev, B. Trzebicka, and W. Walach, *Polymer Bull.* **40**, 461 (1998).
11. T. Basinska, S. Slomkowski, A. Dworak, I. Panchev, and M. Chehimi, *J. Colloid Interf. Sci.*, in print.
12. A. Dworak, G. Baran, B. Trzebicka, and W. Walach, *React. and Funct. Polym.* **42**, 31 (1999).
13. A. Dworak, I. Panchev, B. Trzebicka, and W. Walach, *Macromol. Symp.* **153**, 233 (2000).

BLENDS OF WASTE POLY(ETHYLENE TERE-PHTHALATE) WITH POLYSTYRENE AND POLYOLEFINS

K. P. Chaudhari and D. D. Kale*

1. INTRODUCTION

The usage of plastics is increasing very rapidly and hence problems associated with its waste management are becoming increasingly important. Due to the nonbiodegradable nature of plastics, the landfilling cannot be an attractive solution. In addition, cost of landfill and the stringent legislation[1,2] make it more difficult to use waste plastics in landfill. In the packaging industry and many other end-use applications the waste generated is a mixture of different polymers, which are mostly noncompatible in nature. The wastes from car dismantling units[3], municipal waste,[4,5] PVC bottles and pipe,[6,7] or refrigerator door lining[8] can be cited as examples, where commingled wastes are generated.

Poly(ethylene terephthalate) (PET), high impact polystyrene (HIPS), polystyrene (PS), and polypropylene (PP) are popular commodity plastics which are used very widely in the manufacture of commercial consumer products. The plastic wastes collected at public amusement parks or resorts consist of PET bottles for softdrinks or mineral water, polystyrene (or HIPS) tea cups, polypropylene caps or straws in softdrink packs, and these products are being carried in polyethylene carry bags. Although it is possible to separate different plastics goods from such waste for mechanical recycling, it would be more attractive if the commingled waste is processed together.

---

* PPV Division, Department of Chemical Technology, University of Mumbai (U.D.C.T.), Matunga, Mumbai 400 019, India.

Advanced Macromolecular and Supramolecular Materials and Processes
Edited by K. Geckeler, Kluwer Academic/Plenum Publishers, 2003

To overcome the noncompatible nature of plastic wastes, the use of compatibilizers becomes essential for mechanical recycling. There are some studies on neat blends of PET-PS,[9-11] PET-PP,[12] PS-PP,[13-14] HIPS-PP,[15] HIPS-PA1010,[16] or PS-PE.[17] Most of these studies have recommended the use of poly(styrene-*co*-maleic anhydride) (SMA) for PET-styrenic blends[9-11,18] as well as poly(propylene-*g*-maleic anhydride) (MA-*g*-PP) or poly (ethylene-*co*-acrylic acid) as compatibilizers for PET-PP or HIPS-PP blends.

Although SMA has been used as a compatibilizer for blends of styrenics with PET [9, 12, 19, 20] or polyamides, [21, 22] Ju and Chang [9] have argued that PET carboxyl terminal groups may not react readily with the anhydride groups of styrene maleic anhydride (SMA), while the reaction with terminal hydroxyl groups of PET may require a suitable catalyst. MA-*g*-PP can react with PET in a similar fashion. Therefore, a catalyst like the multifunctional epoxy will enhance the reaction between reactive groups from SMA and MA-*g*-PP and terminal hydroxyl as well as carboxyl groups of PET. The generalized scheme is shown in Figure 1.

Although blends of virgin PET with PS have received some attention,[9-11] blends of waste PET (bottles) with HIPS, PS, or ternary blends of waste PET with HIPS and PP have not been studied. Similarly, the suitability of a trifunctional epoxy resin instead of tetra glycidyl ether of diphenyl diamino methane monomer can be of interest, as such resins are more readily available and melt blending of resin would be easier as compared to handling the monomer.

The present work deals with the binary blends of waste PET bottles with PS or HIPS and ternary blend of waste PET-HIPS and PP using multifunctional epoxy resin in combination with SMA and MA-*g*-PP as compatibilizers respectively. The mechanical, rheological, and morphological properties of these blends are reported in this work.

2. EXPERIMENTAL

2.1. Materials

Waste, discarded PET mineral water bottles were procured from the Essel World Amusement Park, Mumbai, India. The labels on the PET bottles were removed first and the bottles were then shreaded into small strips, which were washed thoroughly with water and detergent to remove dust and other impurities. After thorough drying at 120°C for 6 hours, the small strips were extruded and pelletized. The processing temperature profile for extrusion was 180, 200, 220, 230, and 245 °C for four zones and the die respectively with screw speed of 60 rpm. The intrinsic viscosity of the waste PET bottles was found to be 0.54 and the number average molecular weight, M_n, was estimated[24] as 31,500 g mol^{-1}.

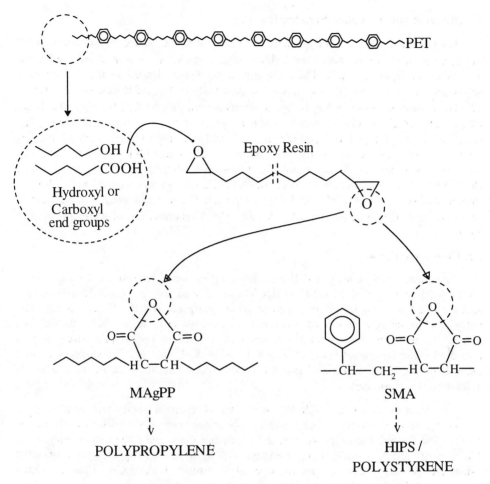

Figure 1. Schematic representation of possible action of compatibilizers.

Commercial grades of HIPS (LG H306) from LG Polymers, Mumbai, and PP (REPOL H100EY) from Reliance Industries Ltd. were used in this work. The melt flow index (MFI) values for these polymers were 6.5 g 10 min at 200°C and 5 kg load, and 9.7 g 10 min at 190°C and 2.16 kg load, respectively.

SMA containing 30.8% maleic anhydride was procured from Supreme Industries Ltd. Mumbai. Trifunctional epoxy resin (Lapox P-304) having an epoxy equivalent of 870-1000 g/eq was procured from Ciba-Atul Ltd., Mumbai. MA-g-PP (OREVAC CA100) from Elf Atochem, France, was also used.

2.2. Blending and Specimen Preparation

Prior to blending, waste PET pellets were dried at 120°C-130°C for six hours, HIPS was dried at 100°C for six hours and SMA as well as epoxy resins were dried at 80°C for four hours in separate ovens. The components were dry blended in the desired ratio before feeding to co-rotating twin screw extruder (MP-19 PC, APV Baker, U.K.; L/D = 29). The compounding extruder had the temperature profile of 190, 210, 230, 245 °C for zones and 248°C for the die while a screw speed of 80 rpm was maintained. The extrudate strand of blend was pelletized, dried, and then the injection molded samples were prepared using a microprocessor based injection moulding machine. The test specimens pertained to ASTM standards, [dumb bell shaped for tensile test (ASTM D638M-91)], notched Izod impact strength (ASTM 256) and rectangular bar for flexural strength measurement (ASTM 790M-92) specimens]. The injection moulding was carried out using the temperature profile 240, 255, and 268 °C and the injection pressure was 85 kg cm^2.

2.3. Characterization

The mechanical (tensile and flexural) properties were determined a using tensile testing machine (LR 50K, LLOYD, U.K.). The cross head jaw speed was 50 mm min for tensile and 2.8 mm min for flexural measurements using a load cell of 5 kN capacity. The notch was cut using a motorized notch cutting machine (Ray Ran U.K.) over the face parallel to the direction of injection moulding. The notched impact strength was measured using an impact tester (AVERY DENISON, U.K.). A 2.7 J striker was used and the striking velocity was 3.46 ms. All results reported in this work are the average of at least eight measurements.

The viscous behavior at 260°C was studied using a rotational, parallel plate viscometer. (RT 10, HAAKE, Germany) The shear rate range was varied from 0.01 s^{-1} to 100 s^{-1}. The fractured samples were etched with chloroform and dried in an oven at 80°C for 12 hours and then these were sputtered with gold and micrographs were obtained using the scanning electron microscope SEM Probe. (CAMECA, France). Other experimental details are reported by Chaudhari.[25]

3. RESULTS AND DISCUSSION

3.1. Binary PET-PS Blends

The mechanical properties of binary blends of waste PET with PS and with HIPS are given in Table 1. Waste PET has better mechanical properties than PS. It is interesting to note that PET rich blends (composition up to 60 %) showed better mechanical properties than those of virgin PS. Properties of these blends were comparable or better than those of waste PET. The % elongation at break was less than that of PS.

Figure 2 shows the rheological properties of waste PET-PS blends. Low viscosity of waste PET can help to obtain better dispersion of two polymers.

Table 1. Mechanical Properties of Binary Blends of waste PET-PS and waste PET-HIPS

Sr. no.	Compositions	Tensile strength (MPa)	Elonga-tion at break (%)	Impact strength (J/m)	Flexural strength (MPa)	Flexural modulus(MPa)
1	Waste PET	48	--	13	--	2413
2	Virgin PS	42	14.8	14	47.	5328
3	Virgin HIPS	23	38.6	87	32	1870
Binary blends of waste PET-PS						
4	Waste PET : PS (60:40)	52	4.3	19	52	2518
5	Waste PET : PS (40:60)	42	2.9	10	39	2725
6	Waste PET : PS (60:40) + 2 phr SMA	46	10.4	13	49	2763
7	Waste PET : PS (60:40) + 2 phr SMA + 0.5 phr Epoxy	55	14.3	18	52	2619
8	Waste PET : PS (40:60) + 2 phr SMA + 0.5 phr Epoxy	42	3.8	13	45	2586
Binary blends of waste PET-HIPS						
9	Waste PET : HIPS (60:40)	38	17.4	16	51	2208
10	Waste PET : HIPS (40:60)	35	14.4	17	37	2165
11	Waste PET : HIPS (60:40) + 2 phr SMA	32	24.8	12	53	2178
12	Waste PET : HIPS (40:60) + 2 phr SMA	31	10.6	15	39	2010
13	Waste PET : HIPS (60:40) + 2 phr SMA + 0.5% Epoxy	38	16.2	15	55	2298

The viscosity of PET-rich, noncompatible blends was comparable with that of waste PET. It is interesting to note that the viscosity of a PS-rich blend was slightly higher than that of PS at lower shear rate (up to 0.5 s^{-1}).

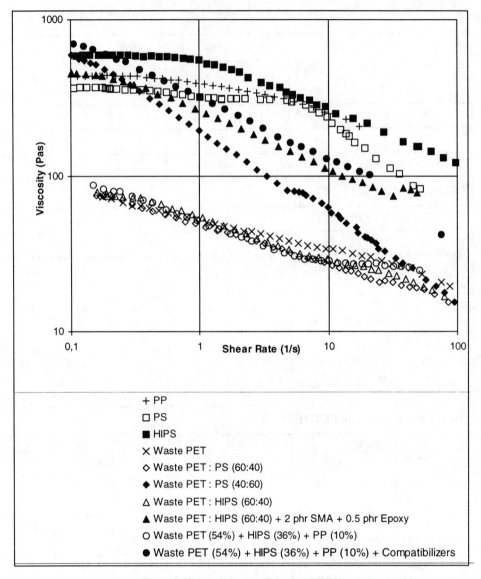

Figure 2. Viscous behavior of blends at 260 °C.

At higher shear rates above 100 s⁻¹, however, the viscosities of all blends were comparable with that of waste PET. These rheological properties, therefore, suggest that processing of blends will not be different from processing of either PS or PET. Extruder employment in blending generates sufficiently high shear rates in the range of 100–500 s⁻¹, which influence the dispersion of one polymer into the other. The processing conditions can also influence the dispersion.[24]

The geometry of the twin screw elements employed in present work provides initial distributive mixing followed by high shearing/stretching. On the other hand, for the PS rich (60 % by weight) noncompatibilized blend, the tensile strength, impact strength and % elongation at break were all lower than that for PET-rich blends. There was a reduction in flexural properties as compared to those for virgin PS. Thus, properties of PS-rich blends were inferior. In PS-rich blends, dispersion of low viscosity PET into high viscosity PS phase would be necessary and processing parameters same as those employed as for PET-rich blends, may not be suitable or effective for a fine dispersion of PET into PS.

Since PS-PET are noncompatible, the effect of SMA and epoxy resin was studied. Addition of 2 phr SMA to a blend containing 40 % PS seems to have a slight increase in the flexural modulus with almost comparable flexural strength, higher % elongation at break but much more reduced impact strength. The combination of SMA and epoxy have given improved properties, for PS-rich blends also.

Figure 3 shows the morphology of these blends. The co-continuous structure having some degree of orientation is evidently seen for PET rich 60:40 blend. However, for the noncompatible 40:60 blend (Figure 3-b) the large droplet size of the dispersed phase was observed and the fibrous structure also seems to be present when SMA and epoxy were added to the PS-rich blend.

3.2. Binary PET-HIPS Blends

Table 1 also compares the properties of PET-HIPS blends. The trend of properties is similar to that for PET-PS blends. PET-rich blends showed better mechanical properties than HIPS except for impact strength. Although HIPS has superior impact strength, the PET-rich PET-HIPS blends did not show a high impact strength. As a matter of fact, PET-PS blends showed a better impact strength than PET-HIPS blends at a given composition. Combination of SMA and epoxy seem to be acting as better compatibilizer than SMA alone. These results are similar to those reported by Ju and Chang[9] for the PET-PS system.

The morphological results shown in Figure 3 elucidate that the effect of combined compatibilizers seem to have resulted in a reduced droplet size of the dispersed phase. The fibrilar morphology present in the PET-PS system was not observed for the PET-HIPS blends. This may be due to the presence of rubber in HIPS.

The rheological characteristics of PET-HIPS blends are also depicted in Figure 2. The trend of rheological changes are similar to that for PET-PS blend. The viscosity of HIPS is higher that that of PS and therefore the viscosity of PET-HIPS blend is higher than that of the corresponding PET-PS system.

3.3 Ternary PET-HIPS-PP Blends

Table 2 compares the properties of ternary blends of PET, HIPS, and PP. The ratio of PET:HIPS was maintained at 60:40. The tensile strength of the blends does not seem to be affected much by the concentration of PP. The tensile strength of the blend seems to be higher than that of HIPS and comparable with that of PP, but less than that of waste PET. The flexural strength of the blend was higher than for the PP and HIPS.

Similarly, the flexural modulus was higher than that of PP and HIPS but lower than that of waste PET. The impact strength of the blend was lower than that for HIPS and PP but better than that for waste PET. The %-elongation at break for all the blends was considerably lower, indicating that the blends were not fully compatible.

Table 2. Mechanical Properties of Ternary Blends of Waste PET, HIPS, and PP *

Sr. no.	Compositions	Tensile strength (MPa)	Elonga-tion at break (%)	Impact strength (J/m)	Flexural strength (MPa)	Flexural modulus (MPa)
1	Waste PET	48	--	13	--	2413
2	Virgin HIPS	23	38.6	87	32	1870
3	Virgin PP	34	553	75	29	843
4	Waste PET : HIPS (60:40)	38	17.4	16	51	2208
5	Waste PET : HIPS (54:36) + 10 % PP	39	6.50	17	41	1863
6	Waste PET : HIPS (48:32) + 20 % PP	37	5.99	18	34	1679
7	Waste PET : HIPS (42:28) + 30 % PP	35	5.65	14	29	1566
8	Waste PET : HIPS (54:36) + 10 % PP + 2 phr SMA + 2 phr MA-g-PP + 0.5 phr Epoxy	45	5.6	26	42	1187
9	Waste PET : HIPS (48:32) + 20 % PP + 2 phr SMA + 2 phr MA-g-PP + 0.5 phr Epoxy	42	4.48	21	37	1238
10	Waste PET : HIPS (42:28) + 30 % PP + 2 phr SMA + 2 phr MA-g-PP + 0.5 phr Epoxy	39	4.42	17	33	1461

* Waste PET: HIPS ratio was maintained at 60:40.

The effect of compatibilizers seems to have improved the tensile strength, impact strength, and flexural strength. However, the flexural modulus has decreased to some extent. Interestingly, the flexural modulus of the compatibilized blend increased as the %-PP increased.

The morphology of ternary blends shows that the noncompatible blends had a higher droplet size of the dispersed phase as compared to compatibilized blends, although both compatible and noncompatible blend showed a fibrous structure of PET. The viscosity of compatibilized ternary blend seems to have increased considerably as compared to that of noncompatibilized blends. The viscosity of ternary blend is still comparable with that of HIPS or PP under normal processing conditions.

4. CONCLUSIONS

Binary blends with PS showed better mechanical properties over blends with HIPS at a given composition. SMA was effective when used in combination with epoxy for the binary blends, especially when either PS or HIPS was a major phase. The fibrilar morphology was observed for PET-PS blends but not for PET-HIPS systems. The compatibilizers were effective for ternary compositions and resulted in an increase of mechanical properties over noncompatibilized ternary blends although these ternary blends showed reduced flexural properties over corresponding binary waste PET-HIPS blends. The viscous behavior of binary or ternary blends was comparable with the polymer of highest viscosity amongst blend constituents, indicating that processability of these blends was similar to individual polymers of the blends. The blends could be injection moulded as well as extruded easily.

5. ACKNOWLEDGEMENT

K. P. Chaudhari is thankful to the University Grant Commission (UGC), India, for providing a Senior Research Fellowship (SRF) during the investigation.

6. REFERRENCES

1. J. Brandrup in *Recycling and Recovery of Plastics*, 1st ed, editor Brandrup J., Hanser Publisher, Munich (1996).
2. J. Schiler, in *Polymer Recycling*, ed 1st editor J. Schiler, John Wiley and Sons, U.K. (1998).
3. X. Liu and H. Bertilsson, *J. Appl. Polym. Sci.* **74**, 510 (1999).
4. C. S. Ha, H. D. Park and W. J. Cho, *J. Appl. Polym. Sci.* **74**, 1531 (1999).
5. M. Xantos, A Patel, Deys, S.S Dagli, C. Jacob, T. J. Nasker, R.W. Ranfree, Adv. *in Poly. Tech.*, **13**, 231 (1994).
6. J. C.Arnold and B. Maund, *Polym. Eng. Sci.* **39**, 1234 (1999).
7. M.A. Wenguang, *J. Appl. Polym. Sci.* **59**, 759 (1996).
8. C. R. Lindsey, J. W. Barlow and D. R. Paul, *J. Appl. Polym. Sci.* **26**, 995 (1981).
9. M. H. Ju and F. C. Chang, *J. Appl. Polym. Sci.* **73**, 2029 (1999).
10. J. S. Lee, K. Y. Park, D.J. Yoo, and K.D. Suh *J. Polym. Sci., Part B: Polym, Phys.*, **38**, 1396 (2000).
11. M. H. Ju and F. C. Chang, *Polymer.* **41**, 1719 (2000).
12. K. H. Yoon, H. W. Lee and O. O. Park, *J. Appl. Polym. Sci.*, **70**, 389 (1998).
13. G. Radonjic, V. Musil and I. Smit, *J. Appl. Polym. Sci.*, **69**, 2625 (1998).

14.   O. O. Santana. and A. J. Muller, *Polym. Bull.*, **32**, 47 (1994).
15.   Z. Horak, V. Fort, D. Hlavata, F. Lednicky, and Vecerka, *Polymer*, **30**, 597 (1994).
16.   G. Chen and J. Liu, *J. Appl. Polym. Sci.*, **74**, 857 (1999).
17.   R. Fayt, R. Jerome, and Ph. Tegssie, *J. Polym. Sci., Part B: Polym. Phys.*, **27**, 775 (1989).
18.   S. Paul and D. D.Kale, *J. Appl. Polym. Sci.*, **80**, 2593 (2001).
19.   K. H. Yoon, H. W. Lee, and O. O Park, *Polymer*, **41**, 4445 (2000).
20.   O. Fukushima, S. Onishi, R. Otsubo, and H. Hayanami , U.S. Patent 3,334,153 (1967).
21.   I. Park, J.W. Barlow, and D.R. Paul, J. *Polym. Sci., Part B: Polym. Phys.*, **30**, 1021 (1992).
22.   F. C. Chang and Y. C. Hw, Polym. *Eng. Sci.*, **31**, 1509 (1991).
23.   W. L. Hergenrother and C. J. Nelson, *J. Polym. Sci., Part : Polym. Chem.*, **12**, 2951 (1974).
24.   L. A. Utracki, Polymer Alloys and Blends – Thermodynamics and Rheology, Hanser, München, Wien, New York (1989).
25.   K. P. Chaudhari., *Studies in Waste Polyesters*, Ph.D. Thesis, University of Mumbai (2001).

THE CHEMICAL PROGRESS OF MULTICOMPONENT REACTIONS

Ivar Ugi[*], Günther Roß, and Christoph Burdack

1. INTRODUCTION

An increasing collection of chemical compounds is now prepared by MultiComponent Reactions (MCRs), since a synthesis by a MCR can be accomplished just by mixing their educts and their yields are usually much higher than by the multistep syntheses that correspond to sequences of many steps. The latter require much preparative work and their yield decreases with each step. With the exception of Passerini's work, in the first century of the MCRs and chemistry of the isocyanides were not combined. These two parts of chemistry were combined in 1959 when the four component reaction of the isocyanides (U-4CR) was introduced.[1] In the usual chemistry, their MCRs are less used than its normal reactions, whereas in the chemistry of the isocyanides more MCRs and their libraries are carried out than those of one or two components.

Recently, the stereoselective one-pot syntheses of peptide derivatives and polycyclic ß-lactam derivatives by the U-4CR were accomplished.[2] The chemistry of the isocyanides, the MCRs and their libraries was for many decades of moderate interest, but since 1995 this chemistry became industrially one of its most active fields, and now there more new compounds are formed by the MCRs than by any other methods. Using automating instruments, up to 20,000 and more compounds of MCRs can be produced by a single person in one day. Recently further progress was achieved, when Weber et al. introduced a new mathematically oriented data based computer program by which it is possible to predict preferred products of MCRs and their libraries, and thus new desirable products can even more efficiently be prepared.

* Technical University of Munich, Department of Organic Chemistry, Chair 1, Lichtenbergstr. 4, D-85747 Garching, Germany

Advanced Macromolecular and Supramolecular Materials and Processes
Edited by K. Geckeler, Kluwer Academic/Plenum Publishers, 2003

2. THE EARLY MCR CHEMISTRY

In 1850 Strecker[3] introduced the first three component reaction (3CR) of type I by converting amines, carbonyl compounds and hydrogen cyanide into the α-aminoalkyl cyanides. These were often hydrolysed into the α-aminoacids. Mannich's 3CR[4] are usually reactions of dialkylamines, formaldehyde and carbon atoms of the aldehydes and ketones. Hellmann and Opitz[5] described in their *α-Aminoalkylierung* book that all of such α-aminoalkylating 3CRs and related reactions belong together. They of them are a part of type I.

The classical heterocycles forming MCRs belong to type II. They all begin with α-aminoalkylating 3CRs are closely related, and their resulting intermediate products react further with bi-functional educts. In 1882 Hantzsch[6] and Radziszewski[7] introduced such syntheses of heterocycles. Their last step is always a practically irreversible ring formation. One decade later Biginelli[8] introduced the related synthesis of heterocycles.

In 1929 Bucherer and Bergs[9] introduced the formation of the heterocyclic hydantoin derivatives in high yields from ammonia, carbonyl compounds, hydrogen cyanide and carbon dioxyde. The hydrolysis of these hydantoins is nowadays the preferred method of preparing the α-aminoacids. The Asinger reaction,[10] the last heterocycle forming MCR, was introduced in 1956. It seems that depending on the structures of educts and their reaction conditions these MCRs belong sometimes to type I, and in other cases to type II.

3. THE FIRST CENTURY OF THE ISOCYANIDE CHEMISTRY

The chemistry of the isocyanides began in 1859[11] when Lieke[12] formed the allylisocyanide from allyliodide and silver cyanide. Eight years later Gautier[13] prepared some further alkylisocyanides. At the same time Hoffmann[14] began to produce a variety of isocyanides from primary amines with chloroform and strong bases. The chemistry of the isocyanides remained as a rather empty part of chemistry. For a whole century only 12 isocyanides were prepared, and whole chemistry of the isocyanides remained a rather empty part of chemistry.[11]

In 1921 Passerini[15-17] introduced the first 3CR of the isocyanides, and for a decade he thoroughly investigated this reaction. Later it was realized that the Passerini reaction is one of the most important reactions of the isocyanides.[17] In 1948 Rothe[11,18] found the first naturally occurring isocyanide in the *Penicillium notatum* and in the *Penicillium chrysogenum*, which was shortly later used as the efficient external antibiotic *Xanthocillin*. In 1956 Hagedorn and Tönjes[19] prepared the dimethyl ether of this di-isocyanide by dehydrating its di-formylamine by phenylsulfochloride in pyridine.

4. A NEW ERA OF THE ISOCYANIDE CHEMISTRY

A new chemistry of the isocyanides began in 1958, when the isocyanides became generally well available by dehydrating the formylamines by arylsulphonylchlories,[19] phosphorus oxychloride,[20,21] phosgene[11] diphosgene,[22] triphosgen[23] in the presence of bases. In the *Isonitrile Chemistry* volume of 1971[11] already 325 isocyanides were mentioned.

In January 1959 the Four Component Reaction (4CR) was introduced. The amines, carbonyl compounds, nucleophilic components (which come usually from acidiccompounds) and the isocyanides react together into their products.[24] Since 1962 this 4CR was increasingly often quoted as the Ugi reaction,[16a] which was later abbreviated as the 4CC,[11] U-4CC[25a] or the U-4CR.[25b]

Usually each type of a chemical reaction forms the products with their characteristic skeletons, and only their external substituents are varied, whereas in this 4CR not only the substituents can differ but also intermediate α-adducts can rearrange into many skeletally different types of products. Their characteristic structural features are primarily determined by the different types of acid components, and also by their amines as educts. Nevertheless, very many products of the U-4CR have the same basic structure, since primary amines and carboxylic acids are their most often used types of educts.

The educts and products of the U-4CR are never sterically totally hindered, particularly when their amines and carbonyl compounds are precondensed.[16,26] Therefore they are more variable than any other chemical reactions which usually have their 'scope and limitations'. The purities and yields of products of the U-4CR and their stereoselektivity depend very much on their reaction conditions, however, in some cases also different products can result if different reaction conditions are used.[27] Usually the U-4CRs, including their stereoselective courses, can only be formed in optimal yields if certain ratios and concentrations of educts are used in suitable solvents, and if reactions are accomplished under well selected thermal conditions.

An U-4CR proceeds faster and forms a higher yield of its products, if the amine and the carbonyl compound are precondensed, and subsequently the other components are added,particularly,if sterically hindered educts are sprecondensed.[26, 28]

In the last few years the chemical industry has realized that many chemical compounds can be be prepared easier and in higher yields by the MCRs than by the usual multistep syntheses. Half of the 20 most often used pharmaceutical products could also be prepared by the U-4CR. Recently, Schreibe[29] has demonstrated how advantageous the syntheses can be if the number of steps is reduced by the U-4CR.

This is illustrated by the recent synthesis of the HIV protease inhibitor Crixivan[30] of the Merck whose synthesis included an U-4CR(Scheme 1). This synthesis proceeds much better than any other way of preparing Crixivan.

5. THE SYNTHESIS OF PEPTIDE DERIVATIVES BY THE U-4CR

When the U-4CR was introduced, it was also realized that thereby also derivatives of chiral α-aminoacids and peptide could thereby be prepared. Already in the 1966 introduced peptide volune of Bodanszky and Ondetti[31] the preparative advantages of the syntheses of peptides by the U-4CR were mentioned.

It was soon realized that this methodology of preparing α-aminoacid derivatives cannot be easily accomplished, but this method would have greaz preparative advantages over the usual multistep syntheses, particularly if non-available α-are needed.

It took almost four decades till the synthesis of peptide derivatives could sufficiently well be prepared by a suitable stereoselective U-4CRs and the necessary subsequent selective deprotection of its product.

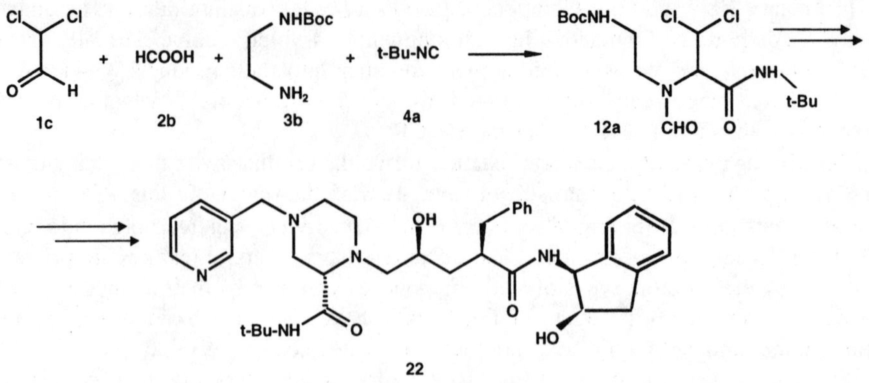

Crixivan (MK 639)

Scheme 1

The first achiral peptide derivatives were prepared by the U-4CR in 1961,[32b] and shortly later stereoselective syntheses of diastereoisomeric products could be formed by the U-4CR from chiral primary amines.[33] The reaction mechanism of an U-4CR in methanol at 0°C was investigated in 1964-1967.[34] There the equilibrium between the chiral Schiff base isobutyraldehyde-(S)-α-phenylethylimine and benzoic acid as educts and their cations and anions was determined, and subsequently the stereoselectivity of the U-4CR of those components the tert.-butylisocyanide was investigated. It was then found that there four different reaction mechanisms compete. Depending on the reaction conditions one or the other chiral product can preferentially be formed.

It was also early recognized that the preparation of peptides from chiral primary amines requires also the removal of the corresponding N-alkyl group of the amine component. The prestudies of the non-trivial cleavage of the corresponding group began in 1964 when alkyl- and alkenyl-groups were used as models.[35]

In 1969 it was recognized that peptide syntheses can potentially be prepared by the U-4CR with a suitable chiral α–ferrocenylalkylamine the amine component.[36] The necessary ferrocen derivatives were then not yet well available, and therefore the necessary chemistry was developed.[26-37]

The U-4CR products of suitable chiral ferrocene derivatives can so be cleaved that chiral ferrocenyl alkyl cations result, which can react with ammonia and form again the initial chiral α–ferrocenylalkylamine.[26]

This chemistry was more than two decades of intensely investigated,[38] and it was ultimately realized that peptide syntheses by the U-4CR with α–ferrocenylalkylamine had too many preparative disadvantages, and therefore this methodology was given up.

It was later realized that the 1-amino-carbohydrates can be prepared by letting the carbohydrates and ammonia them react for a few days.[39] The resulting chiral 1-amino-carbohydrates of xylose, glucose and 2-acetylamino-glucose were thus prepared, however the products of the U-4CR formed the desired products together with hardly separable by-products.[40]

In 1988 Kunz et al.[41] introduced a formation of α-aminoacid derivatives by U-4CRs of O-pivalyl amino-carbohydrate in the presence of Zinc chloride etherate. Such highly stereoselective U-4CRs form often excellent yields of products. However, its disadvantage was that no peptide derivatives can be formed since only formic acid could be used as the acid component of the U-4CR. Furthermore, their U-4CR products can only be cleaved into the α-aminoacids or their amides if their U-4CR products are hydrolyzed by hot concentrated solutions of hydrochloric acid in methanol.

A few years later Goebel et al.[42] formed peptide derivatives by the U-4CR with tetra-O-alkyl-1-glucopyranosylamines. These reactions had the advantage that thus any carboxylic acids could participate in such U-4CRs. However, the auxiliary groups of their products could also not be removed under sufficiently mild conditions.

Lehnhoff et al.[43] used 1-amino-2-deoxy-2-N-acetyl-3,4,6-tri-O-acetyl-ß-D-glucopyranose as the amine component of the U-4CR. Their corresponding peptide derivatives could be prepared in excellent yields, but their auxiliary group could also not easily be removed under sufficiently mild conditions.

Zychlinski et al.[44] prepared the 1-amino-2-deoxy-2-acetamido-3,4,6-tri-O-acetyl-ß-D-glucopyranose by a synthesis of 11. This amine component can very well form peptide derivatives by the U-4CRs. However, its products are so unstable since they are already in neutral media partly cleaved. Therefore this amine component has not yet suitable properties. An 1-amino-2-acylamino-carbohydrate of more stable products of the U-4CR would be needed. Thus more suitable N-acyl group should be used.

An alternative was introduced by Ross[45] when he formed the 1-amino-5-deoxy-5-thio-2,3,4-tri-O-isobutanoyl-ß-D-xylopyranose from xyline by seven preparative steps (Scheme 2). A variety of α-aminoacid and peptide derivatives were formeds by the U-4CR of this amine in the presence of Zinc chloride etherate. All essential components of educts of 12 different types of stereoselective U-4CRs form their products in good yields.

The structure of the U-4CR product N-benzoyl-N-(5-desoxy-5-thio-ß-D-xylopyranosyl)-R-leucine-tert.-butylamine was confirmed by an X-ray determination.[46] Such products are usually rather stable, but in the presence of Hg(AcO)$_2$ and TFA they can selectively be cleaved into the derivative of α-aminoacid or peptide together with its thiocarbohydrates, which can be re-converted into the initial chiral amine.[45]

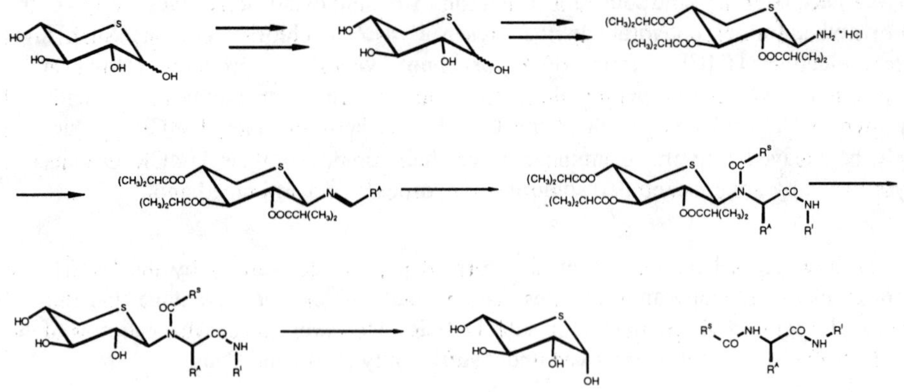

Scheme 2

6. SYNTHESES OF ß-LACTAM DERIVATIVES BY THE U-4CR

In 1961 it was found that the ß-aminoacids, carbonyl compounds and isocyanides form ß-lactam derivatives.[32b] Kehagia, Dömling et al.[41] prepared thus many different types of monocyclic ß-lactams.

The synthesis of polycyclic ß-lactams by the U-4CR began when in 1962 Ugi and Wischöfer[48] prepared various penicillin related bi-cyclic compounds by the U-4CR from thiazole derivatives with carboxylic groups and isocyanides. The resulting intermediate bi-cyclic cis-products with seven-membered rings rearrange into the cis-ß-lactam derivatives.

In 1979 Schütz et al.[49,50] prepared from a new type of multifunctional educts by the U-4CR cis-isomers of penicillin[41] and cephalosporin[50] derivatives. In 1991 Neyer et al.[51] accomplished the synthesis of carbaphenem derivative by an U-4CR with a novel Dieckmann type of reaction and a special version of Wassermann's reaction.[52]

Later several preparative attempts were made to form directly carbaphenems by suitable U-4CR, but this could not successfully be accomplished.[53]

In 2001 Burdack et al.[54] were able to prepare various educts with amino, aldehyde, and carboxylic functional groups that could directly be converted by the U-4CR into bi-cyclic or tri-cyclic carbaphenem derivatives (Scheme 3).

Kametani et al.[55] prepared nitron derivatives and converted these subsequently by further reactions into thienamycin derivatives.

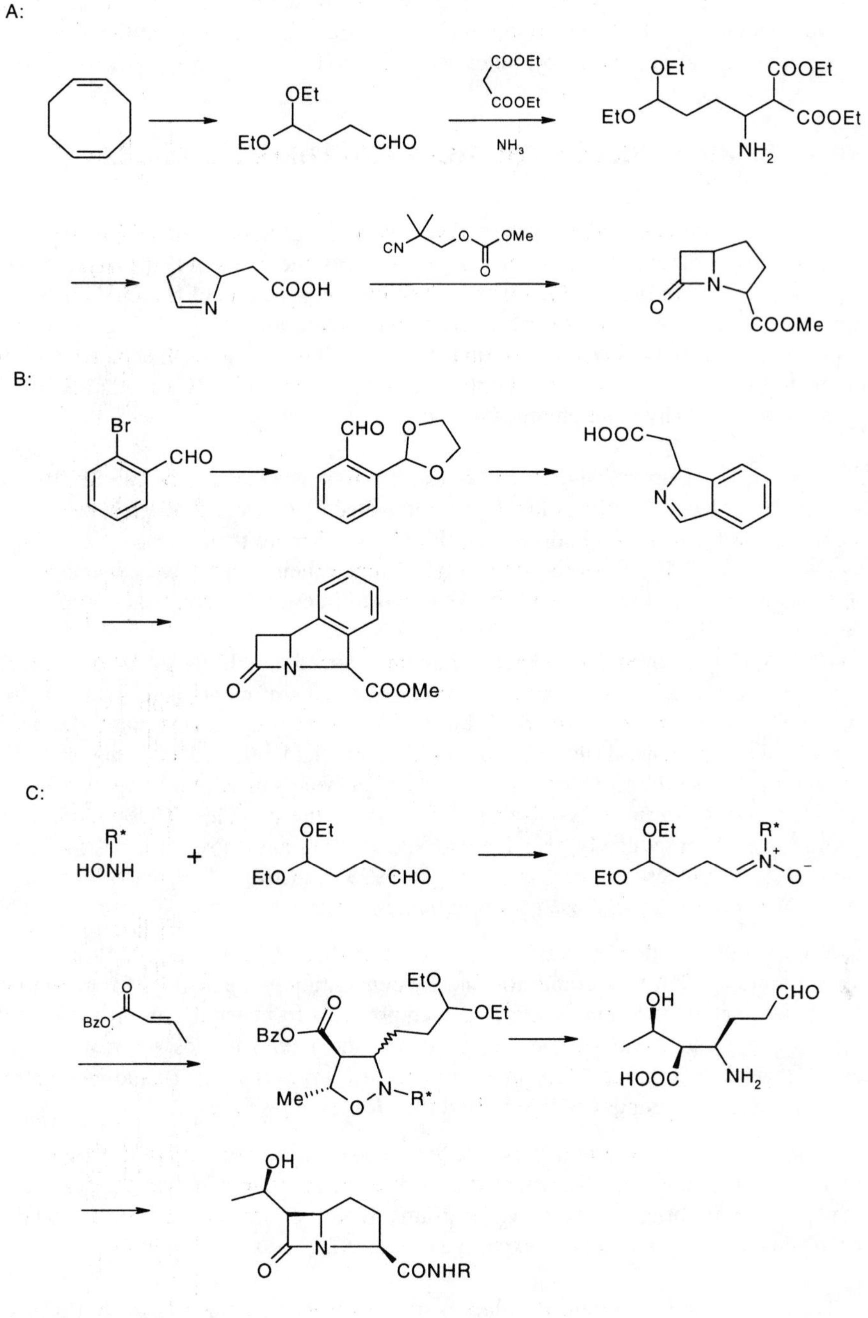
A:
OEt
EtO
CHO
COOEt
COOEt
NH3
OEt
COOEt
EtO
COOEt
NH2
N
COOH
CN
O
OMe
O
N
O
COOMe
B:
Br
CHO
CHO
O
O
HOOC
N
N
O
COOMe
C:
R*
HONH
+
OEt
EtO
CHO
OEt
R*
EtO
N
O
O
BzO
EtO
O
OEt
BzO
Me
N
R*
O
OH
CHO
HOOC
NH2
OH
N
O
CONHR

Scheme 3

Burdack[54] prepared analogously chiral educts with three functional groups via the reductive cleavage of the cycloadduct of benzylcrotonate and the suitable nitron of the (S)-N-(a-phenyl-ethyl)-hydroxylamine with the aldehydes. This multifunctional educt was reacted with the suitable isocyanides according to C.[56]

7. NEW CHEMICAL ERA OF THE MCRS AND THEIR LIBRARIES

For many decades only rather few research groups used chemistry of the isocyanides and their MCR, but in the early 1990s this chemistry and its libraries became one of the most active parts of chemistry.[11] Then many products of the U-4CRs and their unions with further reactions of more than four educts and their libraries were in the industry very intensely formed and investigated, often using sophisticated automatic equipment. Thus more new chemical compounds have been formed by such MCRs of the isocyanides than by any other chemical reaction.

Bossio[57] and Curran[58] introduced a variety of new reactions of the MCRs of the isocyanides. In 1993 Dömling and Ugi[59] published the first MCR of seven different components, and it was then realized that the MCRs of more than four educts are usually unions[60] of the U-4CR and further reactions.[25a] Since their preparative advantages could clearly be recognized, it seems that this event was of interest for many chemists.[61]

In 1961 Ugi and Steinbrückner[32a] had introduced the libraries as collections of different chemical products. These libraries were again mentioned in 1971 in the *Isonitrile Chemistry* volume[11] [*p. 149*], but the importance of this essential idea was for many decades no general interest. In 1963 Merrifield[62] introduced the solid-phase chemistry of the peptides, which was fron then on widely used, and twenty years later Furka[63,64] began to form the solid-phase libraries of the peptides. These libraries were then often used. Subsequently also DNA/RNA libraries and those of products of solid-phase multistep syntheses were produced.[65] In 1995 Armstrong[66] began to form libraries by the U-4CR products and those of subsequent secondary reactions.

For a whole decade a research group of the Hofmann LaRoche A.G. tried to find a thrombin inhibitor, but they could not find by conventional chemistry such a compound. In 1995 Weber et al.[67] began to search for such products from sequences of libraries of U-4CR products. And designed the sequences of the next libraries by mathematically oriented logical reasoning. After three months of investigation, including computer-assisted activity, this group was able to find two desirable products.

The search after optimal new desirable products from MCR libraries by a combination of mathematically oriented and data based computer was one of the most desirable chemistry oriented computer programs. However, this ultimate goals could only be achieved only after a variety of experiences existed.

The mathematically oriented solution of chemical problems began in 1960, when Vledutz[68] proposed the design of the computer program CAOS designing organic syntheses of given from collections of available starting materials.

Around that time, Lederberg et al.[69] developed the mathematically based computer program DENDRAL that determined molecular structural features from spectroscopic data. Somewhat later Corey et al[70] developed the LHASA synthesis design program from data of known molecules and reactions, which was essentially based on a collection of known reactions and their sequences. Corey's previous co-worker Wipke et al.[71] developed the LHASA related more efficient computer program SECS Since it was later realized these such data based computer programs con never produce fundamentally new chemistry, Wipke introduced a modified version of the data base of the SECS[71] into a data delivering computer program.

Gelernter[72] developed a computer program SYNCHEM of synthesis design that was designed in some analogy to the chess-oriented computer programs. This was not only data based, but also on some mathematically oriented logic. The disadvantage of this program was that in the members of this group did not have profound chemical knowledge. In the 1960s Ugi and Kaufhold[73,74] developed a computer program of designing peptide syntheses. This was based on chemical and mathematical information. All of these computer programs that were based on conventional mathematics and chemical data were till the 1980s given up, since none of them was able to solve chemical problems in a non-trivial way.

In 1973 Dugundji and Ugi[75] realized that suitable new mathematical theory can serve as the basis of creative computer programs of constitutional chemistry. There the starting materials and products with n atoms are represented by matrices with nxn numerical entries which correspond to their bonds and free electrons. Their reactions are represented by matrices that convert the educt matrices into the product matrices by suitable additive transformations of matrices. In the reacting matrices the lost bonds of the educt are negative numerical entries, and the newly formed bonds of the products are positive entries. The total sum of these entries is zero since the total number of electrons does not change. Later also molecules with multicenter bonds and delocalised valence electron systems were represented.[76]

On the basis of this mathematical theory a variety of different types of computer program were developed. Thus the search of chemical compounds with desirable properties from large collections of molecules could efficiently be found from the mathematically based computer program CORREL.[74,77] Subsequently a variety of related computer programs are now used like the RESY, KOWIST, HTSS, and S4.[74] The synthesis oriented multipurpose programs IGOR[78] and RAIN[79] were developed, by which e. g. collections of isomeric molecules can be produced, and 12 different. new reactions of different levels of novelty could be found by the IGOR.[78] Thus Forstmeyer et al.[81] produced by the IGOR a reaction of 11 C and N atoms which is not only more related to any known type of chemical reactions.

Wipke et al.[81] generated also a new chemical reaction by the IGOR which they confirm and the capability of this computer program. Since the 1990s also IGOR and RAIN were not any more, essentially used since their application required the learning its methodology.

In the last few years Weber introduced at the Morphochem A.G. a new way of reasoning in the search for promising new pharmaceutical products from libraries of MCR products by the 'molecular diversity analysis and combinatorial library design[82] and the 'practical approaches to evolutionary design[83] based on previous computer-oriented experiences. Recently Weber and Almstetter[84] introduced a new type of a computer program that corresponds to a combination of mathematically oriented logic of this reasoning and practical as well as estimated data of molecular properties which includes the solution of MCR and library problems.

Almstetter[84] demonstrated this computer program for the first time at the *Daylight MUG 2001* conference on March 6-9, 2001, at Santa Fe, New Mexico, USA. This is the first computer program that allows pre-select libraries and to produce new desirable compounds. It can be assumed that the search for desirable products will now be found more by systematic planning than by some companies that still arbitrarily form millions of new compounds per year without systematic reasoning methods.

8. REFERENCES

1.   I. Ugi, *J. Prakt. Chem.* **339**, 499 (1997).
2.   J. Chattopadhyaya, A. Dömling, K. Lorenz, I. Ugi, and Werner, *Nucleosides & Nucleotides*, **16**, 843 (1997).
3.   A. Strecker, *Ann. Chem.* **75**, 27 (1850).
4.   C. Mannich and I. Krötsche, *Arch. Pharm.* **250**, 647(1912); edited by F. F. Blick, and R. Adams, *Organic Reaction*, l. 1, John Wiley & Sons, New York, 303, 1942.
5.   H. Hellmann and G. Opitz, *α-Aminoalkylierung*, Verlag Chemie, Weinheim, 1960.
6.   A. Hantzsch, *Justus Liebigs AnnChem.* **215**, 1; 219, 1 (1982); *Ber. Dtsch. Chem. Ges.* **23**, 1474 (1890); see also: C. Böttinger, *Justus Liebigs Ann.Chem.* **208**, 122 (1981); U. Eisener and J. Kuthan, *Chem. Rev.* **72**, 1 (1972)
7.   B. Radziszewski, *Ber. Dtsch. Chem. Ges.* **15**, 1499, 2706 (1882).
8.   P. Biginelli, *Ber. Dtsch. Chem. Ges.* **24**, 1317, 2962 (1891); **26**, 447 (1893); C. O. Kappe, *Acc. Chem. Res.* **33**, 879 (2000).
9.   H.Bergs, Ger. Pat., 566094 (1929); Chem. Abstr. **27** 1001 (1933); H.T. Bucherer and W. J. Steiner, *Prakt. Chem.* **140**, 291 (1934); H.T. Bucherer, *ibid. 141*, 5 (1934).
10. F. Asinger, *Angew. Chem.,* **68**, 413 (1956); F. Asinger, M. Thiel, and E. Pallas, *Liebigs Ann. Chem,* **602**, 37 (1957); F. Asinger and M. Thiel, *Angew. Chem.,* **79**, 953 (1967); F. Asinger, W. Leuchtenberg, and H. Offermanns, *Chem. Zeitung*, **94**, 6105 (1974); F. Asinger and K. H. Gluzek, *Monats. Chem.*, **114**, 47 (1983)
11. I.Ugi, *Isonitrile Chemistry*, Academic Press, New York, 1971.
12. W. Lieke, *Justus Liebigs Ann. Chem.* **112**, 316 (1859).
13. A. Gautier, *Justus Liebigs Ann. Chem.* **142**, 289 (1867); *Ann. Chim. (Paris)* [4] **17**, 103, 203 (1869).
14. A. W. Hofmann, *Ber. Dtsch. Chem. Ges.* **3**, 63 (1870); see also: W. P. Weber G. W. Gokel, and I. Ugi,. *Angew. Chem.* **84**, 587 (1972); *Angew. Chem., Int. Ed. Engl.* **11**, 530 (1972).
15. M. Passerini, *Gazz. Chim. Ital.,* **51 II**, 126 (1921); **51 II** 181; **56**, 826 (1926); M. Passerini and G. Ragni, *ibid.* **61**, 964 (1931)..
16. I. Ugi, S. Lohberger, and R. Karl, *Comprehensive Organic Synthesis: Selectivity for Synthetic Efficiency*, vol. 2, Chap. 4.6, B. M. Trost, C. H. Heathcock (eds), Pergamon, Oxford 1991, p. 1083; a) p. 1090.
17. A. Dömling and I. Ugi, *Angew. Chem.* **112**, 3300 (2000); *Angew. Chem., Int. Ed. Engl.* **39**, 3168 (2000).
18. W. Rothe, *Pharmazie* **5**, 190 (1950).
19. I. Hagedorn and H. Tönjes, *Pharmazie* **11**, 409 (1956); **12**, 567 (1957); see also: W. R. Hertler and E. J. Corey *J. Org. Chem.* **23**, 1221 (1958).
20. I. Ugi and R. Meyr, *Angew. Chem.* **70**, 702 (1958).
21. R. Obrecht, R. Herrmann, and I. Ugi *Synthesis*, **1985**, 400.
22. G. Skorna and I. Ugi, *Angew. Chem.* **89**, 267 (1977); *Angew. Chem., Int. Ed. Engl.* **16**, 259 (1977).
23. H. Eckert and B. Forster, *Angew. Chem.* **99**, 922 (1987); *Angew. Chem., Int. Ed. Engl.,* **26**, 1221 (1987).
24. I. Ugi, R. Meyr, U. Fetzer, and C. Steinbrückner, *Angew. Chem.* **71**, 386 (1959).

25. a) I. Ugi, A. Dömling, and W. Hörl, *Endeavour* **18**, 115 (1994); b) GIT *Fachzeitschrift für das Laboratorium* **38**, 430 (1994).

26. I. Ugi, D. Marquarding, and R. Urban, *Chemistry and Biochemistry of Amino Acids, Peptides, and Proteins*, Vol. 6, B. Weinstein (ed.), Marcel Dekker, New York 1982, p. 245.

27. B. M. Ebert and I. K. Ugi, *Tetrahedron* **54**, 11887 (1998).

28. T. Yamada, Y. Omoto, Y. Yamanaka, T. Miyazava, and S. Kuwata, *Synthesis* **1998**, 991.

29. S. L. Schreiber, *Science* **287**, 1964 (2000); D. Lee, J. K. Sello, and S. L. Schreiber, *Org. Lett.* **2**, 709 (2000).

30. D. Askin, K. K. Eng, K. Rossen, R. M. Purick, K. M. Wells, and R. P. Voante, *Tetrahedron Lett.* **35**, 673 (1994); K. Rossen, R. J. Pye, L. M. DiMichele, R. P. Voante, and P. J. Reider, *Tetrahedron Lett.* **39**, 6823 (1998).

31. M. Bodanszky and M.A. Ondetti, edited by G. A. Olah, *Peptide Synthesis*, J. Wiley & Sons, New York. (1966).

32. I. Ugi and C. Steinbrückner, *Chem. Ber.* **94**, a) 734; b) 2802 (1961).

33. I. Ugi and K. Offermann, *Angew. Chem.* **75**, 917 (1963); *Angew. Chem., Int. Ed. Engl.*, **2**, 624 (1963); I. Ugi, K. Offermann, and H. Herlinger, *Angew. Chem.* **76**, 613 (1964); *Angew. Chem., Int., Ed. Engl.*, **3**, 656 (1964); I. Ugi, K. Offermann, H. Herlinger, and D. Marquarding, *Justus Liebigs Ann. Chem.* **709**, 1 (1967)

34. I. Ugi and G. Kaufhold, *Justus Liebigs Ann.Chem.* **709**, 11 (1967).

35. I. Ugi and K. Offermann, *Chem. Ber.* **97**, 2996 (1964).

36. I. Ugi, *Rec. Chem. Progr.* **30**, 289 (1969).

37. G. Wagner and R. Herrmann, *Ferrocenes*, A. Togni and T. Hayashi (eds.), VCH Verlag, Weinheim, 1995.

38. A. Demharter and I. Ugi, *J. Prakt. Chem.* **335**, 244 (1993).

39. L. K. Likhosherstov, O. S. Novikova, V. A. Derivitkava, and N. K. Kochetkov, *Carbohydr. Res.* **146**, C1 (1986); L. K. Likhosherstov, O. S. Novikova, V. N. Shbaev, and N. K. Kochetkov, *Russ. Chem. Bull.*, **45**, 1760 (1996).

40. J. Drabik, J. Achats, and I. Ugi, *Tetrahedron*, in preparation.

41. H. Kunz and W. Pfrengle, *J Am. Chem. Soc*, **110**, 651 (1988); *Tetrahedron*, **44**, 5487 (1988); H. Kunz, W. Pfrengle, and W. Sager, *Tetrahedron Lett.* **30**, 4109 (1989); H. Kunz, W. Pfrengle, K. Rück, and W. Sager, *Synthesis* **1991**, 1039; H. Kunz and K. Rück, *Angew. Chem.* **105**, 355 (1991).

42. M. Goebel and I. Ugi, *Tetrahedron Letters* **36**, 6043 (1995); M. Goebel, H.-G. Nothofer, G. Roß, and I. Ugi, *Tetrahedron* **53**, 3123 (1997).

43. S. Lehnhoff, M.Goebel, R. M. Karl, K. Klösel, and I.Ugi, *Angew. Chemie* **107**, 1208 (1995); *Angew. Chem., Int. Ed. Engl.* **34**, 1104 (1995); S. Lehnhoff, Doctoral thesis, Technical University of München, 1994.

44. A. von Zychlinski, edited by Z.Hippe and I Ugi, *MultiComponent Reactions & Combinatorial Chemistry*, p. 31; German-Polish Workshop, Rzeszów, 28.-30. Sept. 1 1998, University of Technology, Rzeszów / Technical University, Munich;; Doctoral thesis, Technical University Munich, 1998.

45. G. Ross, Doctoral thesis, Technical University Munich, 2001; G. Ross and I. Ugi, *Canadian J. Chem.*submitted.

46. G. Ross, I. Ugi, and E. Herdweck, Angew. Chem., submitted.

47. K. Kehagia, A. Dömling, and I. Ugi, *Tetrahaedron* **51**, 139 (1995); K. Kehagia, and I. Ugi, *Tetrahaedron* **51**, 9523 (1995) A. Dömling and I. Ugi, *Angew. Chemie* **107**, 2465 (1995); *Angew. Chem., Int. Ed. Engl.* **34**, 2238 (1995); A. Dömling, K. Kehagia, and I. Ugi, *Tetrahedron* **51**, 9519 (1995).

48. I. Ugi and E. Wischhöfer, *Chem. Ber.* **95**, 136 (1962).

49. A. Schutz, I. Ugi, and J. Kabbe, *J. Chem. Res.* (S) 1979, 157; (M) 1979, 2064.

50. A. Schutz and I. Ugi, *Z. Naturforschung* **34d**, 1159 (1979).

51. G. Neyer and I. Ugi, *Synthesis* **9**, 743 (1991).

52. H. H. Wassermann and M. B. Floyd, *Tetrahaedron Suppl.* **7**, 441 (1966); H. H. Wassermann, F. E. Mac Carthy, and K. S. Prowse, *Chem. Rev.* **86**, 845 (1986).

53. S. Ganslmeier, Doctoral thesis, Technical University of München, 1998.

54. C. Burdack, Doctoral thesis, Technical University Munich, 2001; C. Burdack and I. Ugi, *Angew. Chem.*, submitted.

55. T. Kametani, S. Huang, A. Hakayama, and T. Honda, *J. Org. Chem.* **47**, 2328. (1985); T. Kametani, T. Hagahara, and T. Honda, *J. Org. Chem.* **50**, 2327 (1982); T. Kametani, S.-D. Chu, and T. Honda, *J. Chem. Soc. Perk. Trans.* **1**, 1593 (1988).

56. T. Lindhorst, H. Bock, and I. Ugi, *Tetrahedron* **55**, 7411 (1999).

57. R. Bossio, S. Marcaccini, P. Paoli, R. Pepino, and C. Polo, *Synthsis* **1991**, 999; R. Bossio, S. Marcaccini, P. Paoli, P. Papaleo, R. Pepino, and C. Polo, *Liebigs Ann.Chem.* **1991**, 843; R. Bossio, S. Marcaccini, and R. Pepino, *Justus Liebigs Ann.Chem.* **1991**, 1107; R. Bossio, S. Marcaccini, R. Pepino, and Torroba, *Synthesis* **1993**, 783 R. Bossio, S. Marcaccini, and R. Pepino, *Liebigs Ann. Chem.* **1993**, 1229; R. Bossio,

S. Marcaccini, S. Papaleo, Pepino, *J. Heterocycl. Chem.* **31**, 397 (1994); R. Bossio, S. Marcaccini, P. Paoli, and R. Pepino, *Synthsis* **1994**, 672; R. Bossio, S. Marcaccini, and R. Pepino, *Tetrahedron Letters* **36**, 2325 (1995); *J. Org. Chem.* **61**, 2202 (1996); R. Bossio, S. Marcaccini, R. Pepino, and T. Torroba, *J. Chem. Soc,. Perkin Trans 1*, **1996**, 229; R. Bossio, C. F. Marcos, S. Marcaccini, and R. Pepino, *Heterocycles.* **45**, 1589 (1997); *Synthesis* **1997**, 1389; R. Bossio, S. Marcaccini, and R. Pepino, *Tetrahedron Letters* **38**, 2519 (1997).

58. I. Ryu, N. Sonoda, and D. P. Curran, *Chem. Rev.* **96**, 177 (1996); A. Studer, P. Jeger, P. Wipf, and D. P. Curran, *J. Org. Chem.* **62**, 2917 (1997); A. Studer, S. Hadida, R. Ferrito, S.-Y. Kim, P. Jeger, and D. P. Curran, *Science* **275**, 823 (1997); H. Josien, S.-B. Ko, D. Bom, and D. P. Curran, *Chem. Eur.* **4**, 1043 (1998).

59. A. Dömling and I. Ugi, *Angew. Chem.,* **105**, 634 (1993); *Angew. Chem,. Int. Ed. Engl.,* **32**, 563 (1993).

60. S. MacLane and G. Birkhoff, *Algebra*, MacMillan Company, New York, 1967, p. 3: Given sets of *R* and *S* have the intersection $R \cap S$ with the common elements *R* and *S* . This means $R \cap S = \{x \mid x \subset R \text{ and } x \subset S\}$, where as a *union* $R \cup S$ is $R \cup S = \{x \mid x \subset R \text{ or } x \subset S\}$

61. D. Bradley, *New Scientist,* **16,** July 3 (1993); *C&EN,* **32**, April 19 (1993).

62. R. B. Merrifield, *J. Am. Chem. Soc.*, **85**, 2149 (1963).

63. A. Furka, *Drug Dev. Res.,* **36**, 1 (1995).

64. F. Balkenhohl, C. v. Buschen-Hünnefeld, A. Lanshy, and C. Zechel, *Angew. Chem.*, **108**, 3436 (1996); *Angew. Chem. Int. Ed. Engl.*, **35**, 2288 (1996).

65. N. K Terret, *Combinatorial Chemistry*, Oxford Univ. Press: New York, 1998.

66. R. W. Armstrong, *Combinatorial libraries related to natural products at the ACS National Meeting in Anaheim*, Calif. on April 2, 1995; *J. Am. Chem. Soc.* **117**, 7842 (1995).

67. L. Weber, S. Waltbaum, C. Broger, and K. Gubernator, *Angew. Chem.* **107**, 2452; (1995); Angew. Chem. Int. Ed. Engl. **34**, 2280 (1995); O. Lacke and L. Weber, *Chimia 50*, 445 (1996); L. Weber, *Current Opinion in Chemical Biol.* **2**, 381 (1998).

68. G. Vledutz and K. A. Finn, *Proc. Dept. of Mechanization and Automization of Information Work*, Acad. Sci. USSR, Moscow, 1960, p. 66; G. Vledutz, *Inf. Storage Retr.* **1**, 117 (1963).

69. R. K. Lindsay, B. G. Buchanan, E. A. Feigencaun, and J. Lederberg, *Applications of Artificial Intelligence for Organic Chemistry: The DENDRAL Project*, McGraw-Hill, New York, 1980; J. Lederberg, *Proc. Natl. Acad. Sci. USA*, **53**, 134 (1990).

70. E. J. Corey, *Pure Appl. Chem.* **14**, 19 (1967); E. J. Corey and X.-M. Cheng, *The Logic of Chemical Synthesis*, Wiley, New York, 1986; E. J. Corey and W. T. Wipke, *Science* **166**, 178 (1991).

71. W. T. Wipke and D. Rogers, *J. Chem. Inf. Comput.* **24**, 71 (1984).

72. H. Gelernter, N. S. Sridharan, A. J. Hart, S. C. Yen, F. W. Fowler, and H. J. Shue, *Top. Curr. Chem.* **41**, 113 (1973); H. Gelernter and J. R. Rose, *J. Chem. Inf. Comput. Sci.* **30**, 492 (1990).

73. I. Ugi, *Rec. Chem. Proc.* **30**, 389 (1969).

74. I. Ugi, J. Bauer, K. Bley, A. Dengler, A. Dietz, E. Fontain, B. Grusber, R. Herges, M. Knauer, K. Reitsam, and N. Stein, *Angew. Chem.* **105**, 210 (1993); *Angew. Chem., Int. Ed. Engl.* **32**, 201 (1993).

75. J. Dugundji and I. Ugi, *Top. Curr. Chem.* **39**, 19 (1973).

76. I. Ugi, N. Stein, M. Knauer, B. Gruber, K. Bley, and R. Weidinger, *Top. Curr. Chem.***166**, 199 (1993).

77. J. Friedrich and I. Ugi, *J. Chem Res. Synop.* **1980**, 70; *J. Chem Res. Miniprint* **1980**, 1301.

78. J. Bauer and I. Ugi, *J. Chem Res. Synop.* **1982**, 298; *J. Chem. Res. Miniprint* **1982**, 3101; J. Bauer, R. Herges, E. Fontain, and I. Ugi, *Chimia* **39**, 43 (1985); J. Bauer, *Tetrahedron Comput. Methodol.* **2**, 269 (1989).

79. E. Fontain, J. Bauer, and I. Ugi, *Chem. Lett.* **1987**, 37; *Z. Naturforsch. B* **42**, 297 (1987); E. Fontain, *Tetrahedron Comput. Methodol.* **3**, 469 (1990).

80. D. Forstmeyer, J. Bauer, E. Fontain, R. Herges, R. Herrmann, and I. Ugi, *Angew. Chem.* **100** 1618 (1988); *Angew. Chem. Int. Ed. Engl.***27**, 1558 (1988); J. Bauer E. Fontain, D. Forstmeyer, and I. Ugi, *Tetrahedron Comput. Methodol.* **1**, 129 (1988); I. Ugi, J. Bauer, R. Baumgartner, E. Fontain, D. Forstmeyer, and S. Lohberger, *Pure Appl. Chem.* **60**, 1573 (1988).

81. G. B. Fisher, J. J. Juarez-Brambila, C. T. Goralski, W. T. Wipke, and B. Singaram, *J. Am. Chem. Soc.* **115**, 440 (1993).

82. L. Weber, *Evolutionary Algogorithms in Molecular Design*, edited by D. Clark, Wiley-VCH, Weinheim, 2000, p. 137-158.

83. L. Weber, *Virtual Screening in Bioactive Molecules*, H.-J.Bohm, G. Scheider, Wiley-VCH, Weinheim, 2000, p. 187-206.

84. L. Weber, K. Illgen, and M. Almstetter, *Synlett*, **3**, 366 (1999); L. Weber and M. Almstetter, *Molecular Diversity in Drug Design*, edited by P. M. Dean and R. A. Lewis, Kluwer Academic Publisher, Dordrecht, 1999, p. 93-114.

85. M. Almstetter, *Daylight MUG 2001*, March 6-9, 2001, Santa Fe, New Mexico, USA.

WATER-SOLUBLE POLYMERS
FOR METAL INTERACTION

Bernabé L. Rivas[*]

1. INTRODUCTION

Polymeric supports with complexing groups are widely investigated and applied for the metal recovery from dilute solutions such as industrial fluids and waste waters.[1-10] Liquid-liquid extraction, sorption, precipitation, and other methods based on two-phase distributions are used in most cases for the separation of inorganic species contained in dissolved matrices, industrial fluids, or natural waters.

Although many such methods have been developed and successfully used, their application can cause problems. Some problems can be connected with heterogeneous reactions and interphase transfer. Other problems can arise, if aqueous solutions are preferred for the subsequent procedure rather than organic solvents or solid concentrates. In such cases, additional procedures are needed, e.g., back extraction, desorption, dissolution of solid concentrates, etc., which complicate the analysis and can result in the contamination of the sample due to the reagent added.

Among these processes, one of the most abundant uses of chemically modified polymers is that in the form of synthetic ion-exchange resins.[11-18] There are different natural and synthetic products which show these properties. The organic resins are by far the most important ion exchangers. The main advantages are high chemical and mechanical stability, high ion-exchange capacity, and ion exchange rate. Another advantage is the possibility of selecting the fixed ligand groups and the degree of cross-linking.

Two-phase system can be avoided by application of separation methods based on membrane processes, which are among the most promising techniques for enrichment of various species from solutions.

[*] Polymer Department, Faculty of Chemistry, University of Concepcion, Casilla 160-C, Concepcion, Chile.

Advanced Macromolecular and Supramolecular Materials and Processes
Edited by K. Geckeler, Kluwer Academic/Plenum Publishers, 2003

Thus, a number of soluble and hydrophilic polymers with chelating groups, termed *polychelatogens*, have been prepared basically through addition polymerization and by functionalizing various polymers, and found to be suitable for the separation and enrichment of metal ions in conjunction with membrane filtration. Membrane filtration allows to easily separate metals bound to soluble chelating polymers from non-chelated metals. This method is known as *Liquid-phase Polymer-based Retention* (LPR) technique.[19-20]

A salient feature of the polychelatogen is its high solubility and hydrophilicity which allow to carry out the complexation in aqueous phase. Applications of the water-soluble polymers to the homogenous enrichment of various metal ions from dilute solutions have been reported.[20-67]

Ultrafiltration is a fast-emerging, new, and versatile operation in separation technology, concentration, purification, and separation processes. This technique can also be used for the detection limits and matrix interferences in instrumental determinations. With the development of asymmetric membranes, other polymeric materials have been subject of investigation.

The study of different polymeric materials for the use in membrane filtration involves the selection of suitable polymers, membrane morphology in relation to its use as an ultrafilter and effect of fabrication, operational, and hydraulic variables. Membrane filtration processes can be successfully used for the separation of inorganic species and for their enrichment from dilute solutions. More effective retention of certain inorganic ions and their enrichment in a membrane filtration unit and separation from other solutes is achieved by using water-soluble polymers.

2. LIQUID-PHASE POLYMER-BASED RETENTION (LPR) TECHNIQUE

Hydrophilic polymers with complexing groups have been tested to show the applicability of the method to the separation of various metal cations for analytical and technological purposes. The method, *Liquid-phase Polymer-based Retention* (LPR) is based on the retention of certain ions by a membrane, which separates low-molecular mass compounds from macromolecular complexes of the ions. Thus, uncomplexed inorganic ions can be removed by the filtrate, whereas the water-soluble polymer complexes are retained.

Different modes of separation by LPR can be used for inorganic ions. To separate the components of a small volume sample in analytical chemistry (relative preconcentration), the liquid sample is placed in the polymer containing cell solution and then washed with water (washing method). The pH is adjusted to a value at which the ions of interest are retained and the other species are removed. The washing method can also be applied to purify a macromolecular compound by eliminating the microsolutes, while maintaining a constant volume in the cell.

To achieve enrichment of the metal ions, their solution can be passed from a reservoir to a smaller volume filtration unit in the presence of a complexing polymer. This concentration method (enrichment method) is designed for metal recovery from diluted technological solutions and for absolute preconcentration of elements in analytical chemistry.

However, interfering components of the test solution partly remain in the cell after the filtration run, even they do not interact with the reagent. This can cause difficulties, e.g, in the trace analysis of highly mineralized waters. To avoid that, a combined procedure is applied to both absolute and relative preconcentration.

Figure 1 shows the principles of the procedure used for the preconcentration of metal ions. The main features of a Liquid-phase Polymer-based Retention system are a membrane filtration unit, reservoirs, and a pressure source, e.g., a nitrogen bottle. Conventional stirred filtration cells or a specially designed tangential-flow cell equipped with a pump can be used. Essential parameters are the molecular mass exclusion rate in a wide pH range (1-12), an appropriate permeate flow rate (0.5-12 mL min^{-1}), retentate volume (2-50 mL) and a gas pressure 300 kPa, are suitable in most cases.

A polymer concentration of 0.5-5 wt.% in the cell solution is most appropriate for both retentions of elements and their subsequent determination in the retentate. The usual molecular mass cut-off ranged between 1 and 300 kg mol^{-1}. A nominal exclusion rate of 10 kg mol^{-1} proved to be convenient for polymers having a molecular mass between 30 and 50 kg mol^{-1}.

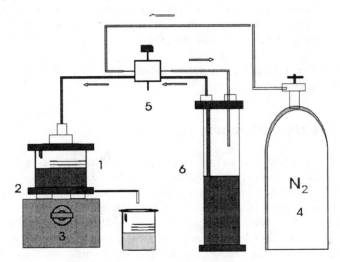

Figure 1. Instrumental arrangement: (1) filtration cell with polymeric and metal ion solution; (2) membrane filtrate; (3) magnetic stirrer; (4) pressure trap; (5) selector; (6) reservoir with water. Adopted from ref. (19), K. Geckeler, et al., Figure 1.

These water-soluble polymers can be synthesized by different routes such as radical, cationic, and spontaneous polymerization, but the usual synthetic procedures are the addition polymerization, especially radical polymerization, and by functionalizing polymer backbones through polymer-analogous reactions. The polychelatogens may be homo and copolymers, and can contain one or more ligand or coordinating groups. These groups are placed at the backbone, or at the side chain, directly or through a spacer group.

The most investigated ligands are amines, carboxylic acids, amides, aminoacids, pyridines, thioureas, imino, ether, phosphoric acids, sulfonic acids, etc. Among them, the amino groups, and particularly the functional poly(ethyleneimine) and derivatives have been extensively studied.[10, 17, 20, 29, 30, 39-41, 44]

This heterochain polymer contains three different types of amino groups: in the main chain secondary and tertiary groups and in the side-chain secondary and primary amino groups. The ratios range are between 1:1:1 and 1:2:1, referred to primary, secondary, and tertiary, but are variable in principle, depending on the degree of branching.

Among the most important requirements for a polychelatogen are the high solubility in water, an easy and cheap route of synthesis, an adequate molecular weight and molecular weight distribution, chemical stability, high affinity for one or more metal ions, and selectivity for the metal ion of interest. It is essential for this technique that the molecular mass of the polymer is higher than the nominal exclusion rate. In practice, an average molecular mass ranging between 20 kg mol^{-1} and 50 kg mol^{-1} was optimum when a 10 kg mol^{-1} ultrafiltration membrane is used.

One versatile route to synthesize polychelatogens is copolymerization, because with a good selection of both comonomers it is possible to improve the properties such as the water-solubility, metal ion binding capability, and selectivity. The copolymer composition is normally determined by ^{1}H-NMR or FTIR spectroscopy by comparison of the characteristic absorption signals for each monomer and by elemental analysis. These polychelatogens, prior to be used by the LPR technique, are purified by ultrafiltration through a membrane of low molecular weight limit exclusion, i.e. 1 kg mol^{-1}, and then lyophilized and characterized.

Table 1 shows some polychelatogens studied by the LPR technique during the last few years in our group.

3. METAL-ION BINDING PROPERTIES

For analytical investigation of the metal uptake, a solution volume of about 200 mL was used. The complexing polymer solution was placed into the membrane filtration cell and the metal salt solution added from the reservoir. The concentration of the polymer in the cell was kept constant, as it is not able to pass through the ultrafiltration membrane. The effect of pH, ionic strength, filtration factor, Z, on the metal ion retention properties of the polychelatogens have been investigated. Z is the ratio between the volume in the filtrate, V_f the volume in the cell, V_o. Mono-, di-, and trivalent metal cations have been studied.

Among the most studied are: Ag(I), Na(I), Cu(II), Co(II), Ni(II), Cd(II), Zn(II), Hg(II), Pb(II), and Cr(III) ions, which have been investigated under different conditions. No precipitation occurred for most polymer complexes with the metals investigated at the low metal concentration normally used. Minor changes of the series of metals were necessary due to the peculiar properties of some metal ions. For example, Cu(II) retention could not be investigated at pH 7 because of precipitation.

After isolation of the complex formed, the proportion of non-bound ions could be determined by atomic absorption spectroscopy of the filtrate, which was collected in fractions. The flow rate depends on the type of membrane, and, above all, on the membrane surface. For example, membrane filtration systems have 300 to 500 mL per hour as average values. In addition, stirred cells, in which the solution is flowing continuously over the membrane, are used.

Table 1. Water-soluble polymers and their interactions with metal cations studied by the LPR technique

Water soluble polymers	Metal cations	pH	Ref.
$-(CH_2-CH_2-N)_x$ $\quad\quad\quad\vert$ $\quad\quad\quad CH_2$ $\quad\quad\quad\vert$ $\quad\quad\quad CH_2$ $\quad\quad\quad\vert$ $\quad\quad\quad OH$ Poly(N-hydroxyethyl)ethyleneimine (PHEI)	Cd^{2+}, Pb^{2+}, Sr^{2+}, Ni^{2+}, Cu^{2+}, Zn^{2+}, Co^{2+}, Fe^{3+}, Cr^{3+}	3, 5, 7	1
$-(CH-CH_2)_x$ $\quad\vert$ $\quad C=O$ $\quad\vert$ $\quad NH_2$ Poly(acrylamide) (PAM)	Cd^{2+}, Pb^{2+}, Ni^{2+}, Cu^{2+}, Zn^{2+}, Co^{2+}, Hg^{2+}, Fe^{3+}, Cr^{3+}	5, 7	62
$-(CH-CH_2)_x$ $\quad\vert$ $\quad CH_2$ $\quad\vert$ $\quad NH_2$ Poly(allylamine) (PALA)	Cu^{2+}, Co^{2+}, Ni^{2+}	1, 3, 5, 7	15
$-(CH-CH_2)_x$ $\quad\vert$ $\quad C=O$ $\quad\vert$ $\quad OH$ Poly(acrylic acid) (PAA)	Cd^{2+}, Pb^{2+}, Ni^{2+}, Cu^{2+}, Zn^{2+}, Co^{2+}, Cr^{3+}	5	44, 46, 50
$\quad CH_3$ $\quad\vert$ $-(C-CH_2)_x$ $\quad\vert$ $\quad C=O$ $\quad/$ HO Poly(methacrylic acid) (PMA)	Ag^+, Cd^{2+}, Hg^{2+}, Ni^{2+}, Cu^{2+}, Zn^{2+}, Co^{2+}, Cr^{3+}	1, 3, 5, 7	61, 65

Table 1 (continued)

 Poly(α-amino acrylic acid) (PAAA)	Ag$^+$, Cd^{2+}, Pb^{2+}, Ni^{2+}, Cu^{2+}, Zn^{2+}, Co^{2+}, Cr^{3+}	3, 5, 7	20
 Poly(N-methyl-N'-methacryloylpiperazine) (PAP)	Ag$^+$, Cd^{2+}, Pb^{2+}, Ni^{2+}, Cu^{2+}, Zn^{2+}, Co^{2+}, Cr^{3+}	3, 5, 7	66
 Poly(sodium 4-styrenesulfonate) (PSS)	Cd^{2+}, Pb^{2+}, Ni^{2+}, Cu^{2+}, Zn^{2+}, Co^{2+}, Cr^{3+}, Fe^{3+}	1, 3, 5, 7	55, 56
 Poly(2-acrylamido-2-methyl-1- propanesulfonic acid) (PAPS)	Ag$^+$, Cd^{2+}, Hg^{2+}, Ni^{2+}, Cu^{2+}, Zn^{2+}, Co^{2+}, Cr^{3+}	1, 3, 5, 7	61, 65, 67

The proportionality between concentration and volume for membrane filtration allows the determination of final microsolute content or filtrate volume, if the original and final volumes are known. The flow-rate is inversely proportional to the logarithm of the concentration of the retained solute. The flow-rate diminishes as the concentration of the retained solute is increased.

The retention, R, of any species, not bound to the polymer and using a membrane filtration system can be expressed by the term:

$$R = \exp(-V_f \bullet V_o^{-1})$$

where V_f is the volume of the filtrate and V_o the volume of cell solution. The quotient $V_f \times V_o^{-1}$ is called filtration factor and abbreviated as Z in the diagrams. In all cases the determination of the R values for Z between 1 and 10 was found to be sufficient to show that metal ions are not bound or complexed by the polychelatogen.

The metal ion retention depends strongly on the pH and the nature of the binding site. In general, in strongly acidic solutions no complexation takes place. Thus, for a polymer containing an amino group as poly(N-acryloyl-N-methyl piperzine) PAP,[64] the protonated PAP and aqua-metal solution co-exist in solution (see Figure 2).

In alkaline solutions, hydroxo complexes are formed, or at least there is a competition between HO⁻ and tertiary amino groups for metal coordination and the formation of mixed complexes can not be ruled out. Therefore, the intermediate pH range (3< pH <8) is the most appropriate for the formation of "single" PAP metal complexes for most of the metal ions. Ag(I) is poorly retained (20%). At pH 3, Cr(III) ions are retained by about 50%. It is necessary to consider that at this pH chromium basically exists as Cr^{3+}, which co-exists with basic species such as $Cr(OH)_2^+$ and $CrOH^{2+}$.

The highest values (> 90% at Z = 1) are achieved at pH 5, which is attributed to the electrons of nitrogen atoms available to coordinate with the metal ions. At pH 5 and Z = 10, the retention of Co(II) (62%) is higher than that of aliphatic amines such as BPEI and PHEI. The retention of Zn(II) ions by PAP is higher than that of PHEI, but lower than that of BPEI.[40]

The nature of the interaction of metal ions with both polyelectrolytes and chelating polymers does not affect appreciably the retention profiles of each metal ion considered individually. Numeric analysis of the retention functions associated to the retention profiles shows that only in polyelectrolyte systems the retention parameter k_z is generally found to be greater than 1.

A typical polyelectrolyte behavior, where electrostatic interactions are dominant, implies that all charged metal ions are very dependent on the ionic strength, and low interaction rates are found when the ionic strength is high. Consequently, at pH 1, where the concentration of the monovalent H^+ ions is high, low interaction rates are always found. The interaction rate increases rapidly increasing the pH, and high retention values are found for divalent and trivalent metal ions at pH > 3.

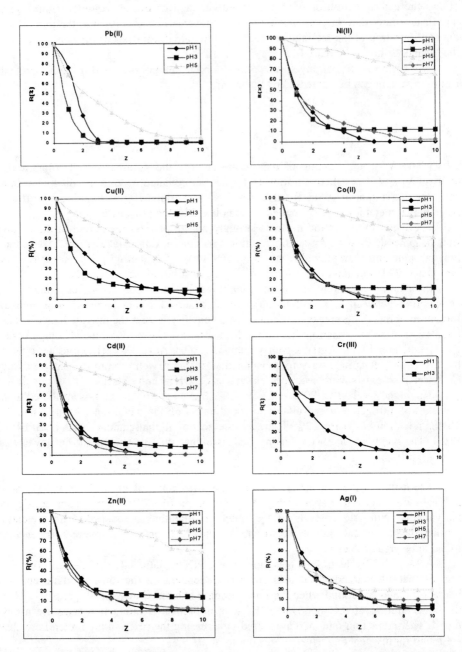

Figure 2. Retention profiles of Cu(II), Cd(II), Co(II), Ni(II), Cd(II), Zn(II), Pb(II), and Cr(III) (0.2 mM) using PAP (0.05 mM) at different pH and Z values. Adopted from ref. (64); B. L. Rivas et al., Figure 1.

On the contrary, when interactions occur by formation of coordinating bonds, the nature of the metal ions and the effect of the pH may induce differences in the retention profiles of like charged metal ions and selectivity is often informed. This has been found for the poly(2-acrylamido-2-methyl-1-propane sulfonic acid) PAMPS,[61] and poly(N-acryloyl-N'-methyl piperazine), PAMP.[64]

In Table 1, the retention of several metal ions is shown when the initial solution has been eluted with a volume of water at two given pH of 10 times the volume of the solution inside the ultrafiltration cell (R(10)). It can be seen that in the case of the polyelectrolyte PAMPS, all the metal ions are similarly retained at pH 5 with high retention values. Differences in the R(10) values are found in the case of chelating polymer PAMP. Regarding to the copolymers, it is found that AMP-AMPS-01, which is rich in sulfonate groups behaves as a polyelectrolyte in ultrafiltration experiments. The other copolymers behave as coordinating polymers applying to the differences found on the retention values for the different metal ions, and their behavior as a function of pH (see Table 2).

Table 2. Amounts of functional groups in the ultrafiltration experiments and values of $R(10)$ for the corresponding metal ions and polymers at pH 3 and 5

Polymer	Amount of Sulfonate groups (mequiv)	Amount of amino groups (mequiv)	pH	Cu^{2+}	Co^{2+}	Cd^{2+}	Zn^{2+}	Pb^{2+}	Ni^{2+}
PAPSA	0.20	0.00	3	0.95	0.93	0.85	0.89	-	0.95
			5	1.00	1.00	1.00	1.00	1.00	1.00
PAPSA	0.00	0.20	3	0.10	0.12	0.08	0.10	0.00	0.12
			5	0.23	0.62	0.48	0.58	0.05	0.68
AMP-APSA-01	0.24	0.10	3	0.79	0.8	0.80	0.82	0.88	0.84
			5	1.00	1.00	0.98	0.98	1.00	1.00
AMP-APSA-02	0.19	0.20	3	0.00	0.00	0.27	0.02	0.00	0.00
			5	0.23	0.09	0.12	0.12	0.29	0.10
AMP-APSA-03	0.09	0.22	3	0.04	0.21	0.13	0.20	0.00	0.13
			5	0.00	0.00	0.00	0.00	0.00	0.25

The dependence of capacity on the concentration of the polymer solution differs from polymer to polymer. Generally, sterical hindrance diminishes the capacity, if a certain concentration of solution is exceeded.

The maximum retention capacity (MRC) is defined as:

$$MRC = M \times V/Pm$$

MRC = mg of metal ion retained per g of polymer
M   = initial concentration of metal salt (mg L^{-1})
V   = metal free volume of the membrane filtrate (L)
Pm   = mass of polymer (g)

A very strong dependency exists between capacity and pH, influenced by both, type of functional group and type of metal ion. Generally, curves are obtained as despicted in Figure 3 for PAPSA. There is a clear difference between the MRC values for the divalent and trivalent metal ions. It is lower for Cr(III) and the MRC for the divalent cations are very similar (see Table 3).

This would demonstrate that the polymer-metal ion interaction is of electrostatic type, explaining why the MRC values for a trivalent cation like Cr(III) are lower than those of the divalent cations. This is due to that one Cr^{3+} ion will interact with three negative charges coming from repeat units PAPSA, whereas a divalent cation as Co^{2+} will interact only with two charges. Hence, to neutralize completely the charges in PAPSA, partially or completely dissociated, it will be necessary a higher number of divalent ions than of trivalent cations, and consequently the MRC will be higher for the former.[67]

Table 3. MRC values of PAPSA for di- and trivalent metal ions under different experimental conditions

Metal ion	Initial Concentration (mg L^{-1})	pH	MRC mg metal/g polymer	mol metal ion/ mol repeat unit
Cr(III)	51.8	3	51.77	0.21
	205.0	3	43.64	0.14
Cu(II)	51.6	3	89.22	0.29
	171.0	3	79.70	0.26
	49.3	5	-	-
	173.0	5	125.20	0.41
Ni(II)	54.3	3	83.35	0.29
	220.0	3	93.97	0.32
	-	5	-	-
	220.0	5	120.50	0.42
	-	7	-	-
	210.0	7	7.00	0.43
Co(II)	51.8	3	78.10	0.28
	211.0	3	95.45	0.34
	49.2	5	-	-
	208.0	5	118.70	0.41
	-	7	-	-
	210.0	7	121.70	0.43

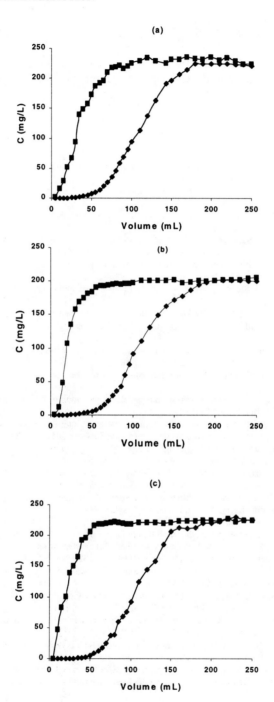

Figure 3. Retention of Co(II) at pH 3 (a), pH 5 (b), and pH 7 (c). Blank (-■-) and PAPSA (-◆-) Adopted from ref. (67), B.L. Rivas, et al., Figure 1.

4. ACKNOWLEDGEMENTS

The author thanks FONDECYT (Grant No. 8990011) and the Dirección de Investigación, Universidad de Concepción (Grant No. 98.24.17-1) for the financial support.

5. REFERENCES

1. A. D. Pomogailo and G. I. Dzhardimalieva, Problems of a unit variability in metal-containinig polymers, *Russ. Chem. Bull.* **47**, 2319-2337 (1998).
2. A. W. Trochimczuk and M. Streat, Novel chelating resins with aminophosphonate ligands, *Reactive Polymers* **40**, 205-213 (1999).
3. E. Tsuchida and H. Nishide, Polymer-metal complexes and their catalytic activity, *Adv. Polym. Sci.* **24**, 1-87 (1977).
4. R. Bogoczek and J. Surowiec, Synthesis of phosphorous-containing of at itcation exchangers and their affinity toward selected cations, *J. Appl.Polym. Sci.* **26**, 4161- 4173 (1981).
5. J. M. Frechet, Synthesis and applications of organic polymers as supports and protecting groups, *Tetrahedron* **37**, 663- 683 (1981).
6. G. F. Vesley and V. I. Stemberg, The catalytic degradation for rapid ester synthesis, *J.Org. Chem.* **36**, 2548-2550 (1971).
7. S. D. Alexandratos and L. A. Hussain, Synthesis of α,- β,- and γ-ketophosphonate polymer-supported reagents: the role of intra-ligand cooperation in the complexation of metal ions, *Macromolecules* **31**, 3235-3238 (1998).
8. T. Soldi, M. Pesavento, and G. Alberti, Separation of vanadium(V) and (IV) by sorption of an iminodiacetic chelating resin, *Anal. Chim. Acta*, **323**, 27-37 (1996).
9. H. Matsuda, Polymers based on divalent metal salts of p-aminobenzoic acid: a review, *Polym. Adv. Technol.* **8**, 616-622 (1997).
10. B. L. Rivas and K E. Geckeler, Synthesis and metal complexation of poly(ethyleneimine) and derivatives, *Adv. Polym. Sci.* **102**, 171-187 (1992).
11. Z. Matejka and Z. Zitkova, The sorption of heavy-metal cations from EDTA complexes on acrylamide resins having oligo(ethyleneimine) moieties, *Reactive Polymers*, **35**, 81-88 (1997).
12. W. H. Chan, S. Y. Lam-Leung, W. S. Fong and F. W. Kwan, synthesis and characterization of iminodiacetic cellulosic sorbent and its application in metal ion extraction, *J. Appl. Polym. Sci.* **46**, 921-930 (1992).
13. J. Lehto, K.Vaaramaa, E. Vesterinen, and H. Tenhu, Uptake of zinc, nickel, and chromium by N-isopropyl acrylamide polymer gels, *J. Appl. Polym. Sci.* **68**, 355-366 (1998).
14. L. Jose and V. N. R. Pillai, Transition metal complexes of polymeric amino ligands derived from tri ethyleneglycol dimethacrylate crosslinked polyacrylamides, *J. Appl. Polym. Sci.* **60**, 1855-1865 (1996).
15. L. G.A. van de Water, F ten Hoonte, W. L. Driessen, J. Reedjik, and D.C. Sherrington, Selective extraction of metal ions by azathia crown ether modified polar polymers, *Inorg. Chim.Acta*, **303**, 77-85 (2000).
16. B. L. Rivas, H. A. Maturana, and S. Villegas, Synthesis, characterization, and properties of an efficient and selective adsorbent to mercury(II), *Polym. Bull.* **39**, 445-452 (1997).
17. B. L. Rivas, in *Polymeric Materials Encyclopedia*, J. C. Salamone (Ed.), Volume **6**, 4137-4143 (1996), CRC Press, Boca Raton, Florida, USA.
18. B. L. Rivas, H. A. Maturana, and S. Villegas, Adsorption behavior of metal ions by an amidoxime chelating resin, *J. Appl. Polym. Sci.* **77**, 1994-1999 (2000).
19. K. Geckeler, G. Lange , H. Eberhardt, and E. Bayer, Preparation and application of water-soluble polymer-metal complexes, *Pure Appl.Chem.* **52**, 1883-1905 (1980).
20. B. Ya. Spivakov, K. Geckeler, and E. Bayer, Liquid-phase polymer based retention- The separation of metals by ultrafiltration on polychelatogens, *Nature* **315**, 313-315 (1980).
21. B. Ya. Spivakov, V. M. Shkinev, and K. E. Geckeler, Separation and preconcentration of trace elements and their physicochemical forms in aqueous media using inert solid membranes, *Pure Appl. Chem.* **66**, 632-640 (1994).
22. G. Asman and O. Sanli, Ultrafiltration of Fe(III) solution in the presence of poly(vinyl alcohol) using modified poly(methylmethacrylate-*co*-methacrylic acid) membranes, *J. Appl. Polym. Sci.* **164**, 1115-1121 (1997).

23. F. Higashi, C. S. Cho, and H. Kakinoki, A new organic semiconducting polymer from Cu^{2+} chelate poly(vinyl alcohol) and iodine, *J. Polym. Sci., Polym. Chem. Ed.* **17**, 313-318 (1979).

24. N. Hojo, H. Shirai, and S. Hayashi, complex formation between poly(vinyl alcohol) and metallic ions in aqueous solution, *J.Polym. Sci. Polym.Symp.* **47**, 299-307 (1979).

25. C. Travers and J. A. Marinsky, The complexing of Ca(II), Co(II), and Zn(II) by polymethacrylic acid and polyacrylic acid, *J. Polym. Sci., Polym. Symp.* **47**, 285- 297 (1974).

26. K. Geckeler, K. Weingartner, and E. Bayer, in *Polymeric Amines and Ammonium Salts*, E. Goethals (Ed.), Pergamon Press, Oxford, p. 227, 1980.

27. E. Bayer, K. Geckeler, and K. Weingartner, Darstellung und derivatisierung von linearem polyvinylamin zur selektiven complex bindung in homogener phase, *Makromol. Chem.* **181**, 585-593 (1980).

28. E. Bayer, H. Eberhardt, and K. Geckeler, Polychelatogene zur anreicherung und abtrenung von metal ionen in homogener phase mit hilfe der membran-filtration, *Angew. Makromol. Chem.* **97**, 217- 230 (1981).

29. E. Bayer, B. Ya. Spivakov, and K. Geckeler, Poly(ethyleneimine) as complexing agent for separation of metal ions using membrane filtration, *Polym. Bull.* **13**, 307-311 (1985).

30. E. Bayer, H. Eberhardt, P. Grathwohl, and K. Geckeler, Soluble polychelatogen for separation of actinide ions by membrane filtration, *Israel J. Chem.* **26**, 40-47 (1985).

31. K. E. Geckeler, E. Bayer, B. Ya. Spivakov, V. M.Shkinev, and G. A. Voroveba, Liquid-phse polymer-based retention, a new method for separation and preconcentration of elements, *Anal.Chim. Acta*, **189**, 285-292 (1986).

32. V. M. Shkinev, G. A. Voroveba, B. Ya. Spivakov, K. E. Geckeler, and E. Bayer, Enrichment of arsenic and its separation from other elements by liquid-phase polymer-based retention, *Sep. Sci. Technol.* **22**, 2165-2173 (1987).

33. V. M. Shkinev, B. Ya. Spivakov, K. E. Geckeler, and E. Bayer, Determination of trace heavy metals in waters by atomic-absorption spectrometry after preconcentration by liquid-phase polymer based retention, *Talanta* **36**, 861-863 (1989).

34. S. Ahamadi, B. Batchelor, and S. S. Koseoglu, The diafiltration method for the study of the binding of macromolecules to heavy metals, *J. Membrane Sci.* **89**, 257-265 (1994).

35. T. Tomida, T. Inoue, K. Tsuchiya, and S. Masuda, Concentration and/or removal of metal ions using a water-soluble chelating polymer and microporous hollow fiber membrane, *Ind. Eng. Chem. Res.* **33**, 904-906 (1994).

36. M. N. Sarbolouk, Properties of asymmetric polyimide ultrafiltration membranes. Pore size and morphology characterization, *J. Appl. Polym. Sci.* **29**, 743-753(1984).

37. A. Bdair, L. Aras, and O. Sanh, Transport of sodium chloride, urea, and creatinine through membranes derived from methylmethacrylate-co-methacrylic acid and its ionomers, *J. Appl. Polym. Sci.***47**, 1497-1502 (1993).

38. K. E. Geckeler, B. L. Rivas, and R. Zhou, Poly [1-(2-hydroxyethyl)aziridine] as polychelatogen for liquid phase polymer retention (LPR), *Angew. Makromol. Chem.* **193**, 195-203 (1991).

39. K. E. Geckeler, R. Zhou, and B.L.Rivas, Metal complexation of poly1-(2-hydroxyethyl)aziridine-co-2-methyl-2-oxazoline in aqueous solution, *Angew. Makromol. Chem.* **197**, 107-115 (1992).

40. K. E. Geckeler, R. Zhou, A. Fink, and B.L.Rivas, Synthesis and properties of hydrophylic polymers.III. ligand effect of the side-chains of poly(aziridines) on metal complexation in aqueous solution, *J. Appl. Polym. Sci.* **60**, 2191-2198 (1996).

41. K. E. Geckeler, V. M. Shkinev, and B. Ya. Spivakov, Interactions of polymer backbones and complexation of polychelatogens with methylthiourea ligand in aqueous solution, *Angew. Makromol. Chem.* **155**, 151-161 (1993).

42. K. E. Geckeler, R. Zhou, A. Novikov, and B. F. Myasoedov, Polymer-supported enrichment for the determination of plutonium applied to natural waters from the Chernobyl area, *Naturwissenschaften*, **80**, 556-558 (1993).

43. V. Palmer, R. Zhou, and K. E. Geckeler, Cetylpyridinium chloride-modified poly(ethylenimine) for the removal and separation of inorganic ions in aqueous solution, *Angew. Makromol. Chem.* **215**, 175-188 (1994).

44. G. del C. Pizarro, B. L. Rivas and K. E. Geckeler, Metal complexing properties of water-soluble poly(N-maleyl glycine) studied by liquid phase polymer-based retention (LPR) technique, *Polym. Bull.* **37**, 525-530 (1996).

45. B. L. Rivas, S. A. Pooley, and M. Soto, Copolímeros de 4-vinilpiridina con acrilamida y N,N'-dimetilacrilamida. Síntesis y caracterización, *Bol. Soc. Chil. Quím.* **41**, 409-414 (1996).

46. G. del C. Pizarro, B. L. Rivas, and K. E. Geckeler, Preparation and characterization of water soluble copolymers of maleyl glycine with acrylic monomers, *J. Macromol. Sci. Pure Appl. Chem.* **A34**, 854-864 (1997).

47. B. L. Rivas, S. A. Pooley, M. Soto, and K. E. Geckeler, Synthesis, characterization and polychelatogenic properties of poly(acrylic acid-*co*-acrylamide), *J. Polym. Sci. Part A. Polymer Chem.* **35**, 2461-2467 (1997).

48. G. del C. Pizarro, O. Marambio, B. L. Rivas, and K. E. Geckeler, Interactions of the water-soluble poly(N-maleylglycine-*co*-acrylic acid) as polychelatogen with metal ions in aqueous solution, *J. Macromol. Sci.-Pure Appl. Chem.* **A34**, 1483-1491 (1997).

49. B. L. Rivas and I. Moreno-Villoslada, Analysis of the retention profiles of poly(acrylic acid) with Co(II) and Ni(II), *Polym. Bull.* **34**, 656-660 (1997).

50. B. L. Rivas, S. A. Pooley, M. Soto, H. A. Maturana and K. E. Geckeler, Poly(N,N'dimethylacrylamide-*co*-acrylic acid): synthesis, characterization and application for the removal and separation of inorganic ions in aqueous solution, *J. Appl. Polym. Sci.* **67**, 93-100 (1998).

51. B. L.Rivas and I. Moreno-Villoslada, Poly(sodium 4-styrene sulfonate) metal-ion interactions, *J. Appl. Polym. Sci.* **70**, 219-225 (1998).

52. B. L. Rivas and I. Moreno-Villoslada, poly[acrylamide-*co*-1-(2-hydroxyethyl)aziridine], an efficient water-soluble polymer for selective separation of metal ions, *J. Appl. Polym. Sci.* **69**, 817-824 (1998).

53. B. L. Rivas and I. Moreno-Villoslada, Chelation properties of polymer complexes of poly(acrylic acid) with poly(acrylamide), and poly(acrylic acid) with poly(N,N-dimethylacrylamide), *Macromol. Chem. Phys.* **199**, 1153-1160 (1998).

54. G. del C. Pizarro, O. G. Marambio, B. L. Rivas and K. E. Geckeler, Application of a synthetic water-soluble poly(N- maleylglycine-co-acrylamide) as polychelatogens for inorganic ions in aqueous solutions, *Polym. Bull.* **41**, 687-694 (1998).

55. B. L. Rivas and I. Moreno-Villoslada, Binding of Cd^{++} and Na$^+$ ions by poly(sodium 4-styrene sulfonate) analyzed by ultrafiltration and its relation with the counterion condensation theory, *J. Phys. Chem. B.* **102**, 6994-6999 (1998).

56. B. L. Rivas and I. Moreno-Villoslada, Evaluation of the counterion condensation theory from the metal ion distributions obtained by ultrafiltration of a system poly(sodium 4-styrene sulfonate)/Cd^{2+}/Na$^+$, *J. Phys. Chem. B* **102**, 11024-11028 (1998).

57. B. L. Rivas, S. A. Pooley, M. Soto, and K. E. Geckeler, Water-soluble copolymers of 1-vinyl-2-pyrrolidone and acrylamide derivatives. Synthesis, characterization, and metal binding capability studied by liquid-phase polymer based retention (LPR) technique, *J. Appl. Polym. Sci.* **72**, 741-750 (1999).

58. B. L. Rivas and I. Moreno-Villoslada, Synthesis and behavior of two copolymers of poly[(acrylamide-*co*-N(1- hydroxymethylacrylamide)) in ultrafiltration experiments, *Polym. Bull.* **44**, 159-165 (2000).

59. B. L. Rivas and E. Pereira, Obtention of poly(allylamine)-metal complexes through liquid-phase polymer based retention LPR) technique, *Bol. Soc. Chil. Quím.* **45**, 165-171 (2000).

60. B. L. Rivas and I. Moreno-Villoslada, Effect of the polymer concentration on the interactions of water-soluble polymers with metal ions, *Chem. Letters,* 166-167 (2000).

61. B. L. Rivas, E. Pereira, E. Martínez and I. Moreno-Villoslada, Metal ion interactions with poly(2-acrylamido-2-methyl-1-propanesulfonic acid-*co*-methacrylic acid), *Bol. Soc. Chil. Quím.* **45**, 199-205 (2000).

62. B. L. Rivas and E. D. Pereira, Viscosity properties of aqueous solution of poly(allylamine)-metal complexes, *Polym. Bull.* **47**, 69-76 (2000).

63. B. L. Rivas and I. Moreno-Villoslada, Prediction of the retention values associated to the ultrafiltration of mixtures ions and high molecular weight water-soluble polymers as a function of the initial strength, *J. Membrane Sci.* **178**, 165-170 (2000).

64. B. L. Rivas, S. A. Pooley, and M. Luna, Chelating properties of poly(N-acryloyl piperazine) by liquid-phase polymer-based retention (LPR) technique, *Macromol. Rapid Commun.* **13**, 905-908 (2000).

65. B. L. Rivas, E. Martínez, E. Pereira, and K. E. Geckeler, Synthesis, characterization, and polychelatogenic properties of poly(2-acrylamido-2-methyl-1-propane sulfonic acid-*co*-methacrylic acid), Polymer International **50**, 456-462 (2001).

66. B. L. Rivas, S. A. Poley, and M. Luna, Poly(N-acetyl-α-acrylic acid). Synthesis, characterization, and chelation properties through liquid-phase polymer-based retention (LPR) technique. *Macromol. Rapid Commun.* **22**, 418-421 (2001).

67. B. L. Rivas, S. A. Pooley, E. D. Pereira, and P. Gallegos, Liquid-phase polymer-based retention (LPR) technique to determine the maximum retention capacity of a strong polyelectrolyte for di- and trivalent cations. Polym. Bull. (submitted).

BOVINE SERUM ALBUMINE COMPLEXATION WITH SOME POLYAMPHOLYTES

Alexander G.Didukh, Gulmira Sh.Makysh, Larisa A.Bimendina, and Sarkyt E.Kudaibergenov*

1. INTRODUCTION

Purified protein preparations, which recently were used only in biochemical laboratories, nowadays are coming more and more into our everyday life as medicine preparations and detergents. They are used in fine organic synthesis and food production as well as in a variety of analytical techniques. One of the effective methods of protein separation and purification is protein precipitation (liquid phase splitting) with the help of polyelectrolytes. The most extensively studied protein complexes are mixtures of lysozyme[1], albumins[2-4], gelatin[5,6], and catalase[7,8] with weak and strong polyelectrolytes of linear and crosslinked structure.

Earlier Dubin[9] has reviewed in detail the interaction of linear anionic and cationic polyelectrolytes with proteins and outlined purification of proteins by selective phase separation. The effects of various relevant factors such as molecular weight and concentration of polymers, charge density, pH, and ionic strength of the solution have been thoroughly considered. It has been shown that the polyelectrolyte precipitation technique is effective to recover some proteins up to 100%.

However, the main problem is to separate the precipitated proteins without the loss of their original functions and to recycle the polyelectrolyte precursors. In principle, protein-polyelectrolyte complexes can be redissolved by pH adjustment and a high molecular weight polymer could then be removed by ultrafiltration. If the polyelectrolyte is relatively inexpensive, it could also be removed by precipitation with a polyelectrolyte of opposite charge.

---

* Institute of Polymer Materials and Technology, Satpaev Str.18a, 480013, Almaty, Kazakhstan.

Advanced Macromolecular and Supramolecular Materials and Processes
Edited by K. Geckeler, Kluwer Academic/Plenum Publishers, 2003

265

Although protein precipitation by polyelectrolytes has been studied for many years, few theoretical works have been directed toward understanding the mechanism of precipitation. Authors[10] developed a molecular-thermodynamic approach for precipitation of charged globular proteins by oppositely charged linear polyelectrolytes.

The complexation of proteins, such as soybean tripsin inhibitor, ovalbumin, ribonuclease and lysozyme with dilute solutions of a random- and blockpolyampholyte DMAEM-MAA-MMA was studied by turbidimetric titration.[11,12] Separation of protein mixture with the help of random triblock polyampholytes $DMAEM_8MMA_{12}MAA_{16}$ was performed by other authors.[13-15] It follows from the turbidimetric titration curves of 1-methyl-4-vinylethynylpiperidinol-4-methacrylic acid (MVEP-MAA) by HSA that the most favorable region of interaction of HSA ($pH_{iep} = 5.4$) and MVEP-MAA ($pH_{iep} = 9.4$) is arranged between their isoelectric points,[16-19] The increase of the ionic strength of the solution makes the association interval of pH narrower.

In this paper the complex formation of bovine serum albumin (BSA) with annealed polyampholyte based on 1,2,5-trimethyl-4-vinylethynylpiperidinol-4-acrylic acid (TMVEP-AA) and a novel betaine-type polyampholyte based on a Schiff base and acrylic acid (SB-AA) was investigated in comparison with PAA.

2. EXPERIMENTAL PART

2.1. Materials

Bovine serum albumin (BSA) with a molecular weight of 68 kg mol^{-1} and an isoelectric point pI =5.4 with 99% purity was purchased from Boehringer Mannheim (Germany) and used without further purification. Commercial poly(acrylic acid) (PAA) with a molecular weight of 260 kg mol^{-1} was received from Aldrich Chemical Co. (USA). Acrylic acid (AA) and Schiff base (SB) were purified twice by distillation. Preparation and purification of 1,2,5-trimethyl-4-vinilethinylpiperidol-4 has been described earlier.[20] HCl and NaOH solutions and NaCl were from Fisher and used as received.

2.2. Procedures

2.2.1. Synthesis of Polyampholyte with Betaine Structure

The linear polybetaine based on SB and AA was synthesized by Michael addition reaction followed by radical polymerization in a water-ethanol mixture (50:50 vol.%).[21] During the mixing of AA and SB at first the formation of an intermediate product via the Michael addition reaction takes place that is accompanied by protonation of nitrogen and internal salt formation (Scheme 1).

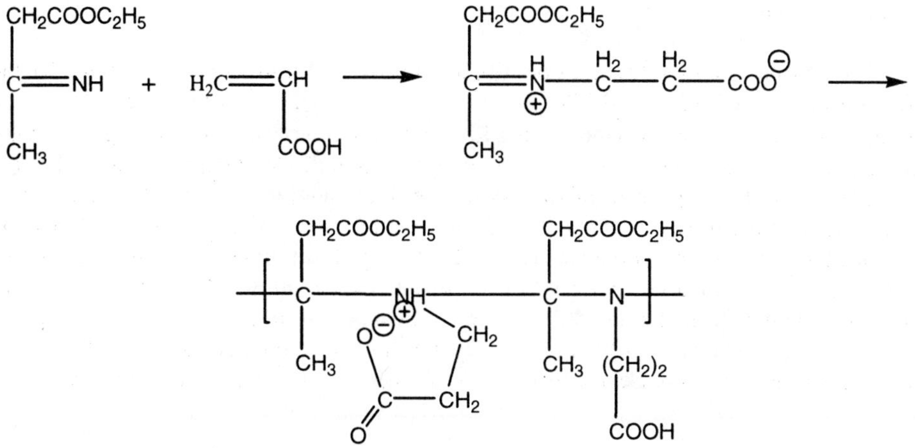

CH$_2$COOC$_2$H$_5$ CH$_2$COOC$_2$H$_5$

C=NH   +   H$_2$C=CH   ⟶   C=N—C—C—COO$^{\ominus}$ ⟶

CH$_3$ COOH   CH$_3$

Scheme 1. Formation of the betaine-type polyampholyte.

At the next stage such intermediates are involved in the polymerization process in the presence of azobisisobutyric acid (c = $5\cdot10^{-3}$ mol·L^{-1}). By adding of AA to a solution of polyiminoethylene or polyiminohexamethylene both a protonation and a Michael addition reaction take place simultaneously.[22] An equimolar monomer mixture was placed into the ampoule from the molybdenum glass and bubbled by argon to remove the oxygen. The ampoule was then sealed and thermostated at 70 °C during one hour. The linear polymer was purified by three-fold precipitation from a water-ethanol mixture into ethyl acetate and washed with acetone several times. The samples were dried *in vacuo* up to constant mass.

The synthesis of copolymer based on the 1,2,5-trimethyl-4-vinylethinylpiperidinol-4 (TMVEP) and acrylic acid (AA) was described earlier.[20]

2.2.2. Potentiometric Proton Titration

The acid-base ratio of functional groups was investigated by potentiometric proton titration. Polyampholyte solutions with concentration 1 g·L^{-1} in 0.01M NaCl were adjusted to pH = 7 by addition of 0.5 N NaOH or HCl. 10 mL aliquots were titrated by 0.1 N NaOH or 0.1 N HCl in nitrogen atmosphere. A 2-point calibration "Mettler Toledo pH meter", equipped with a combined electrode, was used for the registration of the pH values.

For correction and calculation of the results the blank solutions of water with the same ionic strength were also adjusted to pH = 7 and titrated by 0.1 N NaOH and 0.1 N HCl. The composition of polymer samples obtained was determined by elementary analysis, potentiometric, and conductimetric titration. For both polyampholytes the molar ratio of carboxylic and amino groups was equal to 1:1.

2.2.3. Turbidimetric Titration

The dependence of solution turbidity on pH ("type 1 titration") was obtained by the addition of 0.1 M NaOH or 0.1 M HCl to a protein-polymer mixture at constant ionic strength and at constant polymer (0.1 g·L^{-1}) and protein concentrations (1 g·L^{-1}) [8-9]. Solutions of protein and polymer were prepared independently and filtered through Gelman 0.2 μm filters prior to mixing. After the titrant addition, the solution was gently stirred (the stirring time was generally 2-3 min) until a stable turbidity (±0.1% T) reading was obtained. A nitrogen purge was employed during all titrations. A Mettler Toledo pH meter with a combined electrode was used to monitor solution pH. Transmittance was monitored with a Brinkman PC 800 colorimeter, connected to a 2 cm path length optical probe, and equipped with a 420 nm filter. The turbidity was reported as 100-% T, and %T fluctuations (±0.1%) were treated by consistently selecting the highest transmittance.

2.2.4. Measurement of the Isoelectric Point of Polyampholytes

Isoelectric point of TMVEP-AA was measured using Ubellohde viscometer. Polyampholyte solutions were filtered through the 0.45 μm filter and adjusted to different pH by addition of 0.5 N NaOH or HCl. Then these solutions were stirred and relative viscosity values were measured. The electrophoretic mobility of both SB-AA and polyampholyte-protein complexes was measured by Doppler electrophoretic light scattering with the help of the Coulter DELSA 440SX. Electrophoretic light scattering was carried out at 25±0.1 °C at four angles (8.9, 17.6, 26.3, 35,2°). The electric field was applied at a constant current of 8-14 mA. The electrophoretic cell has a rectangular cross section connecting the hemispherical cavities in each electrode. The total sample volume was about 1 mL. The measured electrophoretic mobility, U, was the average value at the upper stationary layer for the four scattering angles. pH of polymer solutions was adjusted by 0.5N NaOH or HCl. After adjusting pH and stirring about 30 min solution was injected into cell through the 0.45 μm filter.

3. RESULTS AND DISCUSSION

The electrophoretic mobility of the betaine-type polyampholyte in dependence of the pH shows that the amphoteric polymer has a zero mobility at pH ≈ 2.1 (Fig. 1). The "asymmetry" in the mobility behavior can be prescribed to differences in the displacement of oppositely charged groups along the chain and condensation of counterions to positive and negative charges.[27] Ionization of nitrogen atoms located in the main chain at low pH probably causes lesser stretching effect with respect to the whole macromolecule than the ionization of carboxylic groups displaced in a side position (Scheme 2).

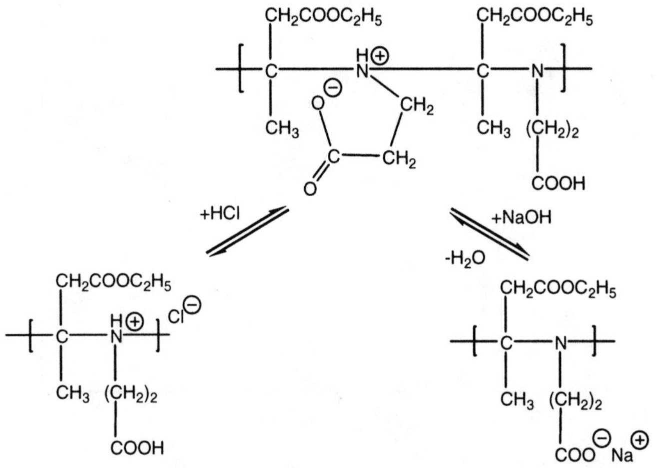

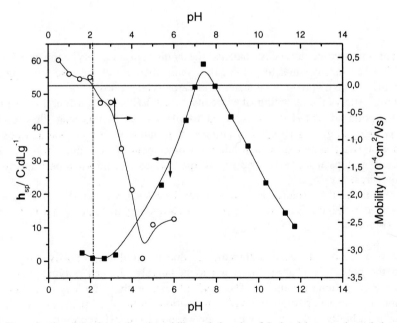

Scheme 2. Ionization of the betaine-type polyampholyte.

Figure 1. Dependencies of the electrophoretic mobility and viscosity of the betaine-type polyampholyte SB-AA on pH.

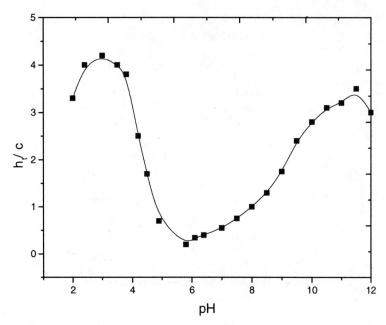

Figure 2. Dependence of the reduced viscosity of annealed polyampholyte TMVEP-AA on pH.

According to viscometric measurements the minimum viscosity corresponds to the isoelectric point that is equal to pH ≈ 2,0. a sharp increase of the viscosity takes place between pH = 4 - 6 owing to the ionization of carboxylic groups.

Figure 2 shows the changing of viscosity of TMVEP-AA as a function of pH. The isoelectric point of TMVEP-AA is equal to pH ≈ 5.9. In this state the number of positively charged groups is equal to number of negatively charged one and oppositely charged groups are attracted to each other. Displacement of pH values in more basic or more acidic regions from isoelectric region (pI) causes to conformation transitions and viscosity increasing.

The results of the turbidimetric titration for the PAA/BSA and SB-AA/BSA systems are presented in Figure 3. Three complex formation regions can be defined for these systems (Figure 4).

The first region of constant turbidity is due to Coulomb repulsive forces action between the positively charged protein and negatively charged polymer that (retard) prohibit the formation of complex. The weak plateau on the curve can be considered as the region of primary or soluble complex formation. In the third part of curve the sharp increase in turbidity indicates to phase separation. Here pH_c represents the boundary between the non-associative and primary phases and pH_ϕ represents the boundary between the primary and aggregate phases. It is seen that TMVEP-AA forms non-stable and soluble complex with BSA.

This can be explained by close pI values for the polyampholyte and protein (pI_{BSA} = 5.4 and $pI_{TMVP-AA}$ = 5.9). In the pH_c = 3.3 they have weak opposite charges and form fast breaking electrostatic bounds, while SB-AA has a pI = 2.1. The protein binding behavior of this polyampholyte is similar to that of PAA.

It can be supposed that the region of complexation will be at weak acidic pH values. As seen from Figure 4, the results of the turbidity titration coincide well with DELSA measurements.

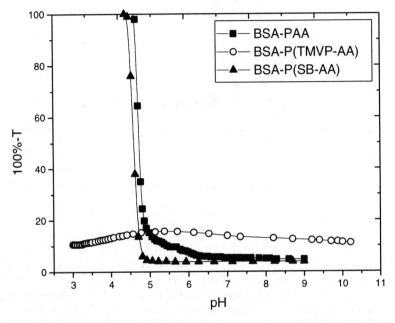

Figure 3. Type 1 turbidity titration of the PAA/BSA and polyampholyte/BSA systems (r = 10).

The comparison of the viscosity change of TMVEP-AA and the complex TMVEP-AA/BSA (Figure 5) confirms the previous turbidimetric titration results on the formation of a soluble complex between BSA and TMVEP-AA.

Figure 6 shows the PAA/BSA complex phase boundary data plotted as a function of pH and ionic strength. An increase of ionic strength leads to a shielding effect of charged groups of both the protein and polyelectrolyte. As a result an increase of pH_c and pH_ϕ is observed. It is shown that the phase boundary separates the non-associative, soluble, and coacervate regions. Inspection of the low ionic strength region reveals the formation of a soluble PAA/BSA complex in a pH range where the net charge of the protein is still weak negative.

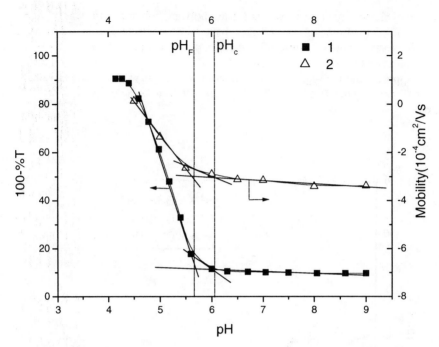

Figure 4. Type 1 turbidimetric titration and electrophoretic mobility data for SB-AA/BSA in 0.01M HCl (r = 10).

The formation of complexes between proteins and polyelectrolytes under conditions at which the protein net charge is of the same sign as that of the polymer has been observed earlier. A possible explanation for the protein-polyanion binding at pH > pI depends on the presence of a negatively charged "patch", thus the complexation at pH_c is a result of electrostatic interactions between the polyelectrolyte and some local region of the huge protein surface. Decreasing of the pH leads to a transition of the "amphoteric" protein system to "cationic". In this case the electrostatic association of the oppositely charged protein and polyanion surface occurred. This is reason of the pH_ϕ location in the low pH region.[9]

Figure 7 represents the phase boundaries of polyampholyte-BSA systems. It is seen that complex formation of SB-AA with BSA is in the region between isoelectric points of polyampholyte and protein. The complexation behavior of this polyampholyte is similar to that of PAA because in both cases the protein binding groups are carboxylic moieties. In the case of the TMVEP-AA/BSA system a weak and soluble complex formation is observed, because the pI values of the polyampholyte and protein are close to each other. Moreover, at the ionic strength > 0.05 a complexation reaction was practically not observed. This is due to screening of macromolecular charges by low molecular weight electrolytes.

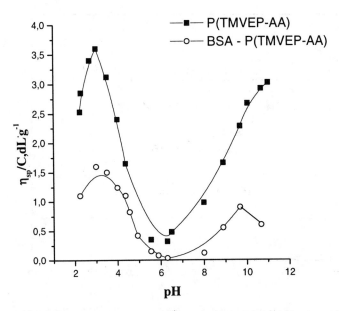

Figure 5. Dependence of the viscosity of annealed polyampholyte TMVEP-AA and its complex with BSA on pH.

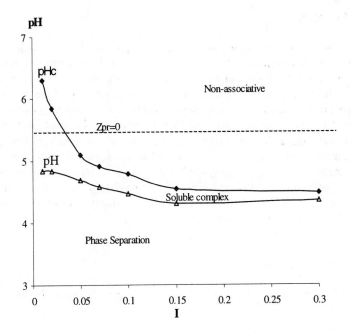

Figure 6. pH$_c$ and phase boundaries for the BS/-PAA complex at c$_{BSA}$= 1 g·L^{-1} and r = 10.

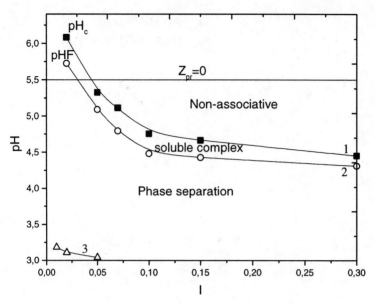

Figure 7. Phase boundary and pH_c for BSA/SB-AA (curve 1,2) and BSA/TMVEP-AA (curve 3) for $c_{protein} = 1$ $g \cdot L^{-1}$ and $r = 10$.

4. CONCLUSION

The complexation of poly(acrylic acid) and polyampholytes with BSA has been studied by the measurement of solution turbidity and electrophoretic mobility. It is observed that for PAA/BSA and SB-AA/BSA systems the primary complex formation started at high pH values ($pH > pI_{BSA}$), when the ionic strength of the solution was rather low. This phenomenon can be interpreted in terms of the presence of positively charged "patch" on the global protein surface. An increase of the ionic strength lead to an increase of pH_c and pH_ϕ. Completed aggregation was observed at low pH when the amphoteric protein becomes positively charged and is attracted to the oppositely charged polyanion. The betaine-type polyampholyte SB-AA has a similar behavior like poly(acrylic acid) due to the presence of carboxylic groups in the side-chain. Zhe polyampholyte TMVEP-AA formed only soluble primary complexes with BSA due to closeness of their isoelectric points.

5. ACKNOWLEDGEMENT

INTAS-00/57 and INTAS-00/113 grants are greatly appreciated for financial support.

6. REFERENCES

1. J. Y. Shieh and C. E. Glatz, Precipitation of proteins with polyelectrolytes: Role of polymer molecular weight, In:*Macromolecular Complexes in Chemistry and Biology*, edited by P. Dubin, J. Block, R. Davis, D. N. Schulz, and C.Thies, (Springer-Verlag, Berlin, 1995). pp.273-284.
2. E. Kokufuta, Complexation of proteins with polyelectrolytes in a salt-free system and biochemical characteristics of the resulting complexes, In: *Macromolecular Complexes in Chemistry and Biology*, edited by P. Dubin, J. Block, R. Davis, D. N. Schulz, and C. Thies, (Springer-Verlag, Berlin, 1995). pp.300-325.
3. K. W. Mattison, I. J. Brittain, and P. L. Dubin, Protein-polyelectrolyte phase boundary, *Biotechnol. Prog.* 11, 632-637 (1995).
4. A. Tsuboi, T. Izumi, M. Hirata, J. Xia, P. Dubin, and E. Kokufuta, Complexation of proteins with a strong polyanion in an aqueous salt-free system, *Langmuir* 12, 6295-6303 (1996).
5. J. Greener, B. A. Contestable, and M. D. Bale, Interaction of anionic surfactants with gelatin: Viscosity effects, *Macromolecules* 20, 2490-2498 (1987).
6. . W. A. Bowman, M. Rubinstein, and J. S. Tan, Polyelectrolyte-gelatin complexation: Light-scattering study, *Macromolecules* 30, 3262-3270 (1997).
7. G. A. Bektenova, E. A. Bekturov, and S.E. Kudaibergenov, Interaction of catalase with cationic hydrogels: Influence of pH, kinetics of process and isotherms of adsorption, *Polym. Adv. Technol.* 10, 141-145 (1999).
8. G. A. Bektenova, E. A. Bekturov, G. K. Sulekeshova, and S. E. Kudaibergenov, Interaction of amphoteric hydrogels with catalase: Influence of pH and ionic strength, *Polym. Prepr.* 41(1), 750-751 (2000).
9. J. Xia and P. L. Dubin, Protein-polyelectrolyte complexes, In: *Macromolecular Complexes in Chemistry and Biology*, edited by P. Dubin, J. Block, R. Davis, D. N. Schulz, and C. Thies, (Springer-Verlag, Berlin, 1995). pp.247-271.
10. C. S. Patrickios, C. J. Jang, W. R. Hertler, and T. A. Hatton, Protein interactions with acrylic polyampholytes, *Polym. Prepr.* 34(1), 954-955 (1993).
11. C. S. Patrickios, C. J. Jang, W. R. Hertler, and T. A. Hatton, Protein interactions with acrylic polyampholytes, in: *Macro-ion Characterization from Dilute Solutions to Complex Fluids*, edited by K. S. Schmitz, (ACS Symposium Series, Washington DC, v.548, Ch.19, p.257-267 1994).
12. C. S. Patrickios, L. R. Sharma, S. P. Armes, and N. C. Billingham, Precipitation of a water-soluble ABC triblock methacrylic polyampholyte. Effects of time, polymer concentration, salt type and concentration, and presence of a protein, *Langmuir* 15, 1613-1620 (1999).
13. S. Nath, C. S. Patrickios, and T. A. Hatton, A turbidimetric titration study of the interaction of proteins with block and random acrylic polyampholytes, *Biotechnol.Prog.* 11(1), 99-103 (1995).
14. S. Nath, Complexation behavior of proteins with polyelectrolytes and random acrylic polyampholytes using turbidimetric titration, *J. Chem. Tech. Biotech.* 62(3), 295-300 (1995).
15. C. S. Patrickios, W. R. Hertler, and T. A. Hatton, Protein complexation with acrylic polyampholytes, *Biotechnol. Bioeng.* 44, 1031-1039 (1994).
16. S. E. Kudaibergenov and E. A. Bekturov, Influence of the coil-globule confromational transition in polyampholytes on sorption and desorption of polyelectrolytes and human serum albumin, *Vysokomol. Soedin. Ser.A.* 31, 2614-2617 (1989).
17. A. K. Tultaev, S. E. Kudaibergenov, and E. A. Bekturov, Study of complexation of synthetic polyampholyte with human serum albumin, *Izv. Akad. Nauk KazSSR, Ser. Khim.* 6, 67-71 (1990).
18. H. Morawetz and W. L. Hughes, The interaction of proteins with synthetic polyelectrolytes. 1. Complexing of bovine serum albumin, *J. Phys. Chem.* 56, 64-69 (1952).
19. L. S. Rodkey and A. Hirata, Studies of ampholyte-protein interactions, *Prot. Biol. Fluides*, 34, 745-748 (1986).
20. O. Sh. Kurmanaliev, E. M. Shaikhutdinov, Sh. S. Tulbaev, and T. M. Mukhametkaliev, Influence of water on radical copolymerization of 1,2,5-trimethyl-4-vinylethynylpiperidinol-4 with methacrylic acid, *Vysokomolek. Soedin. Ser B*, 22, 526-528 (1980).
21. S. E. Kudaibergenov, Synthesis and characterization of Schiff base polyampholytes (to be published).
22. R. C. Schulz, M. Schmidt, E. Schwarzenbach, J. Zoller, Some new polyelectrolytes, *Macromol.Chem., Macromol. Symp.* 26, 221-231 (1989).

FLUORESCENCE QUENCHING TECHNIQUE FOR STUDY OF DNA-CONTAINING POLYELECTROLYTE COMPLEXES

Vladimir A. Izumrudov, Marina V. Zhiryakova, and Natalia I. Akritskaya[*]

1. INTRODUCTION

Polyelectrolyte complexes (PEC) are the products of cooperative coupling reactions between two unlikely charged polyions of high charge density, in particular with ionogenic groups in each monomer unit of the chain. Of late there has been a widespread interest in research of competitive reactions in PEC's solutions mimicking some important regulator processes *in vivo* accompanied by a transfer of charged biopolymers. Data obtained on studying of equilibrium, kinetics and mechanism of the competitive interpolyelectrolyte reactions are summarized in review[1]. These results lead to crucially new consideration of PEC as macromolecular compounds, permanently exchanging by polyions in water-salt solutions. The ability of PEC to combine high stability with the capacity to take part in the interpolyelectrolyte reactions ensures self-assembly of complex particles in the solutions. Perfect selectivity and high rate of the cooperative interpolyelectrolyte reactions endow PEC with sensitivity to external factors (pH, ionic strength, temperature, etc.) making them self-adjustment systems. Both formation of PEC and their transformation are accomplished by the method of trials and errors via polyions transfer until the equilibrium is achieved.

This view on polyelectrolyte complexes and their properties is fundamental and rather common. The principles of competitive binding and chain transfer ascertained for synthetic polyions can be extended to other families of complexes formed by polyelectrolyte and oppositely charged partner of different nature, for review see[2]. Flexible linear polyions with (ir)regular alternating of the charges along the chains, ionogenic polypeptides, branched polyelectrolytes, ionogenic dendrimers of different generations, globular proteins (enzymes, antibodies), conjugates of the proteins with polyelectrolytes, and nucleic acids could be components of the complexes.

[*] Polymer Chemistry Department, Faculty of Chemistry, Moscow State University V-234, Moscow 119899, Russia. E-mail: izumrud@genebee.msu.su, Fax: (095) 9390174.

Advanced Macromolecular and Supramolecular Materials and Processes
Edited by K. Geckeler, Kluwer Academic/Plenum Publishers, 2003

Competitive reactions in solutions of the complexes were shown to obey the same regularities and controlled by similar factors. The features of the chain transfer revealed on studying of equilibrium and kinetics of competitive interpolyelectrolyte reactions form a basis for design of multicomponent polyelectrolyte systems potential for practical use, in particular in biotechnology and bioseparation.

DNA-containing PEC occupy a significant place in the development of this line of inquiry. Being of immense biological importance, DNA is a polyacid that should be attributed to a family of highly charged polyanions. Recently complexing of DNA with polycations were successfully used for increasing an efficiency of transformation of the cells by plasmides and for protection of DNA from splitting by cell nucleases, for review see[3]. The prospects for addressing DNA packed in PEC species to the target cell have motivated extension of study of DNA-containing PEC in order to give precise control over their stability, solubility, and ability to take part in the competitive interpolyelectrolyte reactions.

The informative method of studying both PEC's formation and interpolyelectrolyte reactions in their solutions is fluorescence quenching technique. The approach most extensively employed is based on using of fluorescence labeled polyion and oppositely charged partner-quencher. A pair PMA*–PEVP consisted of pyrene-tagged poly(methacrylate) anion, PMA*, and poly(N-ethyl-4-vinylpyridinium) cation, PEVP, was in most common use[1]. The ability of pyridinium groups of PEVP to quench a fluorescence of the pyrene labels of PMA* allowed one to monitor either formation or destruction of PEC(PMA*–PEVP) by quenching or ignition of the fluorescence, respectively. In turn, this enabled us[4, 5] to reveal the factors affecting the equilibrium of reaction (1) in which native DNA competes with PMA* for binding with PEVP.

$$PEC(PMA*–PEVP) + DNA \Leftrightarrow PEC(DNA–PEVP) + PMA* \tag{1}$$

As might be expected, reaction (1) had much in common with reaction (2) with sodium poly(phosphate) (PPh) as the polyanion-competitor, which was monitored by the same manner[6].

$$PEC(PMA*-PEVP) + PPh \Leftrightarrow PEC(PPh-PEVP) + PMA* \tag{2}$$

Equilibrium of the both reactions proved to be effectively controlled by concentration and nature of added low-molecular-weight electrolyte. However, the ability of alkaline metal cations to shift equilibrium (2) to the right increased as follows[6]: $Li^+ > Na^+ > K^+$, whereas in reaction (1) they exhibited a different order[4], i. e. $Na^+ > K^+ > Li^+$ with Ca^{++} and Mg^{++} being more efficient in this respect[5]. The antipodal position of Li^+ in above series of alkali metal cations is caused by abnormally high affinity of this cation to native DNA. Denatured DNA interacts with Li^+ without any sign of the abnormality and for denatured DNA equilibrium (1) obeys the expected order[4]: $Li^+ \approx Na^+ > K^+$. Equilibrium (1) is controlled by ratio χ of the degrees of polymerization, DP, $\chi = DP(PMA*) / DP(PEVP)$[4]. If PEC(PMA*–PEVP) with $\chi > 1$ is used then it is shifted to the left. Otherwise, at $\chi < 1$ decrease of χ leads to entropy-favorable shift of the equilibrium to the right. These data suggest that binding of DNA with positively charged macromolecules or its release from such complexes can be controlled by the factors of splitting or lengthening of the chains of interacting macromolecular partners.

In the present work another version of fluorescence quenching technique developed specially for investigation of DNA-containing PEC is discussed. The method is based on fluorescence properties of cationic dye ethidium bromide (EB) widely used as fluorescence probe for native DNA. The potential of this approach is demonstrated by the revealed possibilities to estimate complexing of DNA with different positively charged partners, to obtain profiles of the interpolyelectrolyte reactions, and to elucidate factors affecting stability of DNA-containing PEC in water-salt solutions.

2. EXPERIMENTAL PART

2.1. Materials

Ethidium bromide, calf thymus DNA (~10000 base pairs) and poly-L-lysine hydrobromide (DP = 10 ÷ 25) were purchased from Sigma (USA). Sample of histone f_1 from calf thymus was purchased from Sigma too. Samples of other used polycations were synthesized and characterized as described elsewhere[7, 8].

2.2. Measurements

Fluorescence intensity of the solutions was measured using Jobin Yvon-3CS spectrofluorimeter (France) with a water-thermostatic stirred holder. The measurements were performed in a capped quartz cell upon permanent stirring at 25 ^{0}C. The excitation and emission wavelengths were set at 535 and 595 nm, respectively.

The potentiometric titration of the solutions was conducted using MultiLab 540 (WTW, Germany) with a Metler Toledo glass calomel combination electrode in a water-thermostatic stirred cell holder under the argon atmosphere at 25 ^{0}C.

3. RESULTS AND DISCUSSION

Monitoring of either complexing of DNA with different polycations and dissociation of PEC formed were carried out by fluorescence quenching technique using the ability of fluorescence cationic dye ethidium bromide to intercalate into DNA double helix followed by the ignition of ethidium fluorescence.

Ethidium bromide binds strongly with nucleic acids. The (DNA·EB) complexes are very stable (the constant of complex formation, K_C, is of the order of 10^6 M^{-1}) because of the specific interaction of planar ethidium cations with DNA bases and survive even in concentrated solutions of low molecular weight salts[9]. This interaction produces a large increase in the fluorescence quantum yield of EB, for review see[10]. Release of EB from the (DNA·EB) complex causes a decrease in the quantum yield and has been used to study the association of low molecular weight polyamines (short synthetic polypeptides[11] and analogs of spermine[12, 13]) with calf thymus DNA *in vitro*.

Subsequently we reported[14] that the fluorescence spectroscopic assay is applicable for analyzing of complex formation between DNA and different positively charged species, in particular linear synthetic polyamines, basic polypeptides and histones. Moreover, the same approach proved to be appropriate for monitoring a destruction of DNA-containing PEC in water-salt solutions as well[7, 8, 14].

3.1. Complexing of DNA with Different Polyamines

Typical curves of fluorimetric titration of salt-free solution of the (DNA·EB) complex by PEVP (curve 1), 3,6-ionene bromide (curve 2), poly(N,N-dimethylaminoethyl methacrylate), PDMAEM, (curve 3) and histone f_1 (curve 4) are shown in Fig. 1 as a dependencies of fluorescence intensity I on the ratio φ of molar concentration of charged amine groups to molar concentration of phosphate groups, P, of DNA, $\varphi = [+] / [-]$.

It is seen that the curves are sigmoidal with pronounced decrease of the fluorescence intensity at $\varphi \approx 1$. The quenching is caused by abrupt expulsion of EB from the double helix that appears to be conditioned by sudden collapse of DNA molecule in a very narrow range of φ close to unity[15]. Thus, titration of the (DNA·EB) complex by PEVP at charge ratio close to unity was followed by pronounced successive accumulation of free non-fluorescence dye in solution detected on the sedimentation patterns[14].

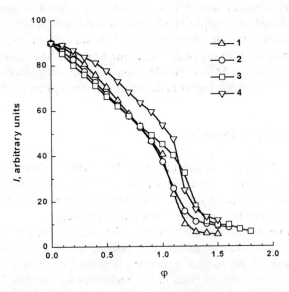

Figure 1. Dependencies of fluorescence intensity I of the mixtures of solution of the (DNA·EB) complex and polycation on the ratio $\varphi = [+] / [-]$ for different polycations: PEVP, DP = 100 (1), 3,6-ionene, DP = 70 (2), PDMAEM, DP = 700 (3), and histone f_1 (4). 0.02 M HEPES, pH 7.0, 25 °C. [P] = 4 × 10^{-5} M, [EB] / [P] = 0.25.

The displacement of EB is quantitatively determined by cooperative electrostatic interaction of DNA with positively charged partner. This is evidenced by data of the fluorescence quenching caused by polycations of different charge density, i. e. partly quaternized poly(N-ethyl-4-vinylpyridinium) bromides[7] and block or graft copolymers of N-(2-hydroxypropyl)methacrylamide and 2-(trimethylammonio)ethyl metha-crylate[15]. The differences among the titration curves for the studied polycations of different composition and structure were very small. In all cases a distinct sudden fall in the fluorescence intensity was observed in the range $\varphi \approx 0.8 \div 1.0$.

The influence of polycation chain length on the fluorescence quenching is evident from the titration curves of the (DNA·EB) complex by exhaustively alkylated PEVP[7] of various DP depicted in Fig. 2. It is seen that shortening of the polycation down to DP = 100 practically does not change the quenching (curves 1 and 2), whereas further cutting of the chain is followed by shift of the curves to the higher φ and their smoothing. The observed decrease of the quenching efficiency is the result of expected weakening of the electrostatic interaction when passing from the polycations with DP $\geq$ 100 to rather short chains, DP $\leq$ 20, that essentially are the positively charged oligomers. Hence low cooperativity of the electrostatic binding between DNA and the charged oligomers is reflected by degeneration of the region of the pronounced quenching. The smooth appearance was also inherent in the fluorimetric curves of association of low molecular weight oligoamines (analogs of spermine) with calf thymus DNA[12, 13]. It suggests that cooperativity of electrostatic interaction between DNA and positively charged species can be, at least qualitatively, estimated from the shape of the fluorimetric curve.

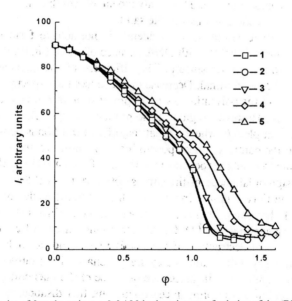

Figure 2. Dependencies of I on the ratio $\varphi = [+] / [-]$ in the mixtures of solution of the (DNA·EB) complex and PEVP of different DP: DP = 1300 (1), 100 (2), 40 (3), 20 (4), and 10 (5). 0.02 M TRIS, pH 9.0. The other conditions are the same as in Fig. 1.

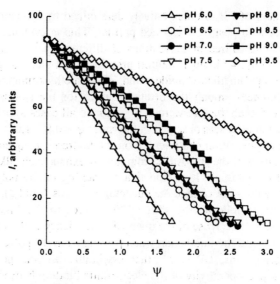

Figure 3. Dependencies of *I* on the ratio ψ = [all amine groups] / [-] obtained on the titration of the complex (DNA·EB) by PEI at different pH. The other conditions are the same as in Fig. 1.

Low cooperativity of the interaction reflected by the absence of the sudden fall in the titration curves can be conditioned not only too small length of the chains but steric hindrances and/or destruction of a part of the salt bonds.

Branched poly(ethyleneimine), PEI, evidently is not able to form highly ordered system of interpolymer salt bonds with DNA because of steric hindrances. The curves of Fig. 3 are accordingly rather smooth. The efficiency of the quenching by PEI in alkaline media, pH 9.5, is minimal. Decrease of pH causes the expected increase of the quenching efficiency of the polycation owing to protonation of primary, secondary, and tertiary amine groups. However, region of the pronounced quenching is not observed even on the titration at pH 6.0, where a great majority of salt bonds between DNA and PEI is formed. On the contrary, linear polyamines containing amine groups of one of above types are able to displace cooperatively the dye from the (DNA·EB) complex. It follows from the sigmoidal shape of the curves obtained at pH ≤ 8 by the addition of PDMAEM with tertiary amine groups in the macromolecule or poly-L-lysine hydrobromide, PLL, bearing primary amine groups in the chain at pH ≤ 9 (Fig. 4).

Data of the quenching by PDMAEM accomplished at different pH provide impressive evidence that destruction of salt bonds might lead not only to decrease of the quenching efficiency, but qualitative change of shape of the curves. Fig. 4 shows that increase of pH results both in successive decrease of the quenching efficiency and degradation of a region of the pronounced quenching that disappeared completely at

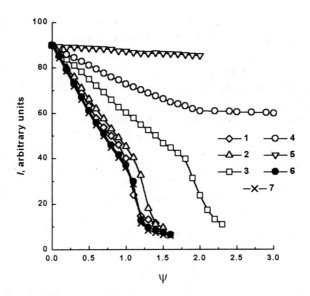

Figure 4. Dependencies of I on the ratio ψ = [all amine groups] / [-] obtained on the titration of the complex (DNA·EB) by PDMAEM (1 – 5) and PLL (6, 7) at different pH: 6.0 (1, 6), 7.0 (2), 8.0 (3), 8.5 (4), and 9.0 (5, 7). The other conditions are the same as in Fig. 1.

8.0 < pH < 8.5. It reflects a lowering of cooperativity of the coupling reaction due to progressive deprotonation of amine groups and consequent destruction of salt bonds in alkaline media. A shape of the curve obtained at pH 8.5 is quite similar to curve of the titration by short polyamines, in particular short cationic peptide KALA[16]. This peptide was not a highly efficient DNA-compacting agent as the fluorimetric titration curve flattened out at charge ratios above 1 / 1 where about a half of the ethidium bromide was still associated with DNA.

3.2. Profiles of the Interpolyelectrolyte Reactions

3.2.1. The pH-Profiles

From comparison of Fig. 3 and Fig. 4 it appears that pH-dependencies of electrostatic binding of PEI and PDMAEM with DNA are distinct. In particular, in alkaline media PDMAEM does not quench the fluorescence at all, whereas the quenching by PEI is noticeable, cf. the curves of the figures obtained at pH 9.0. Furthermore, upon these pH-conditions PLL quenches fluorescence of the (DNA·EB) complex even more effectively than PEI, the corresponding titration curve obtained at pH 9.0 practically coincides with curve of the titration by PDMAEM at pH 6.0 (Fig. 4). In other words, values of a degree of conversion, Θ, in the coupling reactions defined as the ratio of a current number of the salt bonds to the ultimate one[17] are pH-dependent and differ from one another.

Data of the fluorimetric titration performed at different pH allow estimating pH-dependence of Θ, i.e. pH-profiles of DNA complexing with various polycations on the assumption that Θ is proportional to the quenching efficiency.

As a first approximation, assume that $\Theta = 1$ is achieved in neutral media in equimolar mixture of DNA and PLL, i.e. in the mixture of composition $\psi = $ [all amine groups] / [phosphate groups] = 1. This approximation might be crude as a formation of the perfect system of salt bonds between this polypeptide and rigid double helix of native DNA is difficult to perceive. Nevertheless, the degree of conversion in this system should be closed to maximal one due to $pK_b = 10.4$ of PLL. Most likely, it is this relatively high pK_b that provides virtually complete coupling of the polypeptide with DNA even in alkaline media, at pH 9.0. The latter is follows from a coincidence of the titration curves of the (DNA·EB) complex by PLL obtained at pH 6.0 and pH 9.0, cf. curves 6 and 7 of Fig. 4.

From the above reasoning Θ values can be calculated from Eq. (3).

$$\Theta = (I_c - I_n) / (I_c - I_0) \qquad (3)$$

In this formula I_c stands for fluorescence intensity of the (DNA·EB) complex, I_n is fluorescence intensity of equimolar mixture, $\psi = 1$, of DNA with the chosen polyamine at given pH, and I_0 is fluorescence intensity of equimolar mixture of DNA with PLL in neutral media.

Fig. 5 shows pH-dependencies of Θ determined for PEI (curve 2) and PDMAEM (curve 3) by this means from the curves of Fig. 3 and Fig. 4. The pH-profile of DNA coupling reaction with PLL (Fig. 5, curve 1) was constructed by the same procedure from curves 6 and 7 of Fig. 4 and from the analogous curves obtained at different pH in alkaline media (data not shown).

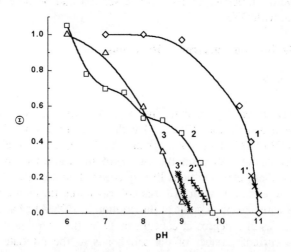

Figure 5. Θ - pH profiles of the reactions of DNA with PLL (1), PEI (2), and PDMAEM (3) derived from the fluorimetric titration curves. The corresponding Θ - pH profiles (1' - 3') obtained by the potentiometric titration are marked by crosses.

In principle, the pH-profiles can be obtained from potentiometric titration curve of the equimolar mixture of DNA and polyamine as it has been detailed in work[17]. However, this approach applies only for studying of the initial stage of the reaction, $\Theta < 0.2$, where the calculation of Θ is of a sufficient accuracy. We conducted the potentiometric titration of mixtures of DNA with the polyamines and calculated values of the conversion degree, as it is described elsewhere[17]. The resulting plots of Θ versus pH are depicted in Fig. 5 by the crosses. It can be seen that these dependencies (curves 1' - 3') fit reasonably to right-hand branches of the corresponding curves 1 - 3. This correspondence is a strong argument in favor of the validity of the fluorescence assay.

The Θ-pH dependencies of Fig. 5 are grouped together in the alkaline region according to rather high pK_b values of the polyamines. In all cases the dependencies are shifted to the alkaline media as compared with the neutralization curve of free polyamine (the neutralization curves are not depicted in Fig. 5 for the sake of simplicity). This is no surprise since the presence of highly charged matrix, DNA in our case, promotes the protonation of the polyamines, or what is the same, formation of interpolymer salt bonds[1]. The extreme right position of curve 1 in Fig. 5 is conditioned by $pK_b = 10.4$ of PLL containing primary amine groups, whereas the left position of curve 3 in the group is in a good agreement with $pK_b = 6.2$ of PDMAEM with tertiary amine groups in the chains.

As would be expected from the irregular structure and chemical composition of branched PEI, pH-profile of the reaction between DNA and PEI, at least its initial portion (curve 2) is intermediate in the position between PLL (curve 1) and PDMAEM (curve 3). PEI contains primary, secondary and tertiary amine groups in roughly equal parts. The basicity of primary and secondary amine groups of polyamines is rather similar and noticeably differs from this one of the tertiary amine groups. Accordingly, right-hand branch of curve 2 reflects protonation of primary and secondary amine groups of PEI on negatively charged DNA matrix in alkaline media, while tertiary amine groups are protonated and form salt bonds with DNA in neutral media, cf. left-hand part of curve 2.

The revealed pronounced sigmoid shape of pH-profile of DNA interaction with PEI as well as smooth pH-profiles of the reactions with PLL and PDMAEM strongly suggest that the fluorescence assay is appropriate for monitoring of DNA coupling reactions at $4 < pH < 11$. This pH region evidently restricted by area of existence of the (DNA·EB) complex[9] allows obtaining the whole pH-profiles of the reactions with a great majority of polyamines.

3.2.2. The μ-Profiles

The study of the coupling polyelectrolyte reactions at different ionic strength, μ, reveals a completely similar behavior. An example presented in Fig. 6a is a quenching of fluorescence of the (DNA·EB) complex by PEVP accomplished at different NaCl concentration. The distinct sudden fall in the fluorescence intensity observed in salt-free solutions at $\varphi \approx 1$ dies out gradually as μ is increased. It suggests a lowering of cooperativity of the interpolyelectrolyte interaction due to progressive destruction of the

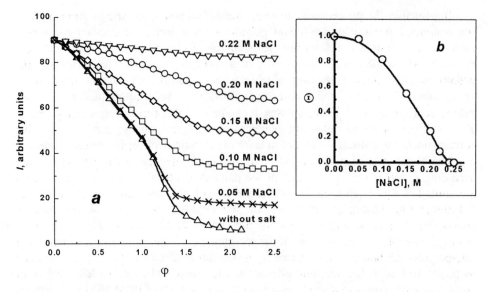

Figure 6. (*a*) Dependencies of I on the ratio φ = [+] / [-] obtained on the titration of the (DNA·EB) complex by PEVP at different concentrations of added NaCl. DP(PEVP) = 10, 0.02 M TRIS, pH 9.0. The other conditions are the same as in Fig. 1. (*b*) The μ-profile determined from *I* values at φ = 1.0, see the text.

salt bonds by added shielding counterions. It is known that PEC can decompose in water solutions dissociating to the original polyelectrolyte components on addition of low-molecular-weight electrolyte. If electrostatic attraction plays a major part in polyelectrolyte coupling the dissociation occurs on exceeding a certain critical concentration of shielding counterions[1]. In the case under consideration, the critical is ca. 0.23 M NaCl, to judge from inability of the relatively short PEVP chains, DP = 10, to quench the fluorescence at [NaCl] ≥ 0.23 M, cf. Fig. 6.

By analogy with described above procedure of pH-profiles determination, one can derive μ-profiles of the reactions from data of the fluorimetric titration performed at different ionic strength on the assumption that Θ is proportional to the quenching efficiency. In this case, too, we are led to assume that Θ = 1 is achieved in salt-free solutions of the equimolar mixture of DNA and the polyamine. On these assumptions the μ-profiles can readily be obtained by the corresponding normalizing.

Fig. 6*b* shows μ-profile of DNA–PEVP polyelectrolyte interaction deduced by this means from data of Fig. 6*a* using Eq. (3). Both I_n corresponding to current μ and I_0 of salt-free solution of the mixture were determined from Fig. 6*a* at φ = 1. The observed sigmoid shape of the profile correlates well with sigmoid shape of Θ-μ dependence for the pair PMA*–PEVP in KI solutions obtained by UV-spectroscopy and reflecting the cooperativity character of PEC destruction[18].

These data clearly demonstrate potential of the fluorescence quenching technique for studying of DNA-containing PEC. The revealed possibility of obtaining pH- and μ-profiles provides a reasonably accurate picture of interpolyelectrolyte interactions between DNA and polycations in a wide range of pH and ionic strength. This, in turn, appears to be crucial for development of monitoring of the positively charged counterparts capable to form DNA-containing PEC under proper conditions, in particular at physiological pH and ionic strength.

3.3. Stability of DNA-containing PEC

It should be borne in mind that intercalation of EB into the DNA helix unwinds the right-handed Watson-Crick helix and consequently weakens electrostatic interaction with polycations, in particular histones[19]. Accordingly, the developed approach is inapplicable in acquisition of quantitative parameters of DNA complexing with polycations. Nevertheless, as it follows from above results this method is well suited for the comparison assessment of affinity of various positively charged species to DNA in salt-free and water-salt solutions. Moreover, fluorescence ignition caused by intercalation of the dye into free sites of DNA is observed at the same salt concentration as the dissociation of PEC(DNA·EB–polycation) on the components, i.e. (DNA·EB) and the polycation[14]. This is an added reason to assume that this approach is appropriate for the estimation of stability of different DNA-containing PEC in water-salt solutions.

In practice, the stability of PEC is conveniently estimated by fluorimetric titration of mixture of the (DNA·EB) and polycation by salt solution. Relying on such experiments, the monitoring of different DNA-containing PEC has been performed[7]. Degree of polymerization, charge density of various polyamines, and structure of N-alkyl substituent of the amine groups were shown to be factors influencing the stability of PEC. The ability of added cations and anions to dissociate PEC decreased in the order $Ca^{++} > Mg^{++} >> Li^+ > Na^+ > K^+ >> (CH_3)_4N^+$ and $I^- > Br^- > Cl^- >> F^-$ which coincided with a decrease of affinity of the same counterions to DNA and positively charged partner.

In subsequent experiments the pH-dependent destruction of DNA-containing PEC formed by polyamines with different type of amine groups has been studied[8]. Structure of the amine groups proved to be a decisive factor of PEC stability. PEC formed by polycations with quaternary amine groups were pH-independent and the least tolerant to destruction by the added salt. Primary amine groups provided the best stability in water-salt solutions under wide pH range. Moderate and pH-dependent stability was revealed for PEC included polyamines with tertiary amine groups in the molecule.

The data obtained[7, 8] appear to be the basis for design of DNA-containing PEC with given and controllable stability. The design may be accomplished not only by proper choice of polyamine of one or another type, but by using of tailor-made polycations with given composition of amine groups of different structure in the chain as well. The latter was illustrated by random copolymer of 4-vinylpyridine and N-ethyl-4-vinylpyridinium bromide[8]. PEC of DNA with this polyamine was pH-sensitive and its destruction could be performed under pH and ionic strength closed to the physiological conditions. This result appears to be particularly promising for addressing DNA packed in PEC species to the target cell.

4. CONCLUSION

The paper illustrates the features and potential of the fluorescence assay developed for investigation of DNA-containing polyelectrolyte complexes.
It should be stressed that

- Use of the assay allows avoiding a covalent attachment of fluorescence tag to DNA, that is costly and labor consuming procedure leading to damage of the double helix on frequent occasions.
- Developed approach makes possible express monitoring and screening of various polycations in respect to their affinity to DNA by mere fluorescence intensity change.
- Revealed possibilities of obtaining a complete pH- and μ- profiles of the coupling polyelectrolyte reactions open a new avenue of attack on the problem of creating self-adjustment gene delivery systems with controllable stability.
- Method could be readily used for prediction and precise control over the equilibrium and kinetics of the competitive reactions in solutions of DNA-containing PEC and polyanions of different chemical nature.

5. ACKNOWLEDGMENTS

The supports of Russian Foundation for Basic Research (grant No. 99-03-33399) and INTAS (grant No. 97-1746) are gratefully acknowledged.

6. REFERENCES

1. V. A. Kabanov, in: *Macromolecular Complexes in Chemistry and Biology*, edited by P. L. Dubin (Springer-Verlag, Berlin-Heidelberg 1994), chapter 10, pp. 151-174.
2. V. A. Izumrudov, I. Yu. Galaev, and B. Mattiasson, Polycomplexes – Potential for bioseparation, *Bioseparation* 7, 207-220 (1999).
3. A. V. Kabanov and V. A. Kabanov, Interpolyelectrolyte complexes of nucleic acids as a means for targeted delivery of genetic material to the cell, *Polymer Science* 36(2), 157-168 (1994).
4. V. A. Izumrudov, S. I. Kargov, M. V. Zhiryakova, A. B. Zezin, and V. A. Kabanov, Competitive reactions in solutions of DNA and water-soluble interpolyelectrolyte complexes, *Biopolymers* 35(5), 523-531 (1995).
5. V. A. Izumrudov, M. V. Zhiryakova, S. I. Kargov, A. B. Zezin, and V. A. Kabanov, Competitive reactions in solutions of DNA-containing polyelectrolyte complexes, *Macromol. Symp.* 106, 179-192 (1996).
6. V. A. Izumrudov, T. K. Bronich, O. S. Saburova, A. B. Zezin, and V. A. Kabanov, The influence of chain length of a competitive polyanion and nature of monovalent counterions on the direction of the substitution reaction of polyelectrolyte complexes, *Macromol. Chem., Rapid Commun.* 9(1), 7-12 (1988).
7. V. A. Izumrudov and M. V. Zhiryakova, Stability of DNA-containing interpolyelectrolyte complexes in water-salt solutions, *Macromol. Chem. Phys.* 200(11), 2533-2540 (1999).
8. V. A. Izumrudov, M. V. Zhiryakova, and S. E. Kudaibergenov, Controllable stability of DNA containing polyelectrolyte complexes in water-salt solutions, *Biopolymers (Nucleic Acid Sciences)* 52(2), 94-108 (1999).
9. J.-B. Le-Pecq and C. Paoletti, A fluorescent complex between ethidium bromide and nucleic acids. Physical-chemical characterization, *J. Mol. Biol.* 27(1), 87-106 (1967).

10. J.-B. Le-Pecq, Use of ethidium bromide for separation and determination of nucleic acids of various conformational forms and measurement of their associated enzymes, *Methods of Biochemical Analysis* **20**, 41-86 (1971).
11. J. Dufoureq, W. Neri, and N. Henry-Toulme, Molecular assembling of DNA with amphipathic peptides, *FEBS Lett.* **421**, 7-11 (1998).
12. K. D. Stewart, The effect of structural changes in a polyamine backbone on its DNA-binding properties, *Biochem. Biophys. Res. Com.* **152**(3), 1441-1446 (1988).
13. H. S. Basu, H. C. A. Schweitert, B. G. Feuerstein, and L. J. Marton, Effects of variation in the structure of spermine on the association with DNA and the induction of DNA conformational changes, *Biochem. J.* **269**, 329-334 (1990).
14. V. A. Izumrudov, A. B. Zezin, S. I. Kargov, M. V. Zhiryakova, and V. A. Kabanov, Competitive displacement of ethidium cations intercalated in DNA by polycations, *Dokl. Phys. Chem.* **342**(4-6), 150-153 (1995).
15. D. Oupický, Č. Koňák, and K. Ulbrich, DNA complexes with block and graft copolymers of N-(2-hydroxypropyl)methacrylamide and 2-(trimethylammonio)ethyl methacrylate: effect of copolymer composition, *J. Biomaterials Sci.* **10**(5), 573-590 (1999).
16. T. B. Wyman, F. Nicol, O. Zelphati, P. V. Scaria, C. Plank, and F. C. Szoka, Design, synthesis, and characterization of a cationic peptide that binds to nucleic acids and permeabilizes bilayers, *Biochemistry* **36**, 3008-3017 (1997).
17. V. A. Kabanov, A. B. Zezin, V. B. Rogacheva, Zh. G. Gulyaeva, M. F. Zansokhova, J. G. H. Joosten, and J. Brackman, Interaction of astramol poly(propyleneimine) dendrimers with linear polyanions, *Macromolecules* **32**(6), 1904-1909 (1999).
18. D. V. Pergushov, V. A. Izumrudov, and A. B. Zezin, Competitive binding of iodide anions by polycation including in water-soluble interpolyelectrolyte complex, *Polymer Science* **39B**, 237-238 (1997).
19. W. H. Strätling and I. Seidel, Relaxation of chromatin structure by ethidium bromide binding: determined by viscometry and histone dissociation studies, *Biochemisry* **15**(22), 4803-4809 (1976).

DNA-CLEAVING AND ADDUCT FORMATION BY FULLERENES

S. Samal, C. N. Murthy, and K. E. Geckeler[*]

1. INTRODUCTION

Since the discovery of C_{60} by Kroto et al.[1] and its subsequent isolation by Krätschmer et al.[2] a number of applications of this molecule and its derivatives have been found. The major areas include electronic applications, which still is a topic of much research effort.[3] Another area that has received considerable attention is to make fullerenes water-soluble to facilitate their use in biological and biomedical applications. These applications emerge from the fullerenes absorbing strongly in the ultraviolet and moderately in the visible range of the spectrum, leading to the formation of singlet oxygen with almost 100% efficiency.[4] The high reactivity of fullerenes towards free radicals has led to these being recognized as a 'free radical sponge'.[5] These two properties (illustrated schematically in Figure 1a and Figure 1b) have been the base of a number of biomedical applications.

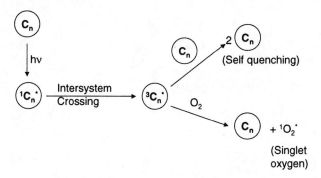

Figure 1a. Illustrative mechanism of singlet oxygen generation by fullerenes.

[*] Laboratory of Applied Macromolecular Chemistry, Department of Materials Science and Engineering, Kwangju Institute of Science and Technology, 1 Oryong-dong, Buk-gu, Kwangju 500-712, South Korea

Advanced Macromolecular and Supramolecular Materials and Processes
Edited by K. Geckeler, Kluwer Academic/Plenum Publishers, 2003

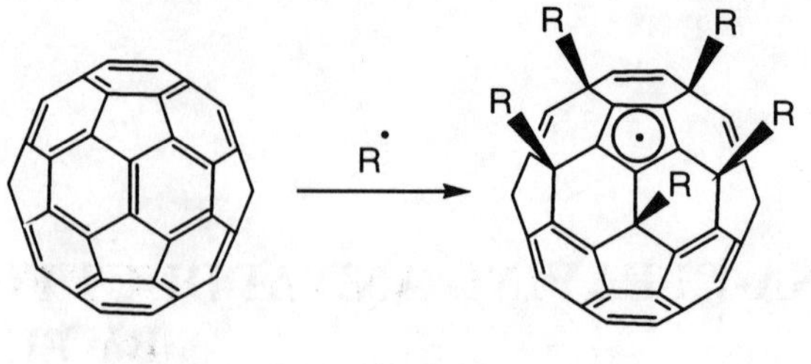

Figure 1b. Radical scavenging by fullerenes.

2. ENZYME INHIBITION ACTIVITY

Based on molecular modeling studies, Friedman and co-workers showed that a water-soluble methano[60]fullerene dicarboxylate (Figure 2) showed a good anti-HIV activity[6] in the μM range, when compared to the peptide-based inhibitors that are effective in the nanomolar range. The reason given was that the fullerene derivative showed good van der Waals interactions with the hydrophobic surface of the enzyme cavity. The water-soluble fullerene derivative N-tris(hydroxymethyl)propylamido methano[60]fullerene (Figure 3) was very active against HIV-1 and exhibited no cytotoxicity.[7] A number of C_{60} and C_{70} derivatives were tested in the form of dimethyl sulphoxide-water emulsions against human cells infected with HIV. Most of the compounds studied (Figure 4) showed antiviral activity in the low micromolar range.[8]

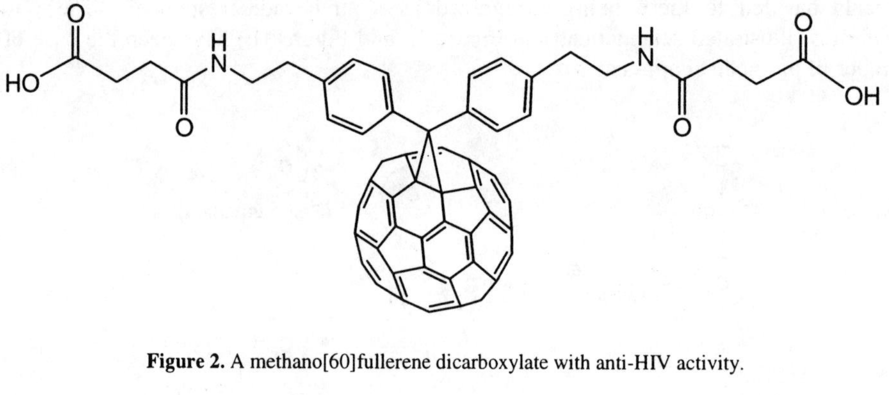

Figure 2. A methano[60]fullerene dicarboxylate with anti-HIV activity.

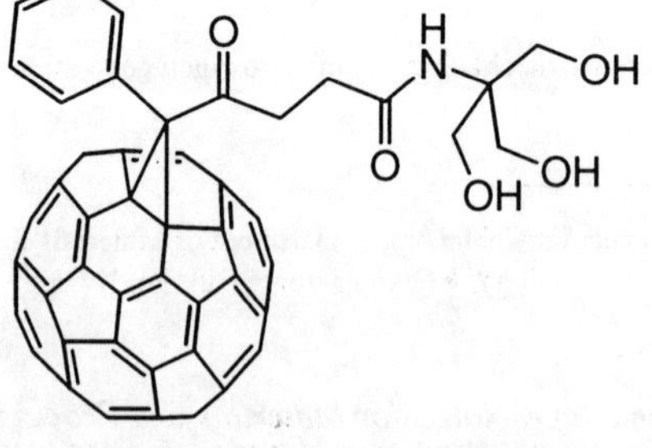

Figure 3. N-Tris(hydroxymethyl)propylamido methano[60]fullerene, active against HIV-1.

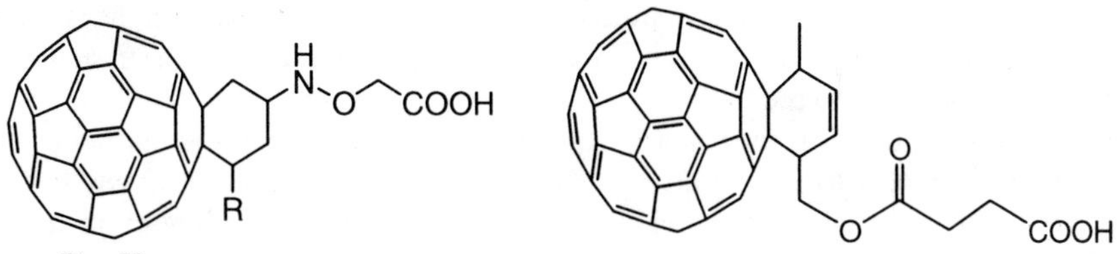

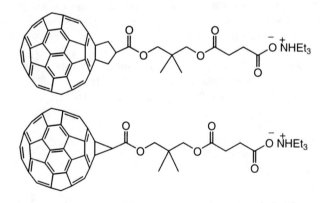

Figure 4. Monocarboxylic acid derivatives of C_{60} with an anti-viral activity.

Enzyme-inhibitory activity is also manifested by a series of amphiphilic fullerene derivatives (Figure 5).[9] These compounds exhibited distinct inhibitory activity against proteases including cysteine proteases and serine proteases, whereas the compounds did not show a high activity against HIV-transcriptase. The inhibitory activity of these amphiphilic fullerene derivatives and fullerols $C_{60}(OH)_n$ and $C_{70}(OH)_n$ was in the micromolar range. The monosubstituted methanofullerene derivative was more potent compared to the disubstituted methanofullerene derivative shown in Figure 2. A kinetic study indicated a typical reversible competitive inhibition, which suggested that the monosubstituted derivative becomes hydrophobically bound inside the active site of the enzyme.

Figure 5. Amphiphilic monosubstituted fullerene derivatives with inhibitory activity against proteases.

3. ANTICANCER ACTIVITY

The water-soluble fullerol $C_{60}(OH)_{24}$ has been studied for its anticancer activity. This compound affected the growth kinetics of human lymphocyte cultures and epidermal carcinoma cell cultures. The compound was found to efficiently prevent cell growth in a manner similar to known anticancer drugs such as taxol.

Generation of singlet oxygen by fullerene in the presence of visible light is the base of photodynamic therapy. A photosensitizer with an affinity for tumor tissue is intra venously administered into the body.

After accumulation of the photosensitizer in the tumor-bearing tissue becomes maximum, the tumor site is selectively light-irradiated. The photosensitizer generates singlet oxygen from oxygen present in the tissue by light irradiation. Singlet oxygen is an extremely reactive species and acts as an effective cytotoxic agent. Fullerene derivatives are excellent photosensitizers. If preferential accumulation in a tumor tissue is achieved, it is possible that a local irradiation of the tumor tissue with visible light would result in tumor necrosis. There are anatomical differences between the tumor and normal tissue. The permeability of blood vessels newly formed in the tumor tissue is generally higher than that of normal blood vessels. In addition, immatured lymph systems in the tumor tissue makes it difficult to excrete large-sized substances from the tissue, enabling them to accumulate to a greater extend and retain for a longer time period in the tumor tissue than in the normal tissue. An increase in accumulation and retention of antitumor drugs in the tumor tissue could be achieved by increasing their apparent molecular size through chemical conjugation with polymers.

When water-soluble polymers were intravenously injected to the tumor-bearing mice, their accumulation in tumor tissue was higher than that in the normal tissue. The apparent molecular size of the water-soluble C_{60} was increased through conjugation with poly (ethylene glycol) (PEG). Besides enhancing the apparent molecular mass of C_{60}, PEG being bioamenable has been widely used for chemical modification of drugs. As expected, the C_{60}-PEG conjugate resulted in enhanced accumulation of C_{60} in tumor tissue after intravenous injection to the tumor-bearing mice.[10]

The experiment revealed that a treatment with the C_{60}-PEG conjugate coupled with light irradiation strongly induced tumor necrosis, whereas the normal skin was not damaged. Conjugate injection alone did not induce any tissue necrosis. The photodynamic effect on tumor greatly depended on the C_{60} dose and the power of light irradiation. All the tumor-bearing mice were cured. The photodynamic effect of the clinically used photosensitizer "photofrin" was less efficient than that of the conjugate, even at 10-times higher doses than the conjugate.

4. DNA-CLEAVING ACTIVITY

4.1. Fullerene Derivatives

Several water-soluble derivatives of fullerenes have been shown to have DNA cleaving activity.[11-15] The photoinduced DNA-cleaving activity of a C_{60} derivative with an acridine group was reported (Figure 6).[14]

An acridine moiety was chosen because of its intercalating activity with DNA double strands. The compound was reacted with a saturated HCl solution in ethanol and then incorporated into a poly(vinyl pyrrolidone) (PVP) micellar system to give a completely transparent aqueous solution that was used for the DNA cleaving study.

The DNA-cleaving activity of the acridine adduct of C_{60} was tested under visible light irradiation using a supercoiled plasmid (pBR322) and the effects were compared under total dark conditions. The acridine adduct showed a strong DNA-cleaving activity at 0-5 °C within 4 h.

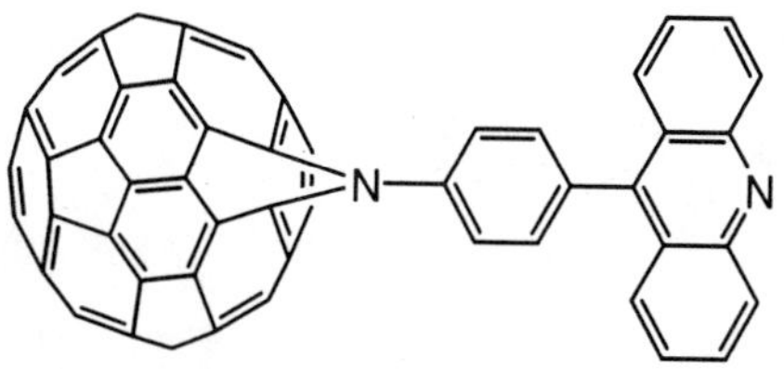

Figure 6. Acridine adduct of C_{60}.

A fullerene-carboxylic acid (Figure 7) cleaved double-stranded DNA with moderate efficiency at a 100 μM concentration upon irradiation with a visible light source. The cleaving took place at the guanine base. The cleaving proceeded much more efficiently in D_2O than in H_2O and was inhibited by the presence of a singlet oxygen quencher. It was also proved that photolysis of the carboxylic acid in solution containing molecular oxygen does indeed generate singlet oxygen. These results were taken as an evidence of singlet oxygen-mediated DNA-cleavage. However, the results do not exclude the possibility of a direct oxidation of DNA by an excited fullerene core.

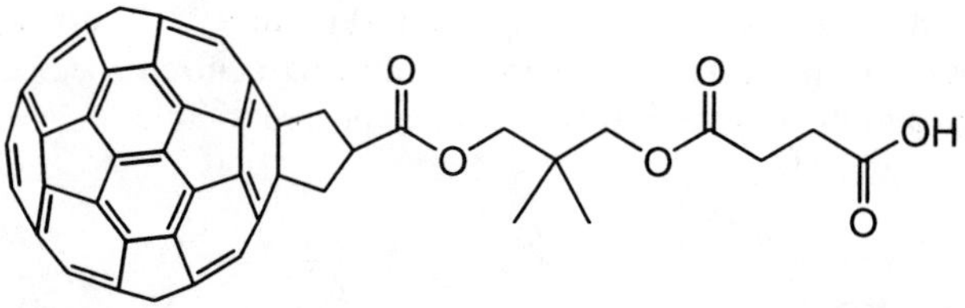

Figure 7. The fullerene-carboxylic acid used for DNA-cleaving studies.

The cleaving activity of the amphiphilic fullerene derivatives was compared with those of the spherically multisubstituted fullerene derivatives such as fullerols and the fullerene amine adducts $C_{60}H_n(NHCH_2CH_2NMe_2)_n$.[9] The amine adduct cleanly cut DNA (supercoiled pBR322) into a nicked circular one under irradiation with visible light.

The efficiency of cleavage was higher than for the carboxylic acid derivative shown in Figure 7, possibly due to the higher water-solubility of the former. Interestingly, the $C_{60}H_n(NHCH_2CH_2NMe_2)_n$ displayed a weak DNA-cleavage even under total darkness, whereas the carboxylic derivative showed no activity in the dark. Both compounds, however, cleaved DNA selectively at the guanine base.

To further demonstrate the site-selective photocleavage of DNA, an appropriate DNA-binding moiety was introduced into a fullerene derivative. This would result in sequence-selective binding to DNA leading to the intended site selective cleavage. An antibiotic netropsin, that contains two *N*-methylpyrrole carboxamides, a known DNA binder, binds to a minor groove of the double helical DNA. Therefore, a fullerene-netropsin conjugate was synthesized (Figure 8).

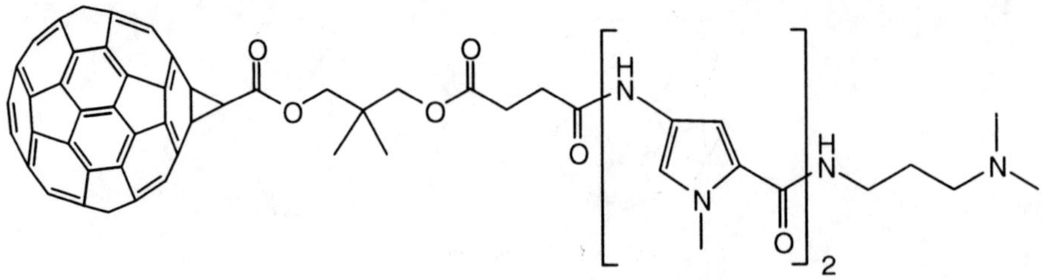

Figure 8. A fullerene derivative containing two *N*-methylpyrrole carboxamides for site-selective DNA binding.

Interestingly, this molecule did not cause photocleavage of DNA. In spite of the electron-donating *N*-methylpyrrole carboxamide units, separated from the electron-accepting fullerene moiety by a 180 pm-long linking group, intramolecular triplet quenching could not be excluded because the linker moiety interconverts between its folded and extended conformation. That the triplet-quenching is very efficient in such a situation was further proved when this molecule did not produce singlet oxygen in test tube experiments. When such folding could be prevented in molecules, the result would be site-selective photocleavage due to efficient DNA binding.

In another approach, DNA-fullerene hybrids were synthesized in which DNA was covalently bound to fullerene *via* linking units (Figure 9).[12] These hybrids could bind single- and double-stranded DNA. Again the cleaving ocurred specifically at guanosine residues proximal to the fullerene moiety upon exposure to light.

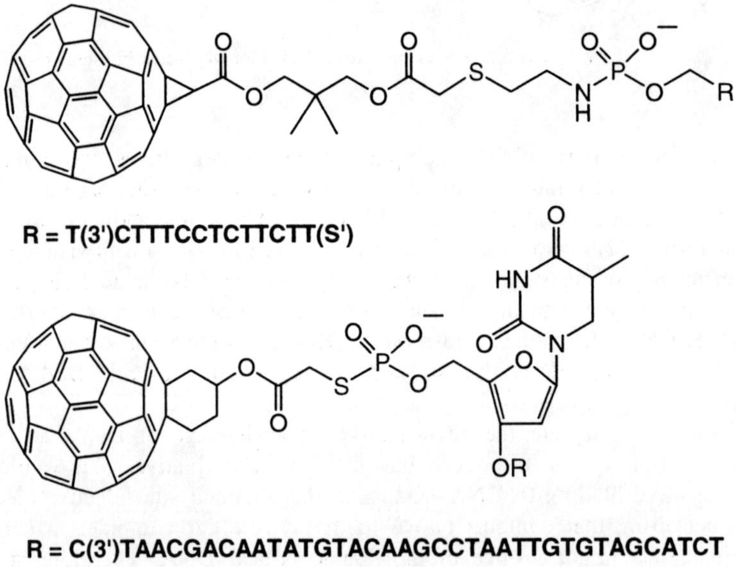

R = T(3')CTTTCCTCTTCTT(S')

R = C(3')TAACGACAATATGTACAAGCCTAATTGTGTAGCATCT

Figure 9. Examples of DNA-fullerene hybrids with linking units.

The immobilization of DNA on a two-dimensional solid surface is of interest both in studies of DNA itself and in various applications such as biosensors. A well-ordered monolayer assembly on a gold surface, that contained cationic groups such as quarternary ammonium salts for the interaction with the phosphate groups of DNA, has been devised.[16] On such a self-assembled monolayer C_{60} has been incorporated and then DNA has been immobilized, so that site-specific cleaving could be studied (Figure 10).

4.2. Mechanism of Cleaving

In most of the experiments cleaving was ascribed to the generated singlet oxygen. However, the hybrid DNA-fullerene compounds (see Figure 9) taken to investigate the cleaving process were shown to bind to the single- or double-stranded DNA. In an effort to show the actual mechanism of the photocleavage process, the fullerene-oligonucleotide-mediated cleaving process was compared with that of an eosin-oligonucleotide. The results suggested that the fullerene-oligonucleotide-mediated cleavage does not involve a singlet oxygen mechanism, but some other mechanism.

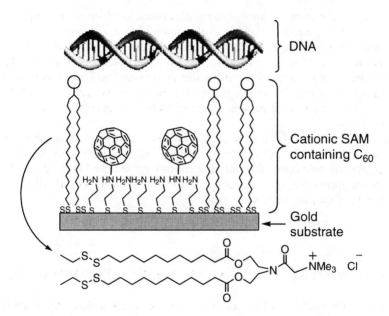

Figure 10. Immobilization of DNA on the cationic self-assembled monolayer (SAM) containing C_{60} on a gold substrate.

In the case of the photocleavage of DNA it has been shown that in the presence of [60]fullerene derivatives, cleavage takes place at the guanosine base site only.[9] This oxidative cleavage at the guanosine site could be possible by two routes: A singlet oxygen ($^1O_2$) generated by the [60]fullerene is the active oxidant or there is a direct transfer of an electron from the guanosine to [60]fullerene. The probable mechanisms are shown schematically in Figure 11.

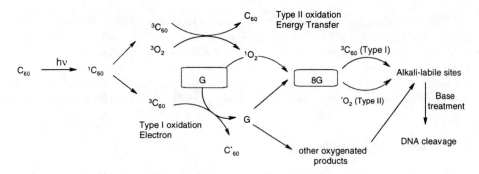

Figure 11. Schematic illustration of mechanism of cleaving action

Recently, it has been shown that the predominant mechanism in the photocleaving process is the direct electron transfer to the [60]fullerene.[17] In the event of the direct electron transfer it is possible that the [60]fullerene may be attached covalently to the cleaved DNA molecule at specific sites. Previous experiments on the photocleavage of DNA were done using water-soluble [60]fullerene derivatives. There was also a report of DNA-cleaving by pristine fullerene, though the molecule was taken in a PVP matrix.[14] However, it was not known if pristine fullerene solubilized in solvents like toluene could also cleave DNA.

Recently, it was found that pristine [60]fullerene dissolved in a suitable solvent like toluene does indeed cleave DNA in the presence of light leading to a quantitative yield of the DNA-fullerene conjugate. The fullerene spheres get attached to the cleaved strand of DNA thus supporting the direct electron transfer mechanism. Details of the study are discussed in the following section.

5. DNA-CLEAVING BY FULLERENE-CYCLODEXTRIN CONJUGATES

In spite of the synthesis of several water-soluble fullerene derivatives and fullerene-bound polymers,[18,19] many are only sparingly soluble in water, and the solubilizing component is often potentially toxic. The synthesis of the first water-soluble fullerene main-chain polymer based on a novel approach of supramolecular masking involving cyclodextrin was reported from our laboratory,[20] and subsequently we synthesized a number of analogs (Figure 12).

All these polymers have shown excellent DNA-cleaving activity.[21] Here, we discuss the cleaving of an oligonucleotide using one of such polymers (**1**). Easy monitoring of the reaction carried out under ambient light condition by UV-Vis spectroscopy, followed by membrane separation led to cleaved DNA in quantitative yields, indicating strong application prospects of this polymer in photodynamic cancer therapy.

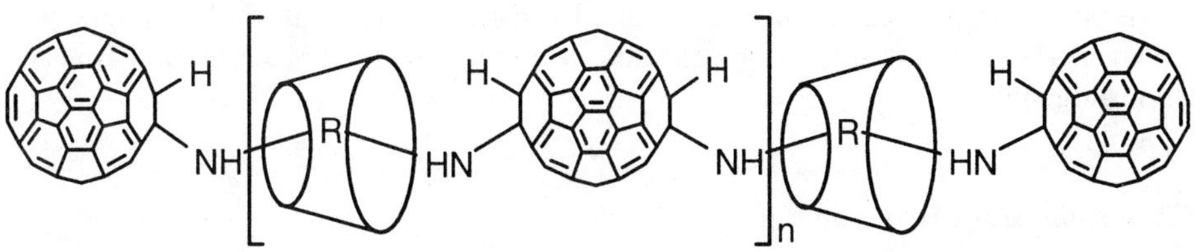

R:

1

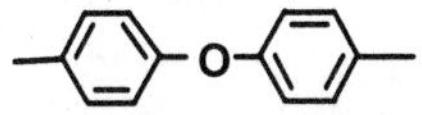

2

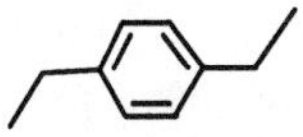

3     $O(CH_2CH_2OCH_2CH_2CH_2-)_2$

Figure 12. Different poly(fullerocyclodextrin)s which are water-soluble fullerene main-chain polymers.

5.1. Experimental

5.1.1. Purification of the DNA Oligonucleotide

DNA used for the present study was a commercial sample isolated from herring sperm (SIGMA). The sample was further purified by membrane filtration of the solution taking 1 g of crude DNA in 100 mL water in the filtration cell and using a membrane of molar mass cut-off (MMCO) of 10 kg mol^{-1}, running through high purity water at pH 7. The volume of solution inside the chamber was always maintained at ~25 mL. Freeze-drying of the retentate solution gave highly pure DNA oligonucleotide with a yield of 87%.

5.1.2. DNA-Cleaving Experiment

A solution of the purified DNA oligonucleotide (100 mg in 10 mL water) was treated with 2.5 mg of **1**, dissolved in an equal volume of the same solvent and the mixture was gently stirred at room temperature (~20°C). The UV-Vis spectra of the mixture were recorded at regular intervals by transferring aliquots of the reaction mixture to a quartz cuvette. After each such measurement, the solution was returned back to the reaction vial. The progress of the reaction was followed by the change of the characteristic fullerene absorbance at 343 nm.

After completion of the reaction, the entire quantity of DNA-polyfullerene solution mixture (20 mL) was taken to the ultrafiltration cell, and diluted to 30 mL. The solution was eluted under pressure (nitrogen atmosphere) with deionized water (pH 7) using the same membrane (MMCO of 10 kg mol^{-1}) used for the purification of the crude DNA. Membrane filtration was continued until the total volume of permeate collected was 600 mL (filtration factor, Z = 20). The UV-Vis spectra of the different portions of collected permeate were recorded and the absorbance at 260 nm was plotted against the permeate volume.

Further, the permeates were freeze-dried and the yields of the DNA samples were recorded. The molar masses of the collected fractions were determined by gel permeation chromatography.

5.2. Results and Discussion

The poly(fullerocyclodextrin) **1** is a short-chain polymer with the molar mass M_n = 18.9, kg mol^{-1}, M_w = 20.0 kg mol^{-1}; polydispersity: 1.06). Figure 13 shows the UV-Vis spectra of the DNA-**1** reaction mixture recorded at different time intervals. The reaction progress is indicated by a gradual decrease of the characteristic fullerene peak at 343 nm. After about 15 hours of reaction time, almost the entire quantity of **1** was used up.

The cleaved DNA fragments were separated using a membrane filtration set-up. The UV-Vis spectra of portions of permeate collected at different time intervals are shown in Figure 14. The characteristic DNA absorbance at 260 nm went on decreasing steadily, indicating a gradual decrease in DNA concentration in the filtration chamber. It is expected that the lower molar mass DNA fragments would be filtered out first followed by components of progressively higher molar masses. The molar mass of purified DNA sample was M_n 9.745, M_w 13.205 kg mol^{-1}; polydispersity: 1.335. From the first 80 mL of permeate the amount of DNA recovered was 43 mg and this portion had the molar mass M_n 4.034, M_w 4.334 kg mol^{-1}; polydispersity: 1.074.

The cleaved DNA content in the next 60 mL permeate was 38 mg, the molar mass slightly increasing to M_n 4.252, M_w 4.800, polydispersity 1.128 (GPC traces shown in Figure 15). It was thus seen that nearly the entire amount of the cleaved DNA passed

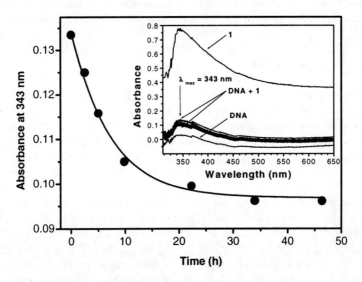

Figure 13. Progressive decrease in absorbance at 343 nm with time for the DNA oligonucletide and poly(fullerocyclodextrin) (**1**) reaction monitored by UV-Vis spectroscopy (inset).

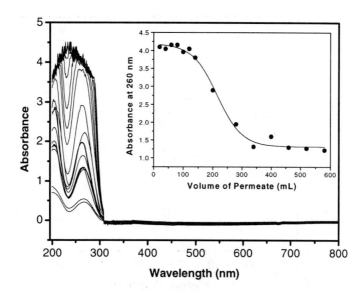

Figure 14. UV-Vis spectra of samples of permeate collected from the membrane filtration of the DNA-**1** reaction. Inset shows a steady decrease in the characteristic DNA absorbance at 260 nm.

through the membrane in the first 140 mL of the permeate volume. Beyond that, the quantities of the materials recovered were very small.

Over 80% of the amount of DNA taken initially got cleaved to nearly half of its original size. Considering that only a very small amount of **1** (2.5 mg) was used to cleave a large excess of the DNA oligonucleotide (100 mg), the cleaving process appears to be highly efficient.

The other poly(fullerocyclodextrin)s (**2**, **3**) subjected to similar cleaving experiments demonstrated a high cleaving efficiency. The water-soluble fullerene derivatives have a long absorption-tail in the visible region starting from ~350 nm, and this was also the observation for all the poly(fullerocyclodextrin)s including **1**.[20]

The ability of these derivatives to promote $^1O_2$-formation using visible region excitation would have significant advantages for *in vivo* photodynamic therapy applications. The reasons are: (i) the typical broad absorption of the fullerene derivatives in the visible region; and (ii) increased optical penetration depth of the longer wavelength radiations.[22]

Tabata and Ikada demonstrated for a water-soluble C_{60}-PEG conjugate that, following its intravenous injection, local irradiation of visible light to the tumor site induced tumor necrosis, in contrast to the conjugate injection alone.[10] That the cleaving process does indeed occur in the presence of visible light was demonstrated by us in an experiment between DNA and pristine fullerene. No cleaving was noticed in the absence of light. In addition, the reaction led to water-soluble DNA-fullerene conjugates in quantitative yields.

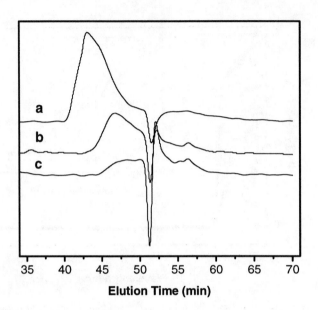

Elution Time (min)

Figure 15. Gel permeation chromatograms of (a) purified commercial DNA oligonucleotide, (b) cleaved DNA in the first 80 mL permeate, (c) cleaved DNA in the following 60 mL permeate.

6. DNA-CLEAVING BY PRISTINE FULLERENE

In the cleaving experiments involving various cyclodextrin derivatives, it is possible that the fullerene moiety would be chemically bonded to the DNA fragments. In some of the DNA-cleaving experiments, a low recovery of DNA due to fullerene-induced aggregation of DNA was observed, and it was suggested that the fullerenes are randomly bound to the DNA chain.

In our experiments involving fullerene-cyclodextrin conjugates, none of the cleaved fragments showed any absorption characteristics of the fullerene moiety in the UV-vis spectra. A possible reason could be that the amount of the fullerene derivative used was extremely small compared to the DNA used for the cleaving experiment. Therefore, it was intended to synthesize a DNA-fullerene conjugate, so that the product itself could be used for biomedical applications.

In order to achieve this target we carried out a direct reaction between pristine fullerene and the DNA oligonucleotide. It was expected that the fullerene moiety, after cleaving the nucleotide, would be incorporated into the small cleaved fragments, thus leading to the desired product. This is schematically presented in Figure 16.

The UV-Vis spectra with progressing reaction showed an interesting trend (Figure 17). There was no change or shift in the absorbance maxima with time when the reaction was continued in dark while the absorption maxima changed when the reaction contents were exposed to light. The absence of any reaction in the dark indicates that [60]fullerene does not react with the DNA under these conditions.

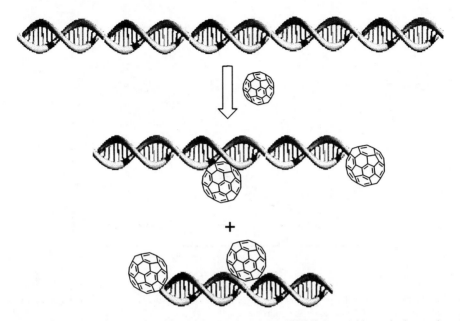

Figure 16. Schematic illustration of the DNA cleaving by pristine [60]fullerene and its anchoring to cleaved fragments leading to DNA-fullerene conjugates.

On exposure to light, however, a significant change in the absorbance maxima within 12 hours of reaction time was observed and the absorption showed a red-shift of about 20 nm from 344 nm to 365 nm. This implies that the excited [60]fullerene in the absence of any dissolved oxygen reacted with the oligonucleotide thus cleaving it. It has been already established previously that the cleaving takes place at the guanosine site only.

However, in this particular situation it shows that along with the cleaving of the DNA moiety there is a simultaneous covalent attachment of the [60]fullerene to the cleaved strand of the DNA. The site of attachment has still not been established unambiguously.

Figure 18 shows the UV-Vis absorbance of the aqueous solution of the DNA-[60]fullerene oligonucleotide. The typical absorbance of the DNA is seen with the additional peak at 345 nm and peak broadening beyond 350 nm assigned to the [60]fullerene. The peak broadening is significant, indicating the complex structure of the [60]fullerene-DNA nucleotide and is due to the aggregation of the [60]fullerenes with the DNA strands coiled around the clusters of C_{60}.

The FTIR spectra of the water-soluble compound shown in Figure 19 confirms that the [60]fullerene is attached covalently to the cleaved nucleotide strand with a typical absorbance band at 527 cm^{-1}. This result taken together with the UV-Vis absorbance studies confirm that the [60]fullerene is actually attached to the DNA strand.

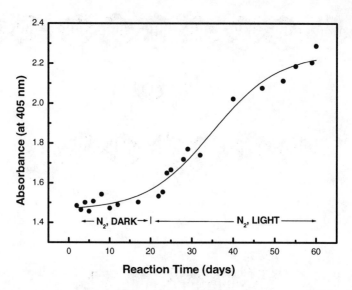

Figure 17. Plot showing the progress of the reaction between fullerene and DNA monitored by UV-Vis absorption maxima of the reaction mixture with time.

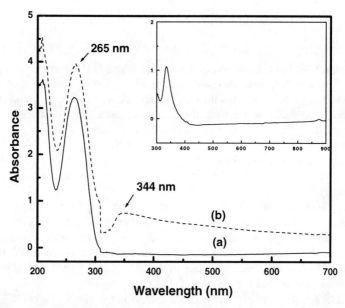

Figure 18. UV-Vis spectra of the fullerene-bound DNA showing the typical absorbance band at 344 nm in water.

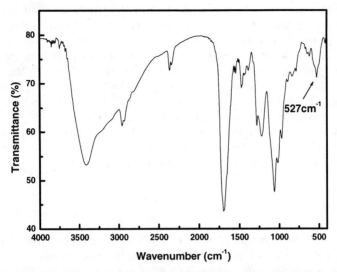

Figure 19. FT-IR spectra of the fullerene-bound DNA showing the typical fullerene bands.

As a further support to this argument we also ran a thermogram of the pure DNA and the product. The heating of the samples was done under inert conditions at a heating rate of 5 °C. The thermogram is shown in Figure 20.

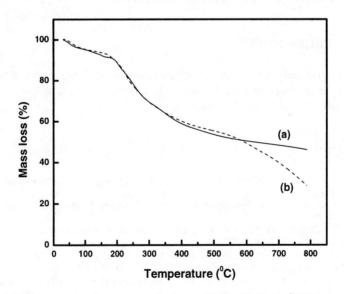

Figure 20. The thermograms of (a) DNA, and (b) the [60]fullerene-DNA conjugate.

The decomposition of both the DNA and the DNA-fullerene conjugate showed a similar trend up to around 400 °C, beyond which the thermal decomposition of fullerene starts. If the product were just a mixture of the DNA and [60]fullerene, then the thermogram would look very different. It is interesting to point out here that the residue content of the pure DNA and the [60]fullerene-bound DNA are different, with the cleaved DNA showing less residue than the pure DNA. Thus, it is possible that the cleaving process resulted in a cleavage of the phosphate di-ester groups linking the sugar units and their removal during the washing and purification steps.

The DNA-fullerene conjugate, being highly water-soluble, is currently being studied further in a number of experiments such as scavenging of a living free radical, to demonstrate properties indicative of possible biomaterial applications.

7. OUTLOOK

Genomic application prospects by way of DNA-cleaving activity hold an immense application potential. Highly efficient DNA-cleaving by small amounts of water-soluble poly(fullerocyclodextrin)s under visible light conditions and also the photo-cleavage of DNA by pristine fullerene in mixed organic solvent system has been found. The cleaving is followed by strong interaction between cleaved DNA fragments and C_{60} leading to DNA-C_{60} conjugates in high yields.

In addition to the use of the novel fullerene derivatives as DNA cleaving reagents, members of this versatile class as well as the DNA-fullerene conjugate should be useful as drugs for the treatment of various DNA-related diseases, e.g., in the targeting of cancers, or in the treatment of virus and retrovirus diseases.

8. ACKNOWLEDGEMENTS

The authors gratefully acknowledge the generous financial support from the Ministry of Science and Technology and the Ministry of Education, Korea, as well as from the Kwangju Institute of Science and Technology.

9. REFERENCES

1. H. W. Kroto, J. R. Heath, S. C. O`Brien, R. F. Curl, and R. E. Smalley, C_{60}-Buckminsterfullerene, *Nature* **318**, 162-163 (1985).
2. W. Krätschmer, L. D. Lamb, K. Fostiropoulos, and D. R. Huffman, Solid C_{60}- a new form of carbon, *Nature* **347**, 354-358 (1990).
3. T. Braun, A. P. Schubert, and R. N. Kostoff, Growth and trends of fullerene research as reflected in its journal literature, *Chem. Rev.* **100**, 23-37 (2000).
4. J. W. Aborgast, A. P. Darmanyan, C. S. Foote, Y. Rubin, F. N. Diedrich, M. M. Alvarez, S. J. Anz, and R. L. Whetten, The photophysical properties of C_{60}, *J. Phys. Chem.* **95**, 11-12, (1991).
5. F. N. Tebbe, J. Y. Becker, R. L. Harlow, D. B. Chase, L. E. Firment, E. R. Holler, B. S. Malone, P. J. Krusic, and E. Wasserman, Multiple, reversible chlorination of C_{60}, *J. Am. Chem. Soc.* **113**, 9900 (1991).
6. S. H. Friedmann, D. L. Decamp, R. P. Sijbesma, G. Srdanov, F. Wudl, and G. L. Kenyon, Inhibition of the HIV-1 protease by fullerene derivatives, *J. Am. Chem. Soc.* **115**, 6506-6509 (1993).
7. R. F. Schinazi, C. Bellavia, R. Gonzalez, C. L. Hill, and F. Wudl, *Proc. Electrochem. Soc.* **95**, 696 (1995).

8. D. I. Schuster, S. R. Wilson, and R. F. Schinazi, Anti-human immunodeficiency virus activity and totoxicity of derivatized buckminsterfullerenes, *Bioorg. Med. Chem. Lett.* **6,** 1253 (1996).

9. E. Nakamura, H. Tokuyama, S.Yamago, T. Shiraki, and Y. Sugiura, Biological activity of water-soluble fullerenes. Structure dependence of DNA cleavage, cytotoxicity, and enzyme inhibitory activities including HIV-protease inhibition. *Bull. Chem. Soc. Jpn.* **69**, 2143-2151 (1996).

10. Y. Tabata and Y. Ikada, Biological function of fullerene, *Pure Appl. Chem.* **71**, 2047-2053 (1999).

11. H. Tokuyama, S. Yamago, E. Nakamura, T. Shiraki, and Y. Sugiura, Photo-induced biochemical activity of fullerene carboxylic acid, *J. Am. Chem. Soc.* **115**, 7918-7919 (1993).

12. A. S. Bourtorine, H. Tokuyama, M. Takasugi, H. Isobe, E. Nakamura, and C. Helene, Fullerene-oligonucleotide conjugates. Photoinduced sequence-specific DNA cleavage, *Angew. Chem., Int. Ed. Engl.* **33**, 2462-2465 (1994).

13. Y.-Z. An, C.-H. B. Chen, J. L. Anderson, D. S. Sigman, C. S. Foote, and Y. Rubin, Sequence-specific modification of guanosine in DNA by a C_{60}-linked deoxyoligonucleotide: evidence for a non-singlet oxygen mechanism, *Tetrahedron* **52**, 5179-5189 (1996).

14. Y. N. Yamakoshi, T. Yagami, S. Sueyoshi, and N. Miyata, Acridine adduct of [60]fullerene with enhanced DNA-cleaving activity, *J. Org. Chem.* **61**, 7236-7237 (1996).

15. J. Hirayama, H. Abe, N. Kamo, T. Shinbo, Y. O. Yamada, S. Kurosawa, K. Ikebuchi, and S. Sekicuchi, Photoinactivation of vesicular stomatitis virus with fullerene conjugated with methoxy polyethylene glycol amine *Biol, Pharm. Bull.* **22**, 1106-1109 (1999).

16. N. Higashi, T. Inoue, and M. Niwa, Immobilization and cleavage of DNA at cationic, self-assembled monolayers containing C_{60} on gold, *Chem. Commun.*, 1507-1508 (1997).

17. R. Bernstein, F. Prat, and C. S. Foote, On the mechanism of DNA cleavage by fullerenes investigated in model systems: electron transfer from guanosine and 8-oxo-guanosine derivatives to C_{60}, *J. Am. Chem. Soc.* **121**, 464-465 (1999).

18. K. E. Geckeler and S. Samal, Macrofullerenes and polyfullerenes: new promising materials, *J. Macromol. Sci., Rev.* **40**, 193-205 (2000).

19. K. E. Geckeler and A. Hirsch, Polymer-bound C_{60}, *J. Am. Chem. Soc.* **115**, 3850-3851 (1993).

20. S. Samal, B.-J. Choi, and K.E. Geckeler, The first water-soluble main-chain polyfullerene, *Chem. Commun.* 1373-1374 (2000).

21. S. Samal and K. E. Geckeler, DNA-cleavage by fullerene-based synzymes, *Macromol. Biosci.* **1**, 329-331 (2001).

22. T. J. Dougherty, C. J. Gomer, B. W. Henderson, G. Jori, D. Kessel, M. Korbelik, J. Moan, and Q. Peng, *J. Natl. Cancer Inst.* **90**, 889 (1998).

AUTHOR INDEX

SUBJECT INDEX